# Mathematical Engineering

Today, the development of high-tech systems is unthinkable without mathematical modeling and analysis of system behavior. As such, many fields in the modern engineering sciences (e.g. control engineering, communications engineering, mechanical engineering, and robotics) call for sophisticated mathematical methods in order to solve the tasks at hand.

The series Mathematical Engineering presents new or heretofore little-known methods to support engineers in finding suitable answers to their questions, presenting those methods in such manner as to make them ideally comprehensible and applicable in practice.

Therefore, the primary focus is—without neglecting mathematical accuracy—on comprehensibility and real-world applicability.

To submit a proposal or request further information, please use the PDF Proposal Form or contact directly: Dr. Thomas Ditzinger (thomas.ditzinger@springer.com)

Indexed by SCOPUS, zbMATH, SCImago.

More information about this series at http://www.springer.com/series/8445

Nita H. Shah · Mandeep Mittal
Editors

# Mathematical Analysis for Transmission of COVID-19

Springer

*Editors*
Nita H. Shah
Satellite Center
Ahmedabad, Gujarat, India

Mandeep Mittal
Department of Mathematics
Amity Institute of Applied Sciences
Noida, Uttar Pradesh, India

ISSN 2192-4732 ISSN 2192-4740 (electronic)
Mathematical Engineering
ISBN 978-981-33-6263-5 ISBN 978-981-33-6264-2 (eBook)
https://doi.org/10.1007/978-981-33-6264-2



This Springer imprint is published by the registered company Springer Nature Singapore Pte Ltd.
The registered company address is: 152 Beach Road, #21-01/04 Gateway East, Singapore 189721, Singapore

Nita H. Shah · Mandeep Mittal
Editors

# Mathematical Analysis for Transmission of COVID-19

Springer

*Editors*
Nita H. Shah
Satellite Center
Ahmedabad, Gujarat, India

Mandeep Mittal
Department of Mathematics
Amity Institute of Applied Sciences
Noida, Uttar Pradesh, India

ISSN 2192-4732 ISSN 2192-4740 (electronic)
Mathematical Engineering
ISBN 978-981-33-6263-5 ISBN 978-981-33-6264-2 (eBook)
https://doi.org/10.1007/978-981-33-6264-2



This Springer imprint is published by the registered company Springer Nature Singapore Pte Ltd.
The registered company address is: 152 Beach Road, #21-01/04 Gateway East, Singapore 189721, Singapore

# Contents

# About the Editors

**Prof. Nita H. Shah** received her Ph.D. in Statistics from Gujarat University in 1994. From February 1990 till now, Prof. Nita is HOD of the Department of Mathematics in Gujarat University, India. She is Postdoctoral Visiting Research Fellow of the University of New Brunswick, Canada. Prof. Nita's research interests include inventory modeling in supply chain, robotic modeling, mathematical modeling of infectious diseases, image processing, dynamical systems and its applications. Prof. Nita has completed 3 UGC sponsored projects. Prof. Nita has published 13 monographs, 5 textbooks and 475+ peer-reviewed research papers. Five edited books are prepared for IGI Global and Springer with Co-editor as Dr. Mandeep Mittal. Her papers are published in high impact Elsevier, Inderscience and Taylor and Francis journals. By the Google Scholar, the total number of citations is over 3278 and the maximum number of citations for a single paper is over 176. The H-index is 25 up to March 12, 2021 and i 10 index is 84. She has guided 28 Ph.D. students and 15 M.Phil. students till now. Eight students are pursuing research for their Ph.D. degree and one Postdoctoral Fellow of D. S. Kothari, UGC. She has travelled in USA, Singapore, Canada, South Africa, Malaysia and Indonesia for giving talks. She is Vice-President of Operational Research Society of India. She is a council member of Indian Mathematical Society.

**Dr. Mandeep Mittal** started his career in the education industry in 2000 with Amity Group. Currently, he is working as Head and Associate Professor in the Department of Mathematics, Amity Institute of Applied Sciences, Amity University Uttar Pradesh, Noida. He earned his post doctorate from Hanyang University, South Korea, 2016, Ph.D. (2012) from the University of Delhi, India, and postgraduation in Applied Mathematics from IIT Roorkee, India (2000). He has published more than 70 research papers in the International Journals and International conferences. He authored one book with Narosa Publication on C language and edited five research books with IGI Global and Springer. He is a series editor of Inventory Optimization, Springer Singapore Pvt. Ltd. He has been awarded Best Faculty Award by the Amity School of Engineering and Technology, New Delhi for the year 2016–2017. He guided three Ph.D. scholars, and 4 students working with him in the area Inventory Control and Management. He also served as Dean of Students Activities at Amity School of

Engineering and Technology, Delhi, for nine years, and worded as Head, Department of Mathematics in the same institute for one year. He is a member of editorial boards of *Revista Investigacion Operacional*, *Journal of Control and Systems Engineering* and *Journal of Advances in Management Sciences and Information Systems*. He actively participated as a core member of organizing committees in the International conferences in India and outside India.

# Chapter 1
# Introduction to Compartmental Models in Epidemiology

**Nita H. Shah and Mandeep Mittal**

**Abstract** In this chapter, we discuss the basics of compartmental models in epidemiology and requisite analysis.

**Keywords** Mathematical model · Dynamics · Reproduction number · Equilibrium points · Stability

## Introduction

The transmission of infections is broadly classified as vector-borne, waterborne or airborne diseases in epidemiology science. The transmission of infections can be epidemic which is a sudden outbreak of a disease, e.g. COVID-19, or endemic in which disease remains in the society, e.g. malaria. Epidemics like the 2002 outbreak of SARS, the Ebola virus, Zika virus and avian flu attracted the research community to study transmission of such diseases. The outbreak of Spanish flu caused huge human life loss. An endemic situation in epidemiology is one in which disease is always prevalent.

Our objective is to discuss mathematical epidemiology, with the formulation of mathematical models for the spread of disease and criteria for their analysis. Mathematical models in epidemiology help to understand the underlying mechanism that stimulates the spread of disease progression which can be used to develop strategies to curtail the transmission of disease. The degree of heterogeneity, usually known as "threshold" or "reproduction number", helps to understand the behaviour of transmission of disease. The term "threshold" or "reproduction number" in epidemiological terms can be defined as follows: if the average number of secondary infections caused by an average infective, called the basic reproduction number, is less than one, then a

N. H. Shah (✉)
Department of Mathematics, Gujarat University, Ahmedabad, Gujarat 380009, India
e-mail: nitahshah@gmail.com

M. Mittal
Department of Mathematics, AIAS, Amity University Uttar Pradesh, Noida, Delhi, India
e-mail: mittal_mandeep@yahoo.com

 
N. H. Shah and M. Mittal (eds.), *Mathematical Analysis for Transmission of COVID-19*, Mathematical Engineering, https://doi.org/10.1007/978-981-33-6264-2_1

disease will die out, while if it exceeds one then disease will be an epidemic, and if it exceeds one and spreads in major part of the globe then it is pandemic. This quantitative quantity can be used to estimate the effectiveness of vaccination programmes, preventive measures and likelihood that a disease may be eliminated or eradicated.

In the mathematical modelling of disease transmission, there is a trade-off between simple models, which neglects most details and is designed to focus on general qualitative behaviour and detailed models usually formulated for specific situations including short-term quantitative predictions. Sometimes, detailed models are difficult to analyse analytically, and hence their utility for theoretical purpose is limited, although their implicative value may be high.

Let us discuss mathematical models to study endemic states. We will also describe age-dependent infectivity. That means, we will try to find explicit solutions of the system of differential equations to visualize the models. We will discuss asymptomatic behaviour of the models, i.e. solutions of the models as $t \to \infty$.

## Simple Epidemic Models

An epidemic can be treated as a sudden outbreak of a disease that infects a substantial fraction of the population in the region.

In compartmental models, the population under consideration is divided into compartments with assumptions about the nature and time rate of motion from one compartment to another. The independent variable in compartmental model is the time $t$, and the rates of transfer between compartments are expressed mathematically as derivatives with respect to time of the sizes of the compartments. Our models are formulated as system of differential equations.

Some basic terminologies required for model constitution are:

- **Susceptible**: An individual who is at a risk of catching infection though still not infected.
- **Exposed**: An individual person who is infected but not capable to transmit infection.
- **Infectious**: An infected individual capable of transmitting infection.
- **Recovered**: After treatment or due to immunity development, infection heals up and the individual does not remain infectious.
- **Epidemic**: Disease is said to be in epidemic stage if the number of cases rapidly increases within a short duration of time for said population.
- **Endemic:** Disease is said to be endemic if it persist for long time among population of particular area.
- **Incubation Period**: The duration between an individual receives an infection and having first apparent symptom of the disease.

The system of nonlinear ordinary differential equation has the form

$$\frac{\mathrm{d}\overline{X}(t)}{\mathrm{d}t} = f\left(\overline{X}\right) \tag{1.1}$$

where $\overline{X}(t) \in \mathbb{R}^n$ is a model compartment and time-derivative, which is also written as $\dot{\overline{X}}$ or $\overline{X}'$. Equation (1.1) is a dynamical system with finite-dimensional state $\overline{X}(t)$ of $n$-dimension. In dynamical models, the problem is divided into compartments with characteristics that variables in each compartment have homogeneous characteristics. Therefore, these models are also known as compartmental models.

In compartmental model, we write

$$\begin{aligned}&\overline{X} = (X_1, X_2, \ldots, X_n)\\ \Rightarrow\ & f\left(\overline{X}\right) = (f_1(X_1, X_2, \ldots, X_n), f_2(X_1, X_2, \ldots, X_n), \ldots, f_n(X_1, X_2, \ldots, X_n))\end{aligned}$$

Using the definition of $f\left(\overline{X}\right)$, the system (1.1) can be replaced by

$$\begin{aligned}\frac{\mathrm{d}X_1(t)}{\mathrm{d}t} &= f_1(X_1, X_2, \ldots, X_n)\\ \frac{\mathrm{d}X_2(t)}{\mathrm{d}t} &= f_2(X_1, X_2, \ldots, X_n)\\ &\vdots\\ \frac{\mathrm{d}X_n(t)}{\mathrm{d}t} &= f_n(X_1, X_2, \ldots, X_n)\end{aligned} \tag{1.2}$$

where $f_1, f_2, \ldots, f_n$ represents the rate of change of compartment with respect to time. Solving this system, realistic solution can be found but at the other end, it is more complex to solve.

Capital letters such as $S(t)$, $E(t)$, $I(t)$, $R(t)$ are state variables used to represent various compartments.

| | |
|---|---|
| $S(t)$ | Number of susceptible individuals at time $t$. |
| $E(t)$ | Number of exposed individuals at time $t$. |
| $I(t)$ | Number of infectious individuals at time $t$. |
| $R(t)$ | Number of recovered individuals at time $t$. |
| $N(t)$ | Total human population at time $t$. |
| $B$ | Recruitment rate. |
| $\beta$ | Transmission rate of infection from an infectious individual to a susceptible one. |
| $\mu$ | Natural death rate. |
| $\alpha$ | Disease-induced death rate. |
| $\sigma$ | Rate of progression from exposed class to infectious class. |
| $k$ | Recovery rate. |
| $\theta$ | Rate of progression from recovered class to susceptible class again. |

Starting with the simplest compartmental model consisting of only two classes,

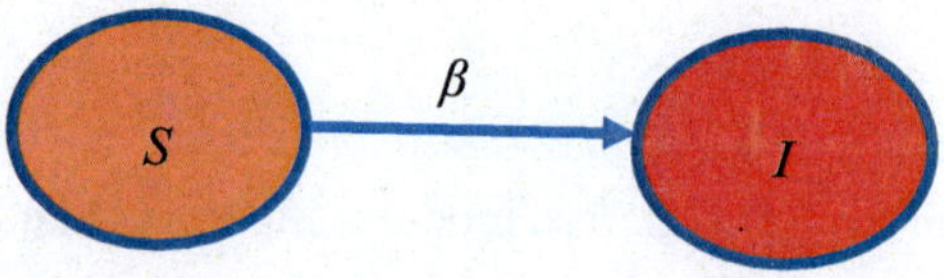


**Fig. 1.1** Transmission in *SI* model

**SI Model**: This model is formulated by dividing total population into two compartments, namely susceptible and infected. An individual from susceptible compartment comes in contact with an infected individual from infectious compartment and joins infectious compartment. It is assumed that an individual entering in infectious compartment will remain in infectious compartment forever. Here, natural birth and natural deaths are not taken into consideration.

Figure 1.1 can be mathematically written as system of differential equations as:

$$\begin{aligned} S' &= -\beta S I \\ I' &= \beta S I - \alpha I \end{aligned}$$

**SIR Model**: In 1927, Kermack and McKendrick had suggested a mathematical model with three compartments, namely susceptible, infected and removed. This model was introduced with natural birth and natural death. SIR model is an extension of SI model with one more added compartment recovered. Infectious individual after getting recovery through natural immunity or by any other sources joins recovered class and remains there forever. Measles, mumps, flu and rubella are few examples of infectious diseases which follows SIR dynamics.

Figure 1.2 can be mathematically written as system of differential equations as:

$$\begin{aligned} S' &= -\beta S I \\ I' &= \beta S I - \alpha I \\ R' &= \alpha I \end{aligned} \tag{1.3}$$

The model is developed based on the following assumptions:

1. An average member of the population makes contact which is sufficient to transmit infection with $\beta N$ others per unit time, i.e. mass action incidence.
2. Infectives leave the infective class at rate $\alpha I$ per unit time.
3. There is no entry into or exit from the population.

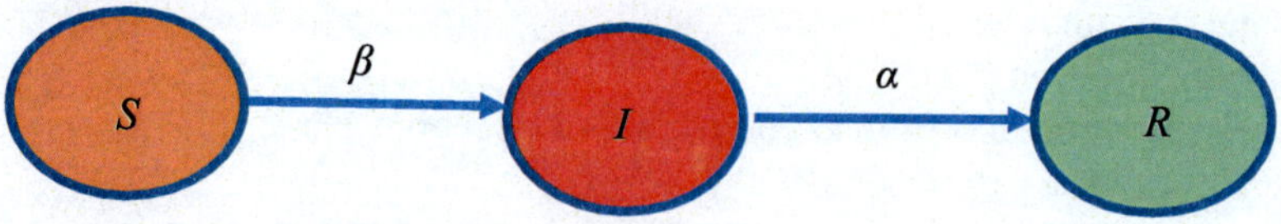


**Fig. 1.2** Transmission in *SIR* model

From (1), the probability that a contact by an infective with susceptible, who can transmit an infection, is $S/N$, the number of new infections in unit time per infective is $(\beta N)(S/N)$, and chance that a rate of new infections is $(\beta N)(S/N)I = \beta SI$. (3) suggests that the timescale of a disease is much faster than the timescale of births and deaths resulting in ignorance of demographic effects on the population. This means we are only interested in analysing the dynamics of a single epidemic outbreak. The assumption (2) can be understood as follows:

Let us consider "cohort" of members who were all infected at one time. Let us denote it by $u(t)$. If a fraction $\alpha$ of this leaves the infective class in unit time, then

$$u' = -\alpha u$$

The solution of this differential equation is

$$u(t) = u(0)e^{-\alpha t}$$

Thus, the fraction of infectives are still infective at $t$ time units after catching infection is $e^{-\alpha t}$. This suggests that the length of the infective period is distributed exponentially with mean $\int_0^{\infty} e^{-\alpha t} dt = 1/\alpha$ which justifies assumption (2).

We can calculate $R$ once $S$ and $I$ are known. Then, Eq. (1.3) reduces to

$$\begin{aligned} S' &= -\beta SI \\ I' &= (\beta S - \alpha)I \end{aligned} \tag{1.4}$$

The nonlinearity hinders to get the closed form of the solution. WLOG, we assume that $S(t)$ and $I(t)$ are non-negative. (Note: If either $S(t)$ or $I(t)$ reaches zero, then system will terminate.) One can see that $S' < 0$ for all $t$ and $I' > 0$ if and only if $S > \alpha/\beta$. That is, $I$ increases till $S > \alpha/\beta$, but since $S$ decreases for all $t$, $I$ eventually decreases to zero (no epidemic), while if $S(0) > \alpha/\beta$, $I$ attains maximum at $S = \alpha/\beta$ and then depletes to zero (epidemic). The quantity $\beta S(0)/\alpha$ is a threshold also known as basic reproduction number and is denoted by $R_0$. It determines whether the disease is an epidemic or not. If $R_0 < 1$, the infection will die out, while if $R_0 > 1$, disease is epidemic. $R_0$ is that threshold number of secondary infections caused by a single infective entry into a wholly susceptible population of size $K \approx S(0)$ over the duration of the infection of this single infective. It means that an infective makes $\beta K$ contacts in unit time, all of which are susceptible, and it produces new infectives with mean infective period $1/\alpha$. Thus, the basic reproduction number is $\beta K/\alpha$ instead of $\beta S(0)/\alpha$.

We divide equations given in (1.4), which gives

$$\frac{I'}{S'} = \frac{dI}{dS} = -1_{-}\frac{\alpha}{\beta S}$$

Integrating this w.r.t $S$ gives

$$I = -S + \frac{\alpha}{\beta}\log S + c \tag{1.5}$$

This can be used to find trajectories in the $(S, I)$-plane. We can write

$$V(S, I) = S + I - \frac{\alpha}{\beta}\log S$$

Then, each trajectory is a curve determined by $V(S, I) = c$ for some constant $c$. This constant $c$ is determined by the initial values $S(0)$, $I(0)$ of $S$ and $I$.

$$c = V(S(0), I(0)) = S(0) + I(0) - \frac{\alpha}{\beta}\log(S(0))$$

Earlier, we had seen that the maximum value of $I$ on each of the trajectories is attained when $S = \alpha/\beta$. Note that these trajectories will never touch $I$-axis, $S > 0$ for all time. In particular, $S_\infty = \lim\limits_{t\to\infty} S(t) > 0$ suggests that fraction of population does not get infection.

- **SIRS Model**: It is further extension of SIR model adding one more situation that a recovered individual loses temporary immunity and rejoins the susceptible class.
- **SEIR Model**: Many time an infection takes a time gap (may vary from disease to disease) to make a susceptible individual infectious after catching infection from an infected individual, called incubation period. During this time period, susceptible individuals who acquired infection but still not become infectious join exposed compartment. SIR model with one more compartment added as exposed compartment suggests the dynamics of *SEIR* model.

Very basic models of mathematical epidemiology are described above. We are going to formulate more complicated models by adding more compartments and variables or parameters associated with quarantine, isolation, treatment, mass media effect and many more in this book with view of current pandemic COVID-19.

Next, we will take a closer look at characteristics of the model.

## What Is Equilibrium Point and Its Existence?

An equilibrium point of a dynamical system is its fixed point which means solution that does not change with time. Equilibrium points occur when each equation of dynamical system is set to be zero, i.e. $\overline{X}_i' = f_i(X) = 0$. Hence, these equilibrium points can be said to be the root of $f_i(X)$. One can also recall that equilibrium points are constant functions which satisfy the dynamical system which means they are time-independent solutions of the model. There are at least two natures of equilibrium points:

1. Problem-free equilibrium point

2. Optimal valued equilibrium point or endemic equilibrium point or interior equilibrium point.

The equilibrium point $X$ exists only when the stability properties are satisfied, i.e. $X > 0$.

## How to Compute Threshold or Basic Reproduction Number?

The threshold is analogous to the most significant concept of the epidemiology world, and this quantity is termed as basic reproduction number and noted as $R_0$. Basic reproduction number measures the maximum reproductive latent for an issue. It aids us to predict the future for an issue that whether it will remain present in atmosphere or die out soon. The concept of basic reproduction number was firstly introduced by Lotka [15] for the disease spread of Malaria. Then after, classical methods were suggested by Dublin and Lotka [7] and Kuczynski [11]. In the current scenario, the method used for calculating a basic reproduction number is through the next-generation matrix method. This was suggested by Diekmann et al. [6], and then this method came into limelight by Van den Driessche and Watmough [20]. Recently, Diekmann et al. [5] have edited this method using simplification of the Jacobian method.

***The next-generation matrix method***

Here, a dynamical system is to be considered which is having $n$-compartments at time $t$. It is a set of $n$ nonlinear ordinary differential equations. The set is further separated into nonlinear and linear systems where nonlinear system $F(\overline{X})$ signifies the rate of affected components and linear system $V(\overline{X})$ represents the transfer rate of components going and coming through the respective compartment. Mathematically, the model representation is

$$\frac{d\overline{X}}{dt} = F(\overline{X}) - V(\overline{X}) \tag{1.6}$$

Now, find the Jacobian matrices of $F(\overline{X})$ and $V(\overline{X})$ about the equilibrium point, say issue-free equilibrium point $E_0$ and then define the next-generation matrix $fv^{-1}$ where $f = [\frac{\partial F(E_0)}{\partial X_j}]_{n\times n}$ and $v = [\frac{\partial V(E_0)}{\partial X_j}]_{n\times n}$. The spectral radius or the largest eigenvalue of the matrix $fv^{-1}$ is the basic reproduction number (threshold) of the model. If the value of threshold is less than 1, then the problem is in controllable state; otherwise, the issue will take epidemic stage in short period of time.

## How to Characterize Nature of Stability?

The fundamental concept for a dynamical system is "stability". An equilibrium point $X$ of a system is said to be stable if for each $\varepsilon > 0$ and $t_0 \in \mathbb{R}$ there exists $\delta = \delta(\varepsilon, t_0) > 0$ such that each solution $x(t)$ having initial conditions within the distance $\delta$, i.e. $\|x(t_0) - X\| < \delta$, then equilibrium point remains within the distance $\varepsilon$, i.e. $\|x(t) - X\| < \varepsilon$ for all $t \geq t_0$. In this thesis, we are focusing on a stronger condition than stability called asymptotic stability. An equilibrium point is said to be asymptotically stable if it is stable and for every $t_0 \in \mathbb{R}$ there exists $\delta_0 = \delta_0(t_0) > 0$ such that whenever $\|x(t_0) - X\| < \delta_0$ then $x(t) \to X$ as $t \to \infty$. We will study two types of stability: (1) local asymptotic stability and (2) global asymptotic stability.

1. ***Local asymptotic stability***

The local asymptotic stability of a model at an equilibrium point $X$ is that the solution of the system must approach an equilibrium point under initial condition close to the equilibrium point, i.e. at $X$ if there is a $\delta > 0$ such that $\|x(t) - X\| < \delta \Rightarrow (x(t)) \to X$ as $t \to \infty$. The local asymptotic stability is established using Jacobian matrix of the system. If all the eigenvalues have negative real parts, then the system is locally asymptotically stable about the equilibrium point [10]. Routh–Hurwitz criteria are also well used for finding the local asymptotic stability [17]. Moreover, this stability about the interior equilibrium point can be done using curl [2–4], second-order additive compound matrix [1, 16].

2. ***Global asymptotic stability***

The global asymptotic stability of a model at an equilibrium point $X$ is the solution of the system must approach to the equilibrium point under all initial conditions; i.e. for every $x(t)$, we have $x(t) \to X$ as $t \to \infty$. Lyapunov function [12] helps to establish that the model is globally asymptotically stable. In addition, graph theory [9] and theory of geometric method [13, 14, 19] deliberate the global asymptotic stability of the model.

## What Is Bifurcation and What It Reflects?

Bifurcation is one of the important concepts for a dynamical system because it helps us to framework the qualitative change of the system. A bifurcation occurs when a small and smooth variation is made to the value of model parameter which leads to a rapid qualitative change in its nature. In other words, it reflects the qualitative change to the state of the system as a model parameter is changed. Hence, we can see the significant impact on the solution while using bifurcation analysis. The parameter which set to be varied is known as bifurcation parameter, and the curve which reflects the qualitative nature of the system is called bifurcation curve.

The dynamical system includes different bifurcations:

1. Saddle-node bifurcation
2. Flip bifurcation
3. Forward bifurcation
4. Backward bifurcation
5. Hopf bifurcation
6. Neimark–Sacker bifurcation
7. Transcritical bifurcation
8. Pitchfork bifurcation.

Backward bifurcation occurs when a branch of the system varies its stability at a point and turns into unstable state, whereas at the same time, a new branch of positive endemic steady sate comes up to coincide with the initial stable state. The point from which state is changing from stable to unstable is called bifurcation point or critical threshold noted as $R_C$. In general, for a dynamical system, when a parameter is permissible to fluctuate within the range then the system may vary. It can result that an equilibrium can attain epidemic state and a periodic solution may happen or a unique stable equilibrium may grow up making earlier equilibrium unstable. The backward bifurcation has a characteristic: it exists only when the threshold of model is less than 1. Moreover, the critical threshold should be less than original threshold, i.e. extended characteristic $R_C < R_0 < 1$ which ensures the expulsion of problem in nearer future. In addition, bifurcation analysis reflects the global stability of a model.

## How to Optimize the Issue?

Optimizing an issue is a fundamental tool in the world of dynamical modelling. Optimization can improve the results. Optimization of dynamical modelling is done through control theory. Control theory suggests that applying sufficient control at appropriate time will try to shatter the problem in sooner time. To achieve this objective, the objective function for the model along with the controls is designed as

$$J(u_i, \Omega) = \int_0^T \left(A_1X_1^2 + A_2X_2^2 + \cdots + A_nX_n^2 + w_1u_1^2 + w_2u_2^2 + \cdots + w_mu_m^2\right)\mathrm{d}t \tag{1.7}$$

where $\Omega$ is the set consisting of all compartmental variables; $A_1, A_2, \ldots, A_n$ (non-negative weight constants for $X_1, X_2, \ldots, X_n$, respectively); and $\omega_1, \omega_2, \ldots, \omega_m$ (non-negative weight constants for $u_1, u_2, \ldots, u_m$, respectively).

The weight constants $\omega_1, \omega_2, \ldots, \omega_m$ normalized the optimal control condition. Now, measure the values of control variables from $t = 0$ and $t = T$ such that

$$J(u_1(t), u_2(t), \ldots, u_m(t)) = \min\left\{J\left(u_i^*, \Omega\right) / (u_1, u_2, \ldots u_m) \in \phi\right\}$$

where $u_i^*; i = 1, 2, \ldots m$ is optimal control, i.e. $\phi = \{u_i(t)/a_i < u_i < b_i; i = 1, 2, \ldots m\}$ is control set with $a_i$ as a lower bound and $b_i$ as a upper bound. Here, $\phi$ is a smooth function on the interval [0, 1]. Before finding optimal controls, some hypothesis should hold:

1. The set of control ($\phi$) and state variable ($\Omega$) are non-empty.
2. The control set $\phi$ is closed and convex.
3. The system is bounded by the linear function made of state and control variable including time-dependent coefficients.
4. The integrand of the objective function is convex on $\phi$ and bounded below by $c_1(|u_1|^2 + |u_2|^2 + \cdots + |u_m|^2)^{\frac{\alpha}{2}} - c_2$ where $c_1$ and $c_1$ are positive constants and $\alpha > 1$.

Now, let us compute the optimal controls $u_i^*, i = 1, 2, \ldots m$ by accumulating all the integrands of objective function (1.7) using the lower and upper bounds described in the results of Fleming and Rishel [8].

Further, consider Pontryagin's principle from Boltyanki et al. Next, construct a Lagrangian function consisting of state equations and adjoint variables $A_i = (\lambda_1, \lambda_2, \ldots, \lambda_n)$ to minimize the cost function given in objective function (1.7). The Lagrangian function is expressed as

$$\begin{aligned} L(\Omega, A_i) = {} & A_1 X_1^2 + A_2 X_2^2 + \cdots + A_n X_n^2 + w_1 u_1^2 + w_2 u_2^2 + \cdots + w_m u_m^2 \\ & + \lambda_1 (X_1') + \lambda_2 (X_2') + \cdots + \lambda_n (X_n') \end{aligned} \tag{1.8}$$

The partial derivative of the Lagrangian function with respect to each compartment gives the adjoint equation corresponding the system, which is followed by

$$\left.\begin{aligned} \dot{\lambda}_1 &= -\frac{\partial L}{\partial X_1} \\ \dot{\lambda}_2 &= -\frac{\partial L}{\partial X_2} \\ &\vdots \\ \dot{\lambda}_n &= -\frac{\partial L}{\partial X_n} \end{aligned}\right\} \tag{1.9}$$

Now, the partial derivative of the Lagrangian function with respect to each control is

$$\left.\begin{aligned} \dot{u}_1 &= -\frac{\partial L}{\partial u_1} \\ \dot{u}_2 &= -\frac{\partial L}{\partial u_2} \\ &\vdots \\ \dot{u}_m &= -\frac{\partial L}{\partial u_m} \end{aligned}\right\} \tag{1.10}$$

Now, to achieve the necessary condition, set each equation of (1.10) equal to zero, we have the values of $u_1, u_2, \ldots, u_m$. Hence, the required optimal control condition is calculated as

$$\left.\begin{aligned} u_1^* &= \max(a_1, \min(b, u_1)) \\ u_2^* &= \max(a_2, \min(b, u_2)) \\ &\vdots \\ u_m^* &= \max(a_m, \min(b, u_m)) \end{aligned}\right\} \tag{1.11}$$

## Inside This Book

### Chapter 2: Modelling the Impact of Nationwide BCG Vaccine Recommendations on COVID-19 Transmission, Severity and Mortality

Coronavirus disease 2019 (COVID-19) is declared as pandemic on 11 March 2020 by World Health Organization (WHO). There are apparent dissimilarities in incidence and mortality of COVID-19 cases in different parts of world. Developing countries in Asia and Africa with fragile health system are showing lower incidence and mortality compared to developed countries with superior health system in Europe and America. Most countries in Asia and Africa have national Bacillus Calmette-Guerin (BCG) vaccination programme, while Europe and America do not have such programme or have ceased it. At present, there is no known Food and Drug Administration (FDA)-approved treatment available for COVID-19. There is no vaccine available currently to prevent COVID-19. As mathematical modelling is ideal for predicting the rate of disease transmission as well as evaluating efficacy of possible public health prevention measures, we have created a mathematical model with seven compartments to understand nationwide BCG vaccine recommendation on COVID-19 transmission, severity and mortality. We have computed two basic reproduction numbers, one at vaccine-free equilibrium point and other at non vaccine-free equilibrium point, and carried out local stability, sensitivity and numerical analysis. Our result showed that individuals with BCG vaccinations have lower risk of getting COVID-19 infection, shorter hospital stays and increased rate of recovery. Furthermore, countries with long-standing universal BCG vaccination policies have reduced incidence, mortality and severity of COVID-19. Further research will focus on exploring the immediate benefits of vaccination to healthcare workers and patients as well as benefits of BCG re-vaccination.

### Chapter 3: Modeling the Spread of COVID-19 Among Doctors from the Asymptomatic Individuals

The present world is in dire straits due to the deadly SARS CoV-2 (coronavirus-2) outbreak, and the experts are trying heart and soul to discover any prevention and/or remedy. The people from all walks of life in the universe are fighting to defeat this novel coronavirus. In this case, doctors are in the front-line fighters who have put themselves at a risk. In this paper, we have formulated a nonlinear mathematical model of COVID-19 based on the tendency of doctors to be infected. The target of this study is to take a look at the transmission of COVID-19 from asymptomatic populations to the doctors. The model is analysed with the determination of the basic reproductive ratio and related stability analysis at the disease-free and endemic equilibrium points. The graph of the basic reproductive ratio for different parameters has been drawn to show the disease behaviour. Finally, numerical simulations have been performed to illustrate the analytic results. Our study shows that the asymptomatic population increases as the disease (COVID-19) transmission rate increases and also the number of infected population increases when the infection rate increases. These increasing asymptomatic and infected populations lead the doctors to get infected by contacting with them. Thus, the whole medical service system is getting down over time.

## Chapter 4: Transmission Dynamics of Covid-19 from Environment with Red Zone, Orange Zone, Green Zone Using Mathematical Modelling

The novel coronavirus or COVID-19 spread had its inception in November of 2019, and in March 2020 it was declared as a pandemic. Since its initial stage, it has now already infected over 5 million people, leading to the lockdown of countries around the world and a halt on global as well as national travel across the globe. Based on this, the research proposes a mathematical COVID-19 model to study the outcome of these classified zones under different control strategies. In the nonlinear mathematical model, the total population has been divided into seven compartments, namely susceptible, exposed, red zone, orange zone, green zone, hospitalized and recovered. The spectral radius is calculated to analyse dynamics of the COVID-19. To control the spread of the virus, the parameters of controls are medical intervention, partial lockdown and strict lockdown. This model has been validated with numerical data. The conclusion validates the implementation of lockdown in curbing COVID-19 cases.

## Chapter 5: A Comparative Study of COVID-19 Pandemic in Rajasthan, India

The treatment of coronavirus diseases is not possible without any vaccine. However, spreading of the deadly virus can be controlled by various measures being imposed by government like lockdown, quarantine, isolation, contact tracing, social distancing and putting face mask on mandatory basis. As per information from the Department of Medical Health and Family Welfare of Rajasthan on 19 September 2020, COVID-19 severely affected the state of Rajasthan, resulting in cumulative positive cases 113,124, cumulative recovered 93,805 and cumulative deaths 1322. Without any appropriate treatment, it may further spread globally as it is highly communicable and because potentially affecting the human body respiratory system, which could be fatal to mankind. Therefore, to reduce the spread of infection, authors are motivated to construct a predictive mathematical model with sustainable conditions as per the ongoing scenario in the state of Rajasthan. Mathematica software has been used for numerical evaluation and graphical representation for variation of infection, recovery, exposed, susceptible and mortality versus time. Moreover, comparative analysis of results obtained by predictive mathematical model has been done with the exact data plotting by curve fitting as obtained from Rajasthan government website. As a part of analysis and result, it is noted that due to the variation of transmission rate from person to person corresponding rate of infection goes on increasing monthly and mortality rate found high as shown and discussed numerically. Further, we can predict that the situation will become worse in the winter months, especially in the month of December due to unavailability of proper vaccine. This model may become more efficient when the researchers, experts from medical sciences and technologist work together.

### Chapter 6: A Mathematical Model for COVID-19 in Italy with Possible Control Strategies

Italy faced the COVID-19 crisis in the early stages of the pandemic. In the present study, a SEIR compartment mathematical model has been proposed. The model considers four stages of infection: susceptible (S), exposed (E), infected (I) and recovered (R). Basic reproduction number R0 which estimates the transmission potential of a disease has been calculated by the next-generation matrix technique. We have estimated the model parameters using real data for the coronavirus transmission. To get a dipper insight into the transmission dynamics, we have also studied four of the most pandemic affected regions of Italy. Basic reproduction number stood differently for different regions of Italy, i.e. Lombardia (2.1382), Veneto (1.7512), Emilia Romagna (1.6331), Piemonte (1.9099) and for Italy at 2.0683. The sensitivity of $R_0$ corresponding to various disease transmission parameters has also been demonstrated via numerical simulations. Besides, it has been demonstrated with the help of simulations that earlier lockdown and rapid isolation of infective individuals would have been helpful in a dual way, by substantially decreasing transmission of COVID-19.

### Chapter 7: Effective Lockdown and Plasma Therapy for COVID-19

COVID-19 is a major pandemic threat of 2019–2020 which originated in Wuhan. As of now, no specific antiviral medication is available. Therefore, many countries in the world are fighting to control the spread by various means. In this chapter, we model COVID-19 scenario by considering compartmental model. The set of dynamical system of nonlinear differential equation is formulated. Basic reproduction number $R_0$ is computed for this dynamical system. Endemic equilibrium point is calculated, and local stability for this point is established using Routh–Hurwitz criterion. COVID-19 has affected more than 180 countries in several ways like medically, economy, etc. It necessitates the effect of control strategies applied by various governments worldwide to be analysed. For this, we introduce different types of time-dependent controls (which are government rules or social, medical interventions) in order to control the exposure of COVID-19 and to increase recovery rate of the disease. By using Pontryagin's maximum principle, we derive necessary optimal conditions which depicts the importance of these controls applied by the government during this epidemic.

### Chapter 8: Controlling the Transmission of COVID-19 Infection in Indian Districts: A Compartmental Modelling Approach

The widespread of the novel coronavirus (2019-nCoV) has adversely affected the world and is treated as a Public Health Emergency of International Concern by the World Health Organization. Assessment of the basic reproduction number with the help of mathematical modelling can evaluate the dynamics of virus spread and facilitate critical information for effective medical interventions. In India, the disease control strategies and interventions have been applied at the district level by categorizing the districts as per the infected cases. In this study, an attempt has been made to estimate the basic reproduction number $R_0$ based on publicly available data at the district level in India. The susceptible-exposed-infected-critically infected-hospitalization-recovered (SEICHR) compartmental model is constructed to understand the COVID-19 transmission among different districts. The model relies on the twelve kinematic parameters fitted on the data for the outbreak in India up to 15 May 2020. The expression of basic reproduction number $R_0$ using the next generating matrix is derived and estimated. The study also employs three time-dependent control strategies to control and minimize the infection transmission from one district to another. The results suggest an unstable situation of the pandemic that can be minimized with the suggested control strategies.

### Chapter 9: Fractional SEIR Model for Modelling the Spread of COVID-19 in Namibia

In this chapter, a fractional SEIR model and its robust first-order unconditionally convergent numerical method are proposed. Initial conditions based on Namibian data as of 4 July 2020 were used to simulate two scenarios. In the first scenario, it is assumed that proper control mechanisms for curbing the spread of COVID-19 are in place. In the second scenario, a worst-case scenario is presented. The worst case is characterized by ineffective COVID-19 control mechanisms. Results indicate that, if proper control mechanisms are followed, Namibia can contain the spread of COVID-19 in the country to a lowest level of 1, 800 positive cases by October 2020. However, if no proper control mechanisms are followed, Namibia can hit a potentially unmanageable level of over 14, 000 positive cases by October 2020. From a mathematical point of view, results indicate that the fractional SEIR model and the proposed method are appropriate for modelling the chaotic nature observed in the spread of COVID-19. Results herein are fundamentally important to policy- and decision-makers in devising appropriate control and management strategies for curbing further spread of COVID-19 in Namibia.

### Chapter 10: Impact of COVID-19 in India and Its Metro Cities: A Statistical Approach

The infectious coronavirus disease is spreading at an alarming rate, not only in India but globally too. The impact of coronavirus disease 219 (COVID-19) outbreak needs to be analysed statistically and modelled to know its behaviour so as to predict the same for future. An exhaustive statistical analysis of the data available for the spread of this infection specifically on the number of positive cases, active cases, death cases and recovered cases and connection between them could probably suggest some key factors. This has been achieved in this paper by analysing these four dominant cases. This helped to know the relationship between the current and the past cases. Hence, in this paper an approach of statistical analysis of COVID-19 data specific to metropolitan cities of India is done. A regression model has been developed for prediction of active cases with 10 lag days in four metropolitan cities of India. The data used for developing the model is considered from 26 April to 31 July (97 days), tested for the month of August. Further, an artificial neural network (ANN) model using back-propagation algorithm for active cases for all India and Bangalore has been developed to see the comparison between the two models. This is different from the other existing ANN models as it uses the lag relationships to predict the future scenario. In this case, data is divided into training, validation and testing sets. Model is developed on the training sets, is checked on the validation set, is tested on the remaining and is implemented for prediction.

**Chapter 11: A Fractional-Order SEQAIR Model to Control the Transmission of nCOVID 19**

The ensuing paper expounds a new mathematical model for a pandemic instigated by novel coronavirus (COVID-19) with influence of quarantine on transmission of COVID-19, using Caputo fractional-order derivative for various fractional orders. Basic reproduction number for the SEQAIR model has been calculated in the study, additionally, proving the existence and uniqueness of the solution using the fixed-point theorem. Furthermore, numerical solution is revealed using the Adams–Bashforth–Moulton method, and its application for real-world data is deliberated.

**Chapter 12: Analysis of Novel Corona Virus (COVID-19) Pandemic with Fractional-Order Caputo–Fabrizio Operator and Impact of Vaccination**

In a very short time period, the coronavirus disease 2019 (COVID-19) has created a global emergency situation by spreading worldwide. This virus has dissimilar effects in different geographical regions. In the beginning of the spread, the number of new cases of active coronavirus has shown exponential growth across the globe. At present, for such infection, there is no vaccination or antiviral medicine specific to the recent coronavirus infection.

Mathematical formulation of infection models is exceptionally successful to comprehend epidemiological models of ailments, just as it causes us to take vital proportions of general well-being interruptions to control disease transmission and spread. This work based on a new mathematical model analyses the dynamic behaviour of novel coronavirus (COVID-19) using Caputo–Fabrizio fractional derivative. A new modified SEIRQ compartment model is developed to discuss various dynamics. The COVID-19 transmission is studied by varying reproduction number. The basic reproduction number $R_0$ is determined by applying the next-generation matrix. The equilibrium points for disease-free and endemic states are computed with the help of basic reproduction number $R_0$ and check the stability property. The Picard approximation and Banach fixed-point theorem based on iterative Laplace transform are useful in establishing the existence and stability behaviour of the fractional-order system. Finally, numerical computations of the COVID-19 fractional-order system are presented to analyse the dynamical behaviour of the solutions of the model. Also, a fractional-order SEIRQ COVID-19 model with vaccinated people has also been formulated and its dynamics with impact on the propagation behaviour is studied.

### Chapter 13: Compartmental Modelling Approach for Accessing the Role of Non-pharmaceutical Measures in the Spread of COVID-19

Epidemic diseases are well known to be fatal and cause great loss worldwide—economically, socially and mentally. Even after around nine months, since the coronavirus disease 2019 began to spread, people are getting infected all over the world. This is one of the areas where human medical advancements fail because by the time the disease is identified and its treatment is figured out, most of the population is already exposed to it. In such cases, it becomes easier to take steps if the dynamics of the disease and its sensitivity to various factors is known. This chapter deals with developing a mathematical model for the spread of coronavirus disease, by employing a number of parameters that affect its spread. A compartmental modelling approach using ordinary differential equation has been used to formulate the set of equations that describe the model.

We have used the next-generation method to find the basic reproduction number of the system and proved that the system is locally asymptotically stable at the disease-free equilibrium for $R_0 < 1$. Stability and existence of endemic equilibrium have been discussed, followed by sensitivity of infective classes to parameters like proportion of vaccinated individuals and precautionary measures like social distancing. It is expected that after the vaccine is developed and is available to use, as the proportion of vaccinated individuals will increase, the infection will decrease in the population which can gradually lead to herd immunity. Since the vaccine is still under development, non-intervention measures play a major role in coping with the disease. The disease generally transmits when the water droplets from an infected individual's mouth or nose are inhaled by a healthy individual. The best measures that should be adopted are social distancing, washing one's hands frequently and covering one's mouth with mask, quarantine and lockdowns. Thus, as more and more precautionary measures are taken, it would gradually reduce the infection which has also been proved numerically by the sensitivity analysis of "w" in our dynamical analysis.

### Chapter 14: Impact of 'COVID-19' on Education and Service Sectors

Coronavirus disease is an infectious disease which is caused by a virus called coronavirus. The people who are infected with this disease will experience respiratory illness. This disease has been declared pandemic by "World Health Organization (WHO)". There are many sectors that have been affected due to the lockdown practised in the entire country, among agricultural sector, manufacturing sector, service sector, education sector, business sector, etc. In our research, we examined the impact of COVID-19 in India vis-a-vis different sectors. For the purposes of this research, we shortlisted two particular sectors, i.e. education and service sectors. These two sectors form the backbone of our country. While the impact on education sector has led to many young minds and vulnerable school kids being affected in an adverse way and have left them to cope with new practices such as online classes during the lockdown period and on the other hand, in the service sector, employees are working from home which in some case has had an impact on the effectiveness and efficiency of their work. In this paper, we assessed the impact on these two sectors on Indian economy by analysing the responses given by our respondents through the questionnaires (Google form) and we then combined the data points to study how students and employees are being affected during the present lockdown period, imposed due to COVID-19. This chapter will help the readers to get to know more about the thinking of the students and employees in lockdown and how much they are affected by this pandemic. Towards the trailing part of our research, we have discussed the possible steps that can be adopted in future, by the employers and educational institutions, in order to limit the damage to the sector and to make recovery in future.

**Chapter 15: Global Stability Analysis Through Graph Theory for Smartphone Usage During COVID-19 Pandemic**

During the pandemic due to coronavirus disease 2019 (COVID-19), technology is regarded as a boon as well as a curse to human life which has a great impact on surroundings, people and the society. One of the innovative, however, perilous (if misused) inventions of humans is the smartphone which is becoming more and more alarmingly common yet an urgent question to be addressed. A wide application of smartphone technology is observed during this pandemic. It has both positive and negative impacts on the prominent areas which include education, business, health, social life and furthermore. Moreover, the impact of such an addiction is observed not only among youngsters but has influenced all age groups. This scenario is modelled in this research through nonlinear ordinary differential equations where individuals susceptible to smartphone use will be either positively or negatively infected/addicted, and may suffer from health issues procuring medication. Threshold is calculated using the next-generation matrix method. Stability analysis is done using graph theory, and for the validation of data, numerical simulation is carried out. This study gives results explaining positive and negative issues on health due to excessive use of smartphone.

**Chapter 16: Modelling and Sensitivity Analysis of COVID-19 Under the Influence of Environmental Pollution**

The ongoing COVID-19 pandemic emerged as one of the biggest challenges of recent times. Efforts have been made from different corners of the research community to understand different dimensions of the disease. Some theoretical works have reported that disease becomes severe in the presence of environmental pollution. In this work, we propose a nonlinear mathematical model to study the influence of air pollution on the dynamics of the disease. The basic reproduction number plays a vital role in predicting the future of an epidemic. Therefore, we obtain the expression of the basic reproduction number and performed a detailed sensitivity and uncertainty analysis. The values of partial rank correlation coefficient (PRCC) have been calculated corresponding to six critical parameters. The positive values of PRCC for pollution-related parameters depict that pollution enhances the chances of a rapid spread of COVID-19.

**Chapter 17: Bio-waste Management During COVID-19**

Since December 2019, coronavirus through human to human hit the world. As this disease is spreading every day, hospitalization of individuals increased. Consequence of this, there is a sudden surge of millions of gloves, masks, hand sanitizers and the other essential equipment in each month. Disposal of these commodities is a big challenge for hospitals and COVID centre, as they may became the reason of creating pollution and infect the surroundings. Increasing hospitalization cases of COVID-19 results in raising bio-waste which creates pollution. Observing the scenario, a mathematical model with four compartments is constructed in this article. The threshold value indicates the intensity of pollution that emerged from bio-waste. Stability of the equilibrium point gave the necessary condition. Optimal control theory is outlined to achieve the purpose of this chapter by reducing pollution. Outcomes are analytically proven and also numerically simulated.

**Chapter 18: Mathematical Modelling of COVID-19 in Pregnant Women and Newly Born**

Enlightened by the coronavirus, the present paper deals with a mathematical model of COVID-19 to investigate the impact of S-I-R-M model on the pregnant women and the newly born due to the influence of availability of suitable conditions. The rates of infection, rate of recovery, rate of mortality for pregnant women before and after delivery and for newly born babies due to the transmission rate have been discussed for the present observed data. The numerical illustrations have been carried out for the parameters and functions and represented graphically by Mathematica software. Moreover, some comparisons have been shown in the figure to estimate the impact of susceptible conditions and represent the particular cases of S-I-R-M model.

**Chapter 19: Sensor and IoT-Based Belt to Detect Distance and Temperature of COVID-19 Suspect**

Owing to the pandemic issue of the coronavirus disease 2019 (COVID-19), it is imperative to keep up more than 1 m of social distancing and 37.5 °C temperature to stop the transmission of COVID-19 from human to human. Therefore, it is utmost requirement to make the smart belt installed with ultrasonic and LM35 sensors for distance and temperature measurements to reduce the transmission of COVID-19, respectively. The embedded sensors with NodeMCU show that once anything come in the proximity of 1 m near to the smart belt or helmet fixed to human body, it automatically makes an alarm for distance contact as well as temperature of incoming/outgoing body and sends an email to the controller with the help of Blynk application through Internet of things (IoT). This data can be stored in the cloud for the future purpose. However, the distance sensor has detected the movement of a person from 3 cm up to around 240 cm. The LM35 temperature sensor measures the actual temperature of the host body, i.e. 35.4 °C with time. With the help of this research, it is possible to interface a camera module which detects the suspects. It could be interfaced with global positioning system (GPS) which can give location-wise data and help us to obtain the probability of suspects at a particular region. It is cost effective, i.e. $14/belt, which can help to control the transmission of coronavirus from human to human.

## References


1. Allen, L., & Bridges, T. J. (2002). Numerical exterior algebra and the compound matrix method. *Numerische Mathematik, 92*(2), 197–232.
2. Awan, A. U., & Sharif, A. (2017). Smoking model with cravings to smoke. *Advanced Studies in Biology, 9*(1), 31–41.
3. Busenberg, S., & Van den Driessche, P. (1990). Analysis of a disease transmission model in a population with varying size. *Journal of Mathematical Biology, 28*(3), 257–270.
4. Cai, L., Li, X., Ghosh, M., & Guo, B. (2009). Stability analysis of an HIV/AIDS epidemic model with treatment. *Journal of Computational and Applied Mathematics, 229*(1), 313–323.
5. Diekmann, O., Heesterback, J. A. P., & Roberts, M. G. (2009). The construction of next generation matrices for compartmental epidemic models. *Journal of the Royal Society Interface, 7*(47), 873–885.
6. Diekmann, O., Heesterbeek, J. A. P., & Metz, J. A. J. (1990). On the definition and the computation of the basic reproduction ratio R0 in models for infectious diseases in heterogeneous populations. *Journal of Mathematical Biology, 28*(4), 365–382.
7. Dublin, L. I., & Lotka, A. J. (1925). On the true rate of natural increase: As exemplified by the population of the United States, 1920. *Journal of the American Statistical Association, 20*(151), 305–339.
8. Fleming, W. H., & Rishel, R. W. (2012). *Deterministic and stochastic optimal control* (Vol. 1). Berlin: Springer Science & Business Media.
9. Guo, H., Li, M. Y., & Shuai, Z. (2008). A graph-theoretic approach to the method of global Lyapunov functions. *Proceedings of the American Mathematical Society, 136*(8), 2793–2802.
10. Johnson, L. (2004). *An introduction to the mathematics of HIV/AIDS modelling*. Centre for Actuarial Research. https://pdfs.semanticscholar.org/7aef/2398872b9f791ab466ee034a633915d384ca.pdf.
11. Kuczynski, R. R. (1931). *The balance of births and deaths* (Vol. 29). New York: Macmillan.
12. LaSalle, J. P. (1976). *The stability of dynamical systems*. Philadelphia, PA: Society for Industrial and Applied Mathematics.
13. Li, M. Y., & Muldowney, J. S. (1996). A geometric approach to global-stability problems. *SIAM Journal on Mathematical Analysis, 27*(4), 1070–1083.
14. Li, Y., & Muldowney, J. S. (1993). On Bendixson's criterion. *Journal of Differential Equations, 106*(1), 27–39.
15. Lotka, A. J. (1923). Contribution to the analysis of malaria epidemiology. II. General part (continued). Comparison of two formulae given by Sir Ronald Ross. *American Journal of Hygiene, 3*(Supp), 38–54.
16. Manika, D. (2013). *Application of the compound matrix theory for the computation of Lyapunov exponents of autonomous Hamiltonian systems* (Master's thesis), Aristotle University of Thessaloniki.
17. Routh, E. J. (1877). *A treatise on the stability of a given state of motion: Particularly steady motion*. New York: Macmillan and Company.

18. Satia, M. H. (2020). *Mathematical models for eco-friendly society* (Ph.D. thesis). Gujarat University, Ahmedabad, India.
19. Smith, R. A. (1986). Some applications of Hausdorff dimension inequalities for ordinary differential equations. *Proceedings of the Royal Society of Edinburgh Section a: Mathematics, 104*(3–4), 235–259.
20. Van den Driessche, P., & Watmough, J. (2002). Reproduction numbers and subthreshold endemic equilibria for compartmental models of disease transmission. *Mathematical Biosciences, 180*(1–2), 29–48.

# Chapter 2
# Modelling the Impact of Nationwide BCG Vaccine Recommendations on COVID-19 Transmission, Severity and Mortality

**Nita H. Shah, Ankush H. Suthar, Moksha H. Satia, Yash Shah, Nehal Shukla, Jagdish Shukla, and Dhairya Shukla**

**Abstract** Coronavirus Disease 2019 (COVID-19) is declared as pandemic on 11 March 2020 by World Health Organization (WHO). There are apparent dissimilarities in incidence and mortality of COVID-19 cases in different parts of world. Developing countries in Asia and Africa with fragile health system are showing lower incidence and mortality compared to developed countries with superior health system in Europe and America. Most countries in Asia and Africa have national Bacillus Calmette-Guerin (BCG) vaccination programme, while Europe and America do not have such programme or have ceased it. At present, there is no known Food and Drug Administration (FDA)-approved treatment available for COVID-19 disease. There is no vaccine available currently to prevent COVID-19 disease. As mathematical modelling is ideal for predicting the rate of disease transmission as well as evaluating efficacy of possible public health prevention measures, we have created

N. H. Shah · A. H. Suthar · M. H. Satia
Department of Mathematics, Gujarat University, Ahmedabad, Gujarat 380009, India
e-mail: nitahshah@gmail.com

A. H. Suthar
e-mail: ankush.suthar1070@gmail.com

M. H. Satia
e-mail: mokshasatia.05@gmail.com

Y. Shah
GCS Medical College, Ahmedabad, Gujarat 380054, India
e-mail: yashshah95aries@gmail.com

N. Shukla (✉)
Department of Mathematics, Columbus State University, 4225, University Avenue, Columbus, GA 31907, USA
e-mail: shukla_nehal@columbusstate.edu

J. Shukla
Department of Medical Education, Family Medicine Residency Program, 1900, 10th Avenue, Columbus, GA 31901, USA
e-mail: Jagdish.shukla@piedmont.org

D. Shukla
Medical College of Georgia, 1120, 15th St, Augusta, GA 30912, USA
e-mail: dshukla@augusta.edu

 
N. H. Shah and M. Mittal (eds.), *Mathematical Analysis for Transmission of COVID-19*, Mathematical Engineering, https://doi.org/10.1007/978-981-33-6264-2_2

a mathematical model with seven compartments to understand nationwide BCG vaccine recommendation on COVID-19 transmission, severity and mortality. We have computed two basic reproduction numbers, one at vaccine-free equilibrium point and other at non-vaccine-free equilibrium point, and carried out local stability, sensitivity and numerical analysis. Our result showed that individuals with BCG vaccinations have lower risk of getting COVID-19 infection, shorter hospital stays and increased rate of recovery. Furthermore, countries with long-standing universal BCG vaccination policies have reduced incidence, mortality and severity of COVID-19. Further research will focus on exploring the immediate benefits of vaccination to healthcare workers and patients as well as benefits of BCG re-vaccination.

**Keywords** COVID-19 · BCG vaccine · Trained immunity · Modelling · Transmission

**MSC Code** 37Nxx

## Introduction

In December of 2019, a novel strain of coronavirus was found in Wuhan, China, and disease caused by it is identified as coronavirus disease 2019 (COVID-19) [1]. Transmission of COVID-19, similar to numerous airborne respiratory viruses including tuberculosis and influenza, occurs through direct contact with an infected person through respiratory droplets when a person coughs or sneezes or indirect contact through interaction with contaminated surfaces with respiratory droplets from infected person [2]. As of 18 April 2020, the disease has spread to 210 countries with 2,324,731 total confirmed new cases reported and 160,434 deaths [4]. Data we used for the current research are most recent as of 18 April 2020. Since currently there is no definitive treatment or an effective vaccine for disease control, there is a crucial need to find measures that cure or reduce morbidity due to COVID-19. There are prominent dissimilarities in how COVID-19 is affecting different countries. Although developing countries like India, Philippines and Sri Lanka reported their initial case in January, they have yet to experience extensive community spread [5]. In contrast, developed countries like Italy and UK have experienced high mortality and widespread infection despite strong curtailing of social interactions. Studies show a disproportionately smaller number of cases have been reported from disadvantaged and low-income countries [6]. It is puzzling that countries with more fragile health systems report less incidence of COVID-19. Figure 2.1 shows a figure from the European Centre for Disease Prevention and Control which indicates that much of the incidence of COVID-19 is in the developing countries such as UK and USA [3]. A possible explanation for lower number of cases detected in these developing countries with extensive travel and trade links to China may stem from the induced heterologous protective activity of the Bacillus Calmette-Guerin (BCG) immunisation [5].

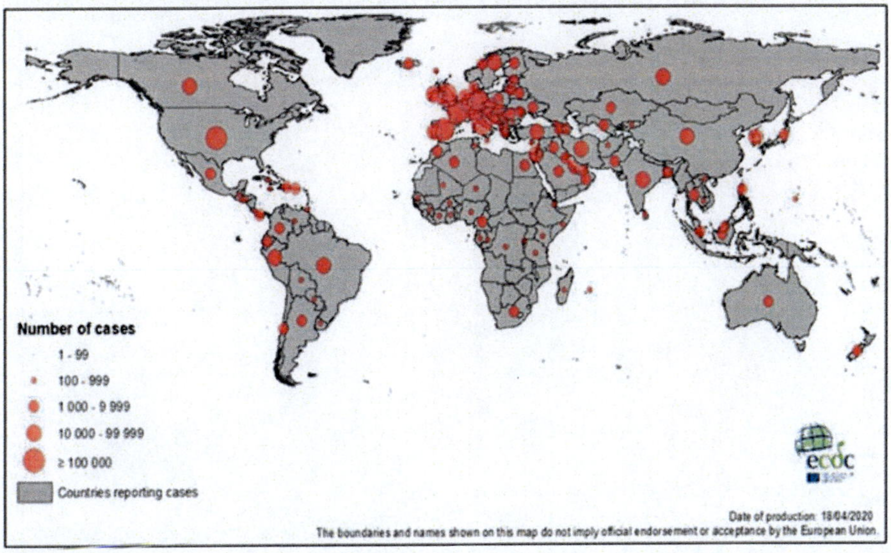


**Fig. 2.1** Geographical distribution of COVID-19 cases worldwide as of 18 April 2020 [3]. Adapted from European Centre for Disease Prevention and Control

Vaccines provide protection from particular pathogens by inducing effector mechanisms directed towards that pathogen. Certain live attenuated vaccines like the BCG vaccine increase immunity against not only a particular pathogen, but also numerous unrelated pathogens that present with similar acute respiratory tract infections [8]. Randomised controlled trials have provided evidence that the BCG vaccine's immunomodulatory properties can protect against respiratory infections. In Guinea-Bissau, a high-mortality setting, BCG Danish reduced all-cause neonatal mortality by 38% (95% CI 17–54), mainly because there were fewer deaths from pneumonia and sepsis [9]. In South Africa, BCG Danish reduced respiratory tract infections by 73% (95% CI 39–88) in adolescents [10]. The BCG vaccine reduced yellow fever vaccine viraemia by 71% (95% CI 6–91) in volunteers in the Netherlands [11]. Previous studies have established that BCG vaccination provided non-specific protection via the induction of innate immune memory [12]. The BCG vaccine induces metabolic and epigenetic changes that enhance the innate immune response to subsequent infections, a process termed trained immunity [13]. The BCG vaccine might therefore reduce viraemia after SARS-CoV-2 exposure, with consequent less severe COVID-19 and more rapid recovery [14]. BCG is used widely across the world as a vaccine for tuberculosis, with many developing nations having a universal BCG vaccination policy for newborn [15]. Studies have indicated that countries with uniform BCG vaccination policy show significantly lower COVID-19 cases and deaths per million people compared to countries where BCG vaccination policy was ceased or was never in place [5, 16]. Figure 2.2 from the BCG world atlas shows that numerous developing countries currently employ a universal BCG vaccination programme, while

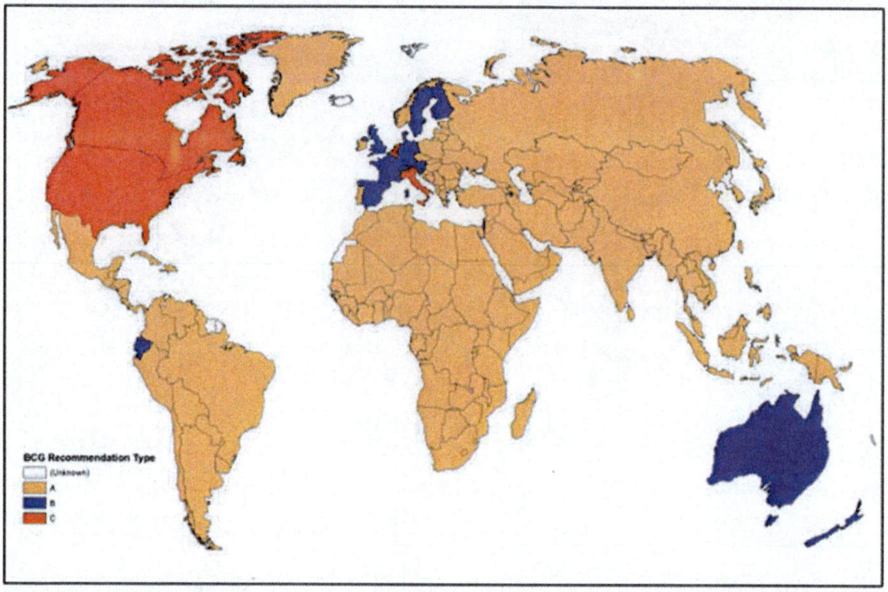


**Fig. 2.2** Map displaying BCG vaccination policy by country. **a** The country currently has universal BCG vaccination programme. **b** The country used to recommend BCG vaccination for everyone, but currently does not. **c** The country never had universal BCG vaccination programmes [7]. Adapted from BCG World Atlas

many developed countries have either ceased their BCG vaccination recommendation or never recommended it [7]. Comparison of Figs. 2.1 and 2.2 also indicates that the largest number of COVID-19 cases is in countries that currently have no BCG vaccination recommendation.

Furthermore, the year that universal BCG vaccination was established within a country has a significant correlation with the mortality rate, indicating that earlier implementation of the policy leads to protection of a larger fraction of the elderly population. For instance, although Iran currently employs a universal BCG vaccination policy, it was implemented as recently as 1984. Consequently, it reports an elevated mortality with 19.7 deaths per million inhabitants [6]. In contrast, Japan started its universal BCG policy in 1947 and found that Japan has around 100 times less deaths per million people, with 0.28 deaths. Similarly, Brazil started universal vaccination in 1920 and also has an even lower mortality rate of 0.0573 deaths per million inhabitants [6].

Exploring this relationship between BCG vaccination and COVID-19 further is crucial in establishing a potential short- or long-term prevention measure that is effective in reducing incidence and mortality of COVID-19. Four clinical trials have been registered with BCG vaccination to prevent or reduce the severity of COVID-19 in the elderly population and healthcare workers [17]. While previous studies have explored a possible link between the BCG vaccine and incidence of COVID-19, they have not modelled the impact of implementing a nationwide BCG immunisation policy [6, 7]. Mathematical modelling is ideal for predicting the rate of disease transmission as well as evaluating efficacy of possible public health prevention measures [18]. This study aims to create a model to evaluate the effect of nationwide BCG vaccine recommendations on incidence, severity and mortality rates of COVID-19. Furthermore, our study aims to model the protentional implications of an immediate implementation of the universal BCG vaccination policy.

## Mathematical Model

A novel coronavirus pandemic is occurring around the world, both in countries with and without the BCG vaccine. To analyse the importance of this vaccine against the fight of COVID-19, we have developed a mathematical model. This model has seven various compartments. First compartment is the class of exposed individuals to COVID-19 noted as $E$. From these individuals, some are vaccinated with BCG vaccine noted as $V$ and those who have not taken it are noted as $N_V$. Now, with the certain rate these individuals transfer into infected stage noted as $I$ and some can become critically infected noted as $C$ compartment. The individuals suffering from COVID-19 that need to be hospitalised are noted as $H$. The infected and hospitalised individuals that recover are noted as $R$.

With the help of transmission shown in Fig. 2.3 and notation of parameter given in Table 2.1, the system of nonlinear ODE is derived as:

$$\frac{\mathrm{d}E}{\mathrm{d}t} = B - \beta_1 EV - (1-\beta_1)EN_V + \beta_9 R - \mu E$$
$$\frac{\mathrm{d}V}{\mathrm{d}t} = \beta_1 EV - \beta_2 V - \mu V$$
$$\frac{\mathrm{d}N_V}{\mathrm{d}t} = (1-\beta_1)EN_V - \beta_3 N_V - \mu N_V$$

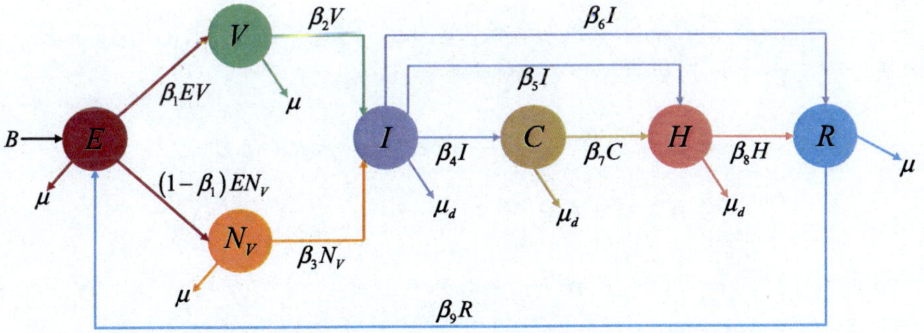


**Fig. 2.3** Transmission diagram of COVID-19 with BCG vaccine

**Table 2.1** Notation and its description

| Parameter | Description |
|---|---|
| $B$ | Birth rate of class of exposed individuals |
| $\beta_1$ | Rate of individual who has taken BCG vaccine |
| $\beta_2$ | Rate of vaccinated individuals who get infection |
| $\beta_3$ | Rate of non-vaccinated individuals who get infection |
| $\beta_4$ | Rate at which infected individual becomes critically infected |
| $\beta_5$ | Rate at which infected individuals get hospitalised |
| $\beta_6$ | Recovery rate of infected individuals |
| $\beta_7$ | Rate at which critically infected individuals get hospitalised |
| $\beta_8$ | Recovery rate of hospitalised individuals |
| $\beta_9$ | Rate at which recovered individuals may get exposed |
| $\mu$ | Natural death rate |
| $\mu_d$ | Diseases-induced death rate |

$$\begin{aligned}\frac{dI}{dt} &= \beta_2 V + \beta_3 N_V - \beta_4 I - \beta_5 I - \beta_6 I - \mu_d I \\ \frac{dC}{dt} &= \beta_4 I - \beta_7 C - \mu_d C \\ \frac{dH}{dt} &= \beta_5 I + \beta_7 C - \beta_8 H - \mu_d H \\ \frac{dR}{dt} &= \beta_6 I + \beta_8 H - \beta_9 R - \mu R \end{aligned} \tag{2.1}$$

with $E > 0$; $V, N_V, I, C, H, R \geq 0$.

Hence, the feasible region of given system is

$$\Lambda = \left\{ (E, V, N_V, I, C, H, R) \in R_+^7 : E + V + N_V + I + C + H + R \leq \frac{B}{\mu} \right\}.$$

Now, consider $\beta_2 + \mu = k_1$, $\beta_3 + \mu = k_2$, $\beta_4 + \beta_5 + \beta_6 + \mu_d = k_3$, $\beta_7 + \mu_d = k_4$, $\beta_8 + \mu_d = k_5$, $\beta_9 + \mu = k_6$; then, we get new system as follows

$$\begin{aligned}\frac{dE}{dt} &= B - \beta_1 E V - (1 - \beta_1) E N_V + \beta_9 R - \mu E \\ \frac{dV}{dt} &= \beta_1 E V - k_1 V \\ \frac{dN_V}{dt} &= (1 - \beta_1) E N_V - k_2 N_V \\ \frac{dI}{dt} &= \beta_2 V + \beta_3 N_V - k_3 I \end{aligned}$$

$$
\begin{aligned}
\frac{\mathrm{d}C}{\mathrm{d}t} &= \beta_4 I - k_4 C \\
\frac{\mathrm{d}H}{\mathrm{d}t} &= \beta_5 I + \beta_7 C - k_5 H \\
\frac{\mathrm{d}R}{\mathrm{d}t} &= \beta_6 I + \beta_8 H - k_6 R
\end{aligned}
\tag{2.2}
$$

The dynamical behaviour of system (2.1) is equivalent to the system (2.2). Hence, every solution of system (2.2) will remain in the region $\Lambda$.

The above system (2.2) has three equilibrium points by setting the equation as zero.

1. Diseases-free equilibrium point $E_P^0(E_0, 0, 0, 0, 0, 0, 0)$ where

$$E_0 = \frac{B}{\mu}$$

2. Vaccine-free equilibrium point $E_P^1\left(E_1, 0, N_{V_1}, I_1, C_1, H_1, R_1\right)$ where

$$E_1 = \frac{k_2}{1-\beta_1}$$

$$N_{V_1} = \frac{k_3 k_4 k_5 k_6 (B\beta_1 + k_2\mu - B)}{(1-\beta_1)[\beta_3\beta_8\beta_9(\beta_4\beta_7 + \beta_5 k_5) + k_4 k_5(\beta_3\beta_6\beta_9 - k_2 k_3 k_6)]}$$

$$I_1 = \frac{\beta_3 k_4 k_5 k_6 (B\beta_1 + k_2\mu - B)}{(1-\beta_1)[\beta_3\beta_8\beta_9(\beta_4\beta_7 + \beta_5 k_5) + k_4 k_5(\beta_3\beta_6\beta_9 - k_2 k_3 k_6)]}$$

$$C_1 = \frac{\beta_3 \beta_4 k_5 k_6 (B\beta_1 + k_2\mu - B)}{(1-\beta_1)[\beta_3\beta_8\beta_9(\beta_4\beta_7 + \beta_5 k_5) + k_4 k_5(\beta_3\beta_6\beta_9 - k_2 k_3 k_6)]}$$

$$H_1 = \frac{\beta_3 k_6 (B\beta_1 + k_2\mu - B)(\beta_4\beta_7 + \beta_5 k_4)}{(1-\beta_1)[\beta_3\beta_8\beta_9(\beta_4\beta_7 + \beta_5 k_5) + k_4 k_5(\beta_3\beta_6\beta_9 - k_2 k_3 k_6)]}$$

$$R_1 = \frac{\beta_3 (B\beta_1 + k_2\mu - B)(\beta_4\beta_7\beta_8 + \beta_5\beta_8 k_4 + \beta_6 k_4 k_5)}{(1-\beta_1)[\beta_3\beta_8\beta_9(\beta_4\beta_7 + \beta_5 k_5) + k_4 k_5(\beta_3\beta_6\beta_9 - k_2 k_3 k_6)]}$$

3. Non-vaccine-free equilibrium point $E_P^2(E_2, V_2, 0, I_2, C_2, H_2, R_2)$

$$E_2 = \frac{k_1}{\beta_1}$$

$$V_2 = \frac{k_3 k_4 k_5 k_6 (k_1\mu - B\beta_1)}{\beta_1[\beta_2\beta_8\beta_9(\beta_4\beta_7 + \beta_5 k_4) + k_4 k_5(\beta_2\beta_6\beta_9 - k_1 k_3 k_6)]}$$

$$I_2 = \frac{\beta_2 k_4 k_5 k_6 (k_1\mu - B\beta_1)}{\beta_1[\beta_2\beta_8\beta_9(\beta_4\beta_7 + \beta_5 k_4) + k_4 k_5(\beta_2\beta_6\beta_9 - k_1 k_3 k_6)]}$$

$$C_2 = \frac{\beta_2\beta_4 k_5 k_6 (k_1\mu - B\beta_1)}{\beta_1[\beta_2\beta_8\beta_9(\beta_4\beta_7 + \beta_5 k_4) + k_4 k_5(\beta_2\beta_6\beta_9 - k_1 k_3 k_6)]}$$

$$H_2 = \frac{\beta_2 k_6 (k_1\mu - B\beta_1)(\beta_4\beta_7 + \beta_5 k_4)}{\beta_1[\beta_2\beta_8\beta_9(\beta_4\beta_7 + \beta_5 k_4) + k_4 k_5(\beta_2\beta_6\beta_9 - k_1 k_3 k_6)]}$$

$$R_2 = \frac{\beta_2 (k_1\mu - B\beta_1)(\beta_4\beta_7\beta_8 + \beta_5\beta_8 k_4 + \beta_6 k_4 k_5)}{\beta_1[\beta_2\beta_8\beta_9(\beta_4\beta_7 + \beta_5 k_4) + k_4 k_5(\beta_2\beta_6\beta_9 - k_1 k_3 k_6)]}$$

The diseases-free equilibrium point is not valid according to current scenario, and therefore we will discuss only two equilibrium points.

## Basic Reproduction Number

In this section, basic reproduction number is derived using next-generation matrix method [19]. The quantity of basic reproduction number helps to understand the behaviour of the spread of COVID-19 among the population. The basic reproduction number is denoted as $R_0$. If the value of $R_0$ is less than 1, then the diseases is in the controllable stage. Otherwise, it has reached the epidemic stage. This is the ratio of affected infected individual by secondary infected individuals within the population. Using next-generation matrix method, $f$ and $v$ are derived as

$$f = \begin{bmatrix} \beta_1 E V \\ (1-\beta_1) E N_V \\ 0 \\ 0 \\ 0 \\ 0 \\ 0 \end{bmatrix} \text{ and } v = \begin{bmatrix} k_1 V \\ k_2 N_V \\ -\beta_2 V - \beta_3 N_V + k_3 I \\ -\beta_4 I + k_4 C \\ -\beta_5 I - \beta_7 C + k_5 H \\ -\beta_6 I - \beta_8 H + k_6 R \\ -B + \beta_1 E V + (1-\beta_1) E N_V - \beta_9 R + \mu E \end{bmatrix}.$$

The Jacobian matrices of $f$ and $v$ are $F$ and $V$ defined as below

$$F = \begin{bmatrix} \beta_1 E & 0 & 0\,0\,0\,0 & \beta_1 V \\ 0 & (1-\beta_1)E & 0\,0\,0\,0 & (1-\beta_1)N_V \\ 0 & 0 & 0\,0\,0\,0 & 0 \\ 0 & 0 & 0\,0\,0\,0 & 0 \\ 0 & 0 & 0\,0\,0\,0 & 0 \\ 0 & 0 & 0\,0\,0\,0 & 0 \\ 0 & 0 & 0\,0\,0\,0 & 0 \end{bmatrix} \text{ and}$$

$$V = \begin{bmatrix} k_1 & 0 & 0 & 0 & 0 & 0 & 0 \\ 0 & k_2 & 0 & 0 & 0 & 0 & 0 \\ -\beta_2 & -\beta_3 & k_3 & 0 & 0 & 0 & 0 \\ 0 & 0 & -\beta_4 & k_4 & 0 & 0 & 0 \\ 0 & 0 & -\beta_5 & -\beta_7 & k_5 & 0 & 0 \\ 0 & 0 & -\beta_6 & 0 & -\beta_8 & k_6 & 0 \\ \beta_1 E & (1-\beta_1)E & 0 & 0 & 0 & -\beta_9 & \beta_1 V + (1-\beta_1)N_V + \mu \end{bmatrix}$$

These Jacobian matrices are used to derive $FV^{-1}$ whose spectral radius is our desired basic reproduction number ($R_0$). Here, we have computed two basic reproduction numbers: one at vaccine-free equilibrium point and other at non-vaccine-free equilibrium point.

The expression of basic reproduction number for non-vaccinated individuals, i.e. at vaccine-free equilibrium point, is denoted by $R_0^{(1)}$ given by

$$\begin{aligned} R_0^{(1)} = {} & ((1-\beta_1)k_3k_4k_5k_6(B\beta_1 + k_2 - B)\big(\beta_1 k_2 k_6 \mu_d^3 + (-\beta_2\beta_6\beta_9\,(1-\beta_1) \\ & + \beta_1 k_2 k_6(\beta_4 + \beta_5 + \beta_6 + \beta_7 + \beta_8))\mu_d^2 \\ & + (-\beta_2\beta_9\,(1-\beta_1)(\beta_5\beta_8 + \beta_6\beta_7 + \beta_6\beta_8) + \beta_1 k_2 k_6(\beta_4\beta_7 \\ & + \beta_4\beta_8 + \beta_5\beta_7 + \beta_5\beta_8 + \beta_6\beta_7 + \beta_6\beta_8 + \beta_7\beta_8))\mu_d \\ & + \beta_7\beta_8(\beta_4 + \beta_5 + \beta_6)\,(-\beta_2\beta_9(1-\beta_1) + \beta_1 k_2 k_6))(\beta_3\beta_8\beta_9\,(\beta_4\beta_7 + \beta_5 k_4) \\ & + k_4k_5\,(\beta_3\beta_6\beta_9 - k_2k_3k_6)))/((1-\beta_1)^2 k_1k_3k_4k_5k_6(\beta_3\beta_8\beta_9\,(\beta_4\beta_7 + \beta_5 k_4) \\ & + k_4k_5(\beta_3\beta_6\beta_9 - k_2k_3k_6))(-Bk_3k_4k_5k_6(1-\beta_1) \\ & + \beta_3\beta_9\mu(\beta_4\beta_7\beta_8 + \beta_5\beta_8k_4 + \beta_6k_4k_5)) \end{aligned}$$

whereas the expression of basic reproduction number for vaccinated individuals, i.e. at non-vaccine-free equilibrium point, is denoted by $R_0^{(2)}$ given by

$$\begin{aligned} R_0^{(2)} = {} & \frac{k_1}{\beta_2 + \mu} - (k_3k_4k_5k_6\,(B\beta_1 - k_1\mu)\big(k_1k_6\mu_d^3 \\ & + (k_1k_6\,(\beta_4 + \beta_5 + \beta_6 + \beta_7 + \beta_8) - \beta_2\beta_6\beta_9)\mu_d^2 \\ & + (-\beta_2\beta_9\,(\beta_5\beta_8 + \beta_6\beta_7 + \beta_6\beta_8) + k_1k_6(\beta_4\beta_7 \\ & + \beta_4\beta_8 + \beta_5\beta_7 + \beta_5\beta_8 + \beta_6\beta_7 + \beta_6\beta_8 + \beta_7\beta_8))\mu_d \\ & + \beta_7\beta_8(\beta_4 + \beta_5 + \beta_6)(-\beta_2\beta_9 + k_1k_6)/(k_2k_3k_4k_5k_6\,(k_4k_5\,(B\beta_1k_3k_6 \\ & - \beta_2\beta_6\beta_9\mu) - \beta_2\beta_8\beta_9\,(\beta_4\beta_7 + \beta_5k_4))) \end{aligned}$$

Moreover, the basic reproduction for the system is $R_0 = \frac{R_0^{(1)} + R_0^{(2)}}{2}$.

## Local Stability Analysis

In this section, we will be discussing the local stability for vaccine and non-vaccine-free equilibrium point utilising the concept of eigenvalues. To find the eigenvalues, the Jacobian matrix of the system (2.2) is calculated as

$$J := \begin{bmatrix} -\beta_1 V - (1-\beta_1)N_V - \mu & -\beta_1 E & -(1-\beta_1)E & 0 & 0 & 0 & \beta_9 \\ \beta_1 V & \beta_1 E - k_1 & 0 & 0 & 0 & 0 & 0 \\ (1-\beta_1)N_V & 0 & (1-\beta_1)E - k_2 & 0 & 0 & 0 & 0 \\ 0 & \beta_2 & \beta_3 & -k_3 & 0 & 0 & 0 \\ 0 & 0 & 0 & \beta_4 & -k_4 & 0 & 0 \\ 0 & 0 & 0 & \beta_5 & \beta_7 & -k_5 & 0 \\ 0 & 0 & 0 & \beta_6 & 0 & \beta_8 & -k_6 \end{bmatrix}$$

Eigenvalues for above Jacobi matrix about the vaccine-free equilibrium point are 0.1225, −0.0664, −0.1375, −0.1375, −0.8463, −0.9431, 0.7019.

Eigenvalues for above Jacobi matrix about the non-vaccine-free equilibrium point are 0.2298, −0.0655, −0.1595, −0.2296, −0.8464, −0.9431, −0.1021.

Hence, both the equilibrium points are not locally asymptotically stable.

## Sensitivity Analysis

Here, sensitivity analysis is carried out to see which parameter is conferring a positive effect on the model. The sensitivity analysis is conducted on the basic reproduction number. It is to be done using Christoffel formula which is $\Gamma_{\alpha}^{R_0} = \frac{\partial R_0}{\partial \alpha} \cdot \frac{\alpha}{R_0}$ where $\alpha$ is the model parameter.

From Table 2.2, it can be observed that the growth rate is producing negative impact. Precisely, growth rate is more negative when there is no vaccinated individual. The rate of individual vaccinated with BCG ($\beta_1$) shows the positive impact among all. Moreover, the disease-induced death rate is lower for $R_0^{(1)}$ when compared to $R_0^{(2)}$. Additionally, considering the sensitivity analysis for both the basic reproduction numbers, we conclude that BCG vaccine is beneficial in preventing COVID-19.

## Numerical Simulation

In this section, results of transmission of COVID-19 outbreak are simulated numerically, helping validate model results.

Figure 2.4 shows the change in the behaviour of individuals. This suggests that exposed individuals quickly progress into infected ones. Infected individuals increase in first 1.5 weeks and then start to decrease, whereas critically infected individuals

**Table 2.2** Sensitivity analysis

| Parameter | $R_0^{(1)}$ | $R_0^{(2)}$ |
|---|---|---|
| $B$ | −2.4260 | −1.6980 |
| $\beta_1$ | 25.09 | −1.6980 |
| $\beta_2$ | −0.9799 | −0.4074 |
| $\beta_3$ | 1.2730 | 0 |
| $\beta_4$ | 0.0116 | 0.0055 |
| $\beta_5$ | 1.2450 | 0.5885 |
| $\beta_6$ | 0.0173 | 0.0080 |
| $\beta_7$ | 0.0124 | 0.0057 |
| $\beta_8$ | 1.224 | 0.5753 |
| $\beta_9$ | 1.224 | 0.5261 |
| $\mu$ | 2.403 | 1.6980 |
| $\mu_d$ | 0.0172 | 0.0261 |

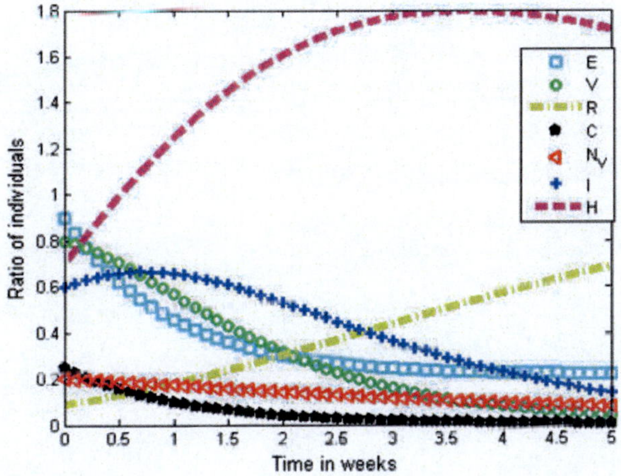


**Fig. 2.4** Transmission of COVID-19

stay in lower numbers. Hospitalisation increases rapidly for at least 3 weeks, and then it flattens. Recovery rate increases but at lower rate.

To get precise results, four clusters of Fig. 2.5 are magnified. Our data indicates that individuals should go into isolation earlier to prevent exposure to COVID-19. BCG-vaccinated individual may get exposed in 0.2 week, while, at the same time, non-vaccinated individual may get critically infected. Moreover, recovered individual also may become exposed and get re-infected after 2 weeks.

Figure 2.6 indicates the cyclic behaviour of recovered individuals. The first graph shows vaccinated individuals, while the other displays non-vaccinated individuals. The graphs show that the cycle of vaccinated individuals is larger than non-vaccinated

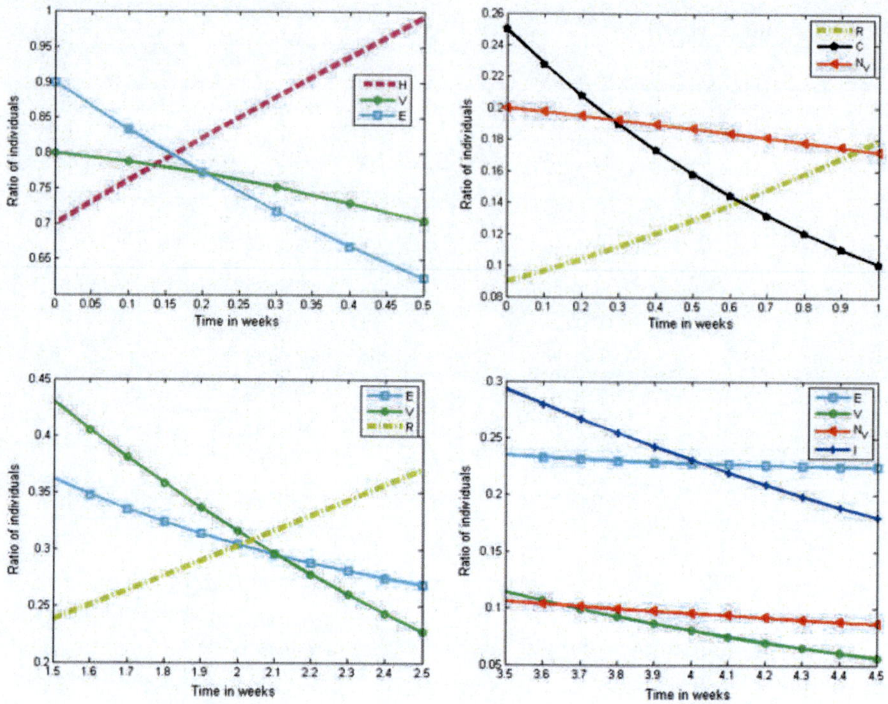


**Fig. 2.5** Magnified plots of transmission of COVID-19

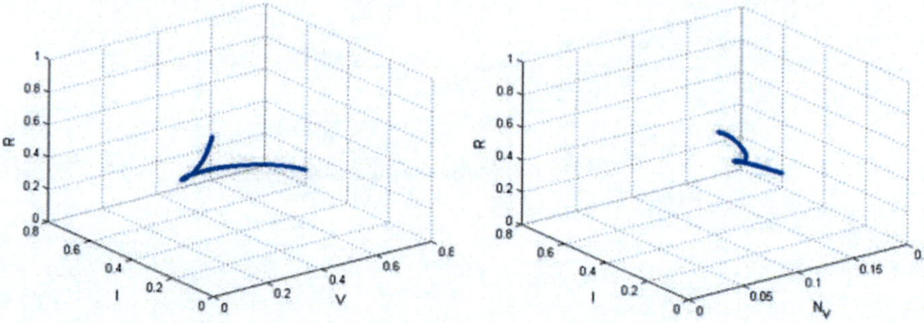


**Fig. 2.6** Cyclic behaviour of recovered individuals

individuals. Furthermore, vaccinated individuals have greater time to reinfection when compared to that of non-vaccinated individuals.

Figure 2.7 compares the behaviour of exposed individuals with vaccinated individuals, non-vaccinated individuals and infected individuals, respectively. The rate of exposure is consecutive for non-vaccinated individuals when compared to vaccinated individuals. Furthermore, the transmission from exposed to infected is quite quick; i.e. almost every individual who is exposed will get infected.

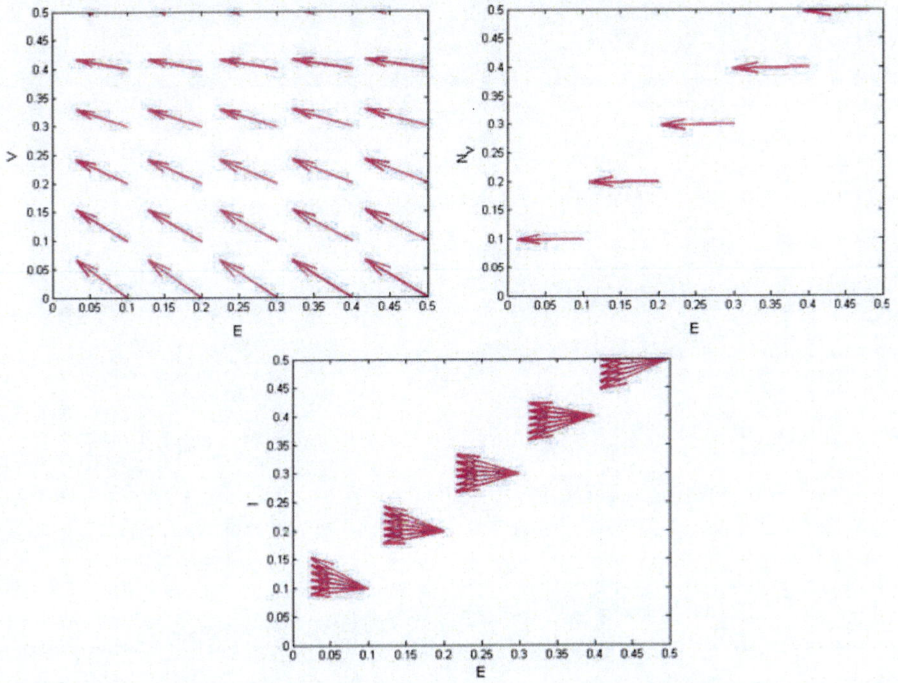


**Fig. 2.7** Behaviour of exposed individuals

Figure 2.8 indicates that vaccinated individuals are infected gradually while non-vaccinated individuals are infected quickly. Consequently, non-vaccinated individuals are at higher risk of becoming infected with COVID-19.

Figure 2.9 shows the behaviour of infected individuals towards critical, hospitalised and recovered individuals, respectively. If individuals are vaccinated early, the chances of becoming critically infection are low. However, this also indicates

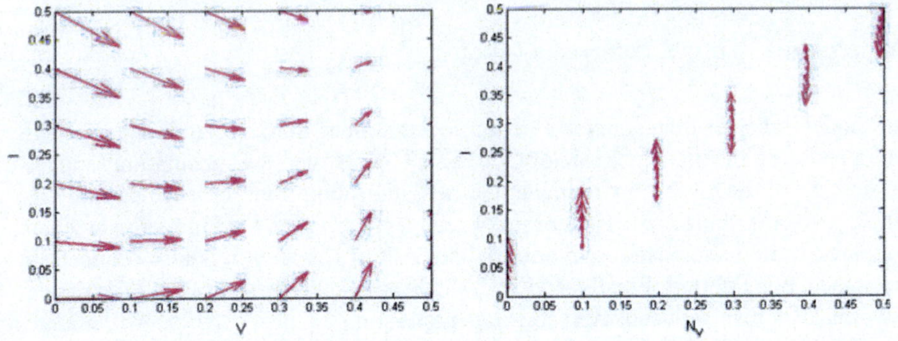


**Fig. 2.8** Behaviour of vaccinated and non-vaccinated individuals

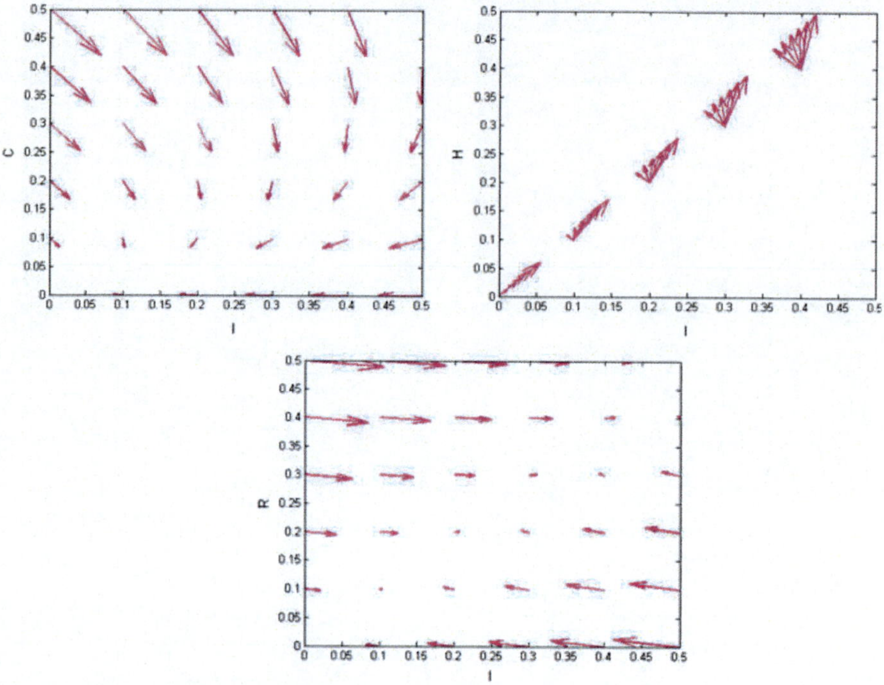


Fig. 2.9 Behaviour of infected individual

that infected individuals must receive treatment in hospitals before the infection progresses to critical stage. This is further elaborated by the last graph, which indicates that the chances of recovering by staying at home are negligible.

Figure 2.10 shows the behaviour of hospitalised individuals and their recovery. This figure further emphasises that individuals recover only after proper hospitalisation is received.

## Conclusion

Our model indicates that a universal recommendation of BCG vaccination is beneficial in reducing the incidence rate of COVID-19. Moreover, the vaccination reduces the severity of the disease by reducing the hospitalisation time for infected individuals. It also contributed to a higher rate of recovery for vaccinated individuals when compared to non-vaccinated individuals. The risk of reinfection is also reduced in vaccinated populations. Finally, the overall disease-specific mortality of infected individuals is also reduced in vaccinated populations. Consequently, we validate the effectiveness of universal BCG vaccination policy in increasing herd immunity against COVID-19.

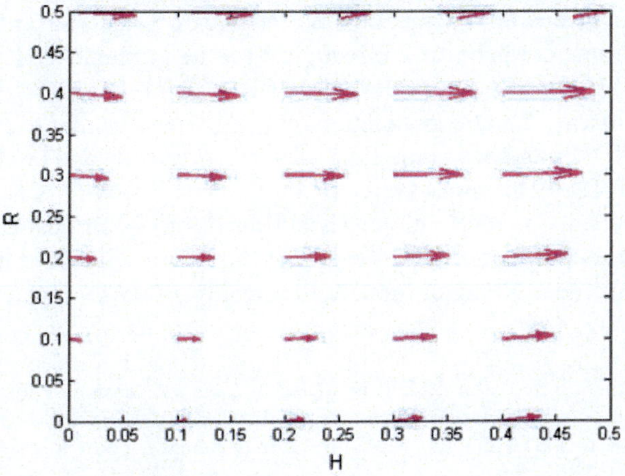


**Fig. 2.10** Behaviour of hospitalised individuals

Considering that the BCG vaccination is established in preventing acute respiratory tract infections even in the elderly [15], vulnerable populations may be immunised with BCG vaccines until a COVID-19-specific vaccine is developed. This strategy would be especially beneficial in the newborns, elderly and front-line essential workers. This study validates the call for clinical trials of BCG vaccine against COVID-19. Given that our model shows a significant reduction in incidence of COVID-19 in countries with universal BCG vaccination, we recommend all countries to conduct more clinical trials to establish and consider policy regarding BCG vaccinations for its citizens.

Our study may be subject to a few limitations. The study does not take into account confounding variables such as the potential differences in geographical and biological factors such as temperature, humidity, life expectancy, average income, social–cultural norms, ethnical genetic background and mitigation between developing and developed countries. Due to these confounding variables, we cannot establish causality [20]. The epidemiological data comparing vaccinated countries and non-vaccinated countries must be monitored throughout this pandemic to establish a causal impact in reducing incidence and mortality. Overall, our study serves to establish that individuals with BCG vaccinations have lower risk of getting COVID-19 infection, shorter hospital stays and increased rate of recovery. Furthermore, countries with long-standing universal BCG vaccination policies have reduced incidence, mortality and severity of COVID-19. Considering the remarkable dissemination capacity and the mortality rates of COVID-19, vaccination capable of conferring even transient protection (e.g. 6–12 months) may be useful in individuals at high risk, such as health workers, first responders and police officers, or those with pre-existing conditions such as obesity, diabetes or cardiovascular disease. Similarly, even enhanced unspecific immunity through BCG vaccination in vulnerable age groups could ameliorate severe COVID-19. Temporarily induced trained immunity could buy time until specific vaccines and/or effective treatments against severe acute respiratory syndrome coronavirus 2 infections become available [21]. Studies done by Moolag et al. [15] and Esobar et al. [21] showed possible benefit of BCG vaccination and suggest that vulnerable population may get benefit from it to ameliorate the severity of COVID-19 and buy some time till specific vaccine and or effective treatment against COVID-19 becomes available. Further research will focus on exploring the immediate benefits of vaccination to healthcare workers and patients as well as benefits of BCG re-vaccination. The results of this study will serve to inform and encourage further research and, ultimately, the creation of policy regarding universal BCG vaccination.

**Acknowledgements** Second author (AHS) is funded by a Junior Research Fellowship from the Council of Scientific and Industrial Research (file no.-09/070(0061)/2019-EMR-I), and first three authors are thankful to DST-FIST file # MSI-097 for technical support to the Department of Mathematics, Gujarat University.

**Conflicts of Interest** The authors declare that they have no competing financial interests or personal relationships that could have appeared to influence the work reported in this paper.

**Data Availability** Our conclusion is based on the data derived from our above mathematical calculations which are available to readers.

## References

1. Cohen, J., & Normile, D. (2020). New SARS-like virus in China triggers alarm. *Science, 367*(6475), 234–235. https://doi.org/10.1126/science.367.6475.234.
2. Holshue, M. L., DeBolt, C., Lindquist, S., Lofy, K. H., Wiesman, J., Bruce, H., et al. (2020). First case of 2019 novel coronavirus in the United States. *The New England Journal of Medicine, 382,* 929–936. https://doi.org/10.1056/NEJMoa2001191.
3. Situation update worldwide, as of 21 April 2020. European Centre for Disease Prevention and Control.
4. Worldometer. (2020). Worldometer for coronavirus cases. Worldometer. https://doi.org/10.1101/2020.01.23.20018549V2.
5. Gursel, M., & Gursel, I. (2020). Is global BCG vaccination-induced trained immunity relevant to the progression of SARS-CoV-2 pandemic? *Allergy European Journal of Allergy and Clinical Immunology, 75*(7). https://doi.org/10.1111/all.14345.
6. Aaron Miller, G. H. O., Reandelar, M. J., Fasciglione, K., Roumenova, V., & Li, Y. (2013). Correlation between universal BCG vaccination policy and reduced morbidity and mortality for COVID-19: An epidemiological study. *Journal of Chemical Information and Modeling,* 1–30. https://doi.org/10.1017/CBO9781107415324.004.
7. Zwerling, A., Behr, M. A., Verma, A., Brewer, T. F., Menzies, D., & Pai, M. (2011). The BCG world atlas: A database of global BCG vaccination policies and practices. *PLoS Medicine, 8*(3), e1001012. https://doi.org/10.1371/journal.pmed.1001012.
8. Kleinnijenhuis, J., Quintin, J., Preijers, F., Benn, C. S., Joosten, L. A. B., Jacobs, C., et al. (2014). Long-lasting effects of BCG vaccination on both heterologous Th1/Th17 responses and innate trained immunity. *Journal of Innate Immunity, 6,* 152–158. https://doi.org/10.1159/000355628.

9. Biering-Sørensen, S., Aaby, P., Lund, N., Monteiro, I., Jensen, K. J., Eriksen, H. B., et al. (2017). Early BCG-Denmark and neonatal mortality among infants weighing <2500 g: A randomized controlled trial. *Clinical Infectious Diseases, 65*(7), 1183–1190. https://doi.org/10.1093/cid/cix525.
10. Nemes, E., Geldenhuys, H., Rozot, V., Rutkowski, K. T., Ratangee, F., Bilek, N., et al. (2018). Prevention of M. Tuberculosis infection with H4:IC31 vaccine or BCG revaccination. *The New England Journal of Medicine, 379*, 138–149. https://doi.org/10.1056/NEJMoa1714021.
11. Arts, R. J. W., Moorlag, S. J. C. F. M., Novakovic, B., Li, Y., Wang, S. Y., Oosting, M., et al. (2018). BCG vaccination protects against experimental viral infection in humans through the induction of cytokines associated with trained immunity. *Cell Host & Microbe, 23*(1), 89–100. https://doi.org/10.1016/j.chom.2017.12.010.
12. Netea, M. G., Quintin, J., & Van Der Meer, J. W. M. (2011). Trained immunity: A memory for innate host defense. *Cell Host & Microbe, 9*(5), 355–361. https://doi.org/10.1016/j.chom.2011.04.006.
13. Netea, M. G., Domínguez-Andrés, J., Barreiro, L. B., Divangahi, T. C. M., Fuchs, E., Joosten, L. A. B., et al. (2020). Defining trained immunity and its role in health and disease. *Nature Reviews Immunology, 20*, 375–388. https://doi.org/10.1038/s41577-020-0285-6.
14. Curtis, N., Sparrow, A., Ghebreyesus, T. A., & Netea, M. G. (2020). Considering BCG vaccination to reduce the impact of COVID-19. *Lancet*. https://doi.org/10.1016/S0140-6736(20)31025-4.
15. Moorlag, S. J. C. F. M., Arts, R. J. W., van Crevel, R., & Netea, M.G. (2019) Non-specific effects of BCG vaccine on viral infections. *Clinical Microbiology Infection, 25*(12), 1473–1478. https://doi.org/10.1016/j.cmi.2019.04.020.
16. Sharma, A. R., Batra, G., Kumar, M., Mishra, A., Singla, R., Singh, A., et al. (2020). BCG as a game-changer to prevent the infection and severity of COVID-19 pandemic? *Allergologia et Immunopathologia (Madr), 48*(5), 507–517. https://doi.org/10.1016/j.aller.2020.05.002.
17. Sala, G., & Miyakawa, T. (2020). Association of BCG vaccination policy with prevalence and mortality of COVID-19. *MedRxiv*. https://doi.org/10.1101/2020.03.30.20048165.
18. Wu, J. T., Leung, K., & Leung, G. M. (2020). Nowcasting and forecasting the potential domestic and international spread of the 2019-nCoV outbreak originating in Wuhan, China: A modelling study. *Lancet*. https://doi.org/10.1016/S0140-6736(20)30260-9.
19. Diekmann, O., Heesterbeek, J. A. P., & Roberts, M. G. (2010). The construction of next-generation matrices for compartmental epidemic models. *Journal of the Royal Society, Interface*. https://doi.org/10.1098/rsif.2009.0386.
20. Kumar, J., & Meena, J. (2020). Demystifying BCG vaccine and COVID-19 relationship. *Indian Pediatrics, 57*, 588–589. https://doi.org/10.1007/s13312-020-1872-0.
21. Escobar, L. E., Molina-Cruz, A., & Barillas-Mury, C. (2020). BCG vaccine protection from severe coronavirus disease 2019 (COVID-19). *Proceedings of the National Academy of Sciences of the United States of America, 117*(30), 17720–17726. https://doi.org/10.1073/pnas.2008410117.

# Chapter 3
# Modeling the Spread of COVID-19 Among Doctors from the Asymptomatic Individuals

**M. H. A. Biswas, A. K. Paul, M. S. Khatun, S. Mandal, S. Akter, M. A. Islam, M. R. Khatun, and S. A. Samad**

**Abstract** The present world is in dire straits due to the deadly SARS coronavirus-2 (CoV-2) outbreak, and the experts are trying heart and soul to discover any prevention and/or remedy. The people from all walks of life in the universe are fighting to defeat this novel coronavirus. In this case, doctors are in the front line fighters who have put themselves at a risk. In this paper, we have formulated a non-linear system of five differential equations of COVID-19 based on the tendency of doctors to be infected. The target of this study is to take a look at the transmission of COVID-19 from asymptomatic populations to the doctors. The model is analyzed with the determination of the basic reproduction number, equilibrium, and related stability analysis at both equilibrium points. The graph of the basic reproductive ratio for different parameters has been drawn to show the disease behavior. Finally, we have simulated our model numerically for visualizing the analytical findings. Our study shows that the asymptomatic population increases as the disease (COVID-19) transmission rate

M. H. A. Biswas (✉) · A. K. Paul · M. S. Khatun · M. A. Islam · M. R. Khatun · S. A. Samad
Mathematics Discipline, Khulna University, Khulna, Bangladesh
e-mail: mhabiswas@yahoo.com

A. K. Paul
e-mail: arindam017@gmail.com

M. S. Khatun
e-mail: shantajsku@gmail.com

M. A. Islam
e-mail: anarul1996@gmail.com

M. R. Khatun
e-mail: reshma161201@gmail.com

S. A. Samad
e-mail: abdussamadku41@gmail.com

S. Mandal · S. Akter
Department of Mathematics, Bangabandhu Sheikh Mujibur Rahman Science and Technology University, Gopalganj, Bangladesh
e-mail: sajibmandal1997@gmail.com

S. Akter
e-mail: soniaaa053@gmail.com

 
N. H. Shah and M. Mittal (eds.), *Mathematical Analysis for Transmission of COVID-19*, Mathematical Engineering, https://doi.org/10.1007/978-981-33-6264-2_3

increases. The number of infected population increases with the infection rate. These increasing asymptomatic and infected populations lead the doctors to get infected by contacting with them. Thus, the whole medical service system is getting down over time.

**Keywords** SARS CoV-2/COVID-19 · Doctors · Mathematical model · Basic reproductive ratio · Numerical simulation

**MSC 2020** Primary: 92Bxx · 37M05 · 34D20 · 92D30 · Secondary: 34A34

## Introduction

This century is experiencing one of the most devastating diseases called COVID-19. The cause of this disease is severe acute respiratory syndrome coronavirus-2 (SARS CoV-2), a newly immerged virus in 2019. COVID-19 is a highly infectious disease that has already spread irrespective of developed, developing, and under-developing countries across the globe. It has become a global health hazard because still now there is no specific medicine or drugs for remedy or even no other prevention like a vaccine that can prevent people from being infected by this disease. The curve of incidence rate and the death rate from coronavirus disease (COVID-19) is becoming skyrocketed day by day. In order to overcome this fatality, our scientists, researchers, government employers, politicians, defenses, experimentalists, intellectuals, health workers are working hand in hand. However, doctors are trying their utmost to rescue people in almost every way. These people are most vulnerable to infection by this disease as they are directly involved in the treatment of COVID-19 patients. Already an alarming number of doctors have died from this fatal disease. And, the death of consultant doctors is an irreversible loss. Because it takes a long time to be a consultant doctor. By this time in the United States, approximately, 600 health workers have died of novel coronavirus disease at the time of performing their duty [1]. In Bangladesh, the coronavirus infection rate among health workers is very high in comparison to other countries in the world. At present, there are about 3301 health workers living with coronavirus infections in Bangladesh, among which 1040 doctors and about 50 doctors have been died by this fatal disease by this time [2]. In the case of doctors' death rate of COVID-19, United States is in the first position followed by Russia (545), UK (540), Brazil (351), Mexico (248), Italy (188), Egypt (111), Iran (91), Ecuador (82), and Spain (63) [3].

Nowadays, mathematical modeling is playing an incredible role in providing quantitative insight into this type of mysterious infectious disease dynamics and maintenance. It has already contributed to a better understanding of the mechanisms of various critical phenomena and has gotten increasing attention because modeling and simulation allow for rapid, cost-effective, and illuminating assessment. In COVID-19, mathematical modeling can be used to better predictions, management, and control strategies.

Our previous work focused on the control policies where we showed that social distancing, educational campaign, and treatment are efficacious for controlling the spread of this horrible disease [4]. Many mathematical models of COVID-19 have been studied as well. Sheffield et al. [5] combined an epidemic model with a cost associated model with a lockdown in comparison to intermittent and moderate lockdown strategies of COVID-19. Bairagi et al. [6] presented an analytical model in order to control the outbreak of COVID-19 by augmenting isolation and social distancing features of populations. Chen and Yu [7] studied a second derivative mathematical model of COVID-19 in order to characterize it in China during the first two months of the outbreak. Chen et al. [8] worked on a Bats-Hosts-Reservoir-People transmission model where they simulated the potential transmission from the infection source to the human infection. Depending on the outbreak of novel coronavirus infections in Thailand, Sookaromdee and Wiwanitkit [9] developed and analyzed a mathematical model of this disease. Zhao et al. [10] proposed a model where they estimated the basic reproductive ratio of COVID-19 on the early detection of its outbreak in China. Ndairoua [11] investigated a compartmental model of COVID-19 by focusing on the transmissibility of super-spreaders populations. We refer Aguilar et al. [12], Carolina et al. [13], Casas-Rojo et al. [14], Chitnis et al. [15], Chen et al. [16], Huo and Feng [17], Kabir et al. [18], Queen Elizabeth Hospital Birmingham COVID-19 airway team [19], Magalhaesa et al. [20], Melliani et al. [21], Nadeem [22], Prompetchara et al. [23], Safi and Garba [24], Zhao et al. [10], Xu et al. [25], Zhang et al. [26], Zhang et al. [27], Biswas [4, 28–34], Khatun and Biswas [35–37] and all other references inside for gathering effective knowledge about this highly infectious disease and a very recent study of modeling and control methods.

The novel coronavirus disease (COVID-19) is a world pandemic. People from all walks of life are taking steps to defend the COVID-19 epidemic but there is no fruitful solution to this disease yet. As per our view, there is no work that highlights on a mathematical model of COVID-19 based on the tendency of doctors to be infected as a result, the medical service system is getting down over the time. Doctors are assets and front line fighters during this tiring situation but most of the countries in the world do not have adequate doctor as many as general people. In that case, the government of those countries has to hire additional doctors and health experts to serve COVID-19 patients. The present paper deals with the five compartmental model: susceptible populations, asymptomatic populations, infected populations, doctor populations, and recovered populations. The novelty of this study is to develop a mathematical model of COVID-19 in terms of a set of non-linear ordinary differential equations showing that doctors are affecting more frequently at the time of serving coronavirus infected patients. Thus, the medical service system has shuttered for the time being.

## Mathematical Model

In our study, we formulate a five compartmental model of coronavirus disease (COVID-19) by showing how frequently doctors are affecting, and thus, the medical service system is getting down over time. Let the compartments are susceptible $S(t)$, asymptomatic $A_s(t)$, infected $I(t)$, doctors $D_r(t)$, and recovered $R(t)$. Here, the first compartment $S(t)$ denotes the susceptible populations and the second compartment $A_s(t)$ is for asymptomatic populations who are actually infected but have no symptoms of COVID-19. The third compartment $I(t)$ denotes the infected populations as those who are actually infected and identified. The fourth compartment $D_r(t)$ denotes the doctor individuals who are serving COVID-19 patients. For that reason, they come to close contact with COVID-19 patients. For this, some doctors are being infected at the time of performing the duty. After being infected, they become unable to provide treatment. While contacting COVID-19 patients, doctors are being infected by a rate of $\xi$. Since there is a non-linear relationship, we can say that if doctors decrease, then more infected people will die. And if the infected people come into contact with susceptible people, the susceptible people can be infected soon at a rate $r$. Another portion of susceptible populations get infected coming into contact with infected people but they do not show any symptoms of COVID-19. This phenomenon is denoted by $\sigma$. For that reason, they can easily contact susceptible people and lead them to become infected faster at a rate $\beta$. When the country demands more doctors for serving the COVID-19 patients, the government starts recruiting the doctors by the rate of $\mu$ from susceptible populations. Some infected populations are dying at a rate $\theta$, and another part is being recovered at the rate $\delta$ after getting proper treatment. $\mu_0$ denotes the natural death of populations. $\xi\gamma$ represents the infection rate of doctors at the time of serving the patients. $(1-\xi)\gamma$ indicates the recovery rate of COVID-19 patients for undertaking treatment at the hospital. The interacting transmissions among different compartments are shown in Fig. 3.1.

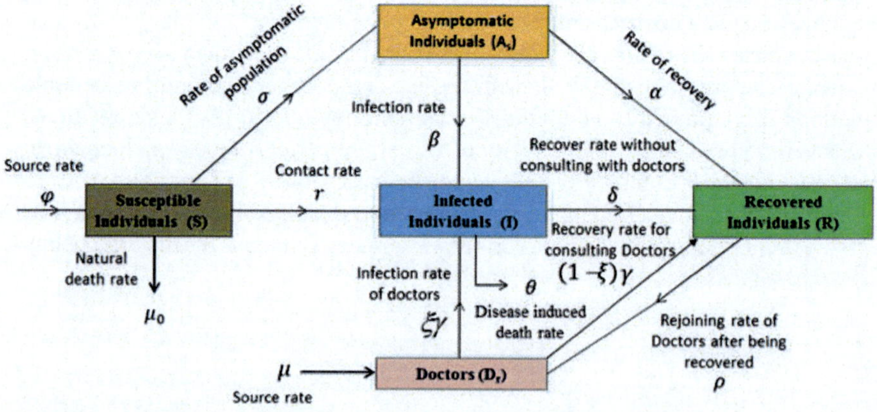


Fig. 3.1 The schematic diagram of the compartmental model of COVID-19

**Table 3.1** Parameter specifications of COVID-19 model (3.1)

| Symbol | Meaning | Values |
|---|---|---|
| $r$ | Contact rate of susceptible and infected populations | 0.002 |
| $\phi$ | Source rate to susceptible populations | 5,000,000 |
| $\gamma$ | Rate of infected cases while consulting with doctors | 0.06 |
| $\theta$ | Death rate from COVID-19 | 0.02 |
| $\delta$ | Recover rate without consulting with doctors | 0.05 |
| $\xi$ | Infection rate of doctors for contacting with COVID-19 patients | 0.00006 |
| $\rho$ | Rejoining rate of doctors after getting recovered | 0.00005 |
| $\beta$ | Infection rate of infected populations | 0.005 |
| $\mu$ | Source rate of doctors | 150 |
| $\sigma$ | Disease transmission rate of asymptomatic individuals for contacting an infected and susceptible population | 0.003 |
| $\alpha$ | Recovery rate of asymptomatic populations | 0.03 |

*Source* Estimated from [1–3]

Now taking all the situations in Fig. 3.1 into consideration, the proposed model can be formulated by the following set of non-linear ODE:

$$\begin{cases} \dfrac{dS(t)}{dt} = \varphi - \mu_0 S(t) - rI(t)S(t) - \sigma I(t)S(t) \\ \dfrac{dA_s(t)}{dt} = \sigma I(t)S(t) - \beta A_s(t)S(t) - \alpha A_s(t) \\ \dfrac{dI(t)}{dt} = \beta A_s(t)S(t) - \theta I(t) - \delta I(t) + rI(t)S(t) + \gamma\xi D_r(t)I(t) \\ \dfrac{dD_r(t)}{dt} = \mu + \rho R(t) - \gamma D_r(t)I(t) \\ \dfrac{dR(t)}{dt} = \alpha A_s(t) + \delta I(t) - \rho R(t) + \gamma D_r(t)I(t)(\xi - 1) \end{cases} \quad (3.1)$$

with boundary conditions

$$S(0) = S_0 > 0,\ A_s(0) = A_{s0} \geq 0,\ I(0) = I_0 \geq 0,\ D_r(0) = D_{r0} \geq 0,\ R(0) = R_0 \geq 0$$

The brief description of the parameters used in the model are shown in Table 3.1.

## Analysis of the Compartmental Model

We have done the positivity analysis of the solutions of the model in order to show the validation and well-posedness of this model. Further, we have calculated different equilibria, basic reproduction ratio and have analyzed the stability at two different equilibrium points.

## *Positivity Analysis*

**Lemma 1** *If* $S(t) > 0$, $A_s(t) \geq 0$, $I(t) \geq 0$, $D_r(t) \geq 0$ *and* $R(t) \geq 0$ *then the solutions* $S(t)$, $A_s(t)$, $I(t)$, $D_r(t)$ *and* $R(t)$ *of the model* (3.1) *are non-negative.*

***Proof*** In order to show the proof of Lemma 1, we use the set of equations (3.1).

$$\frac{\mathrm{d}S(t)}{\mathrm{d}t} = \varphi - \mu_0 S(t) - rI(t)S(t) - \sigma I(t)S(t) \tag{3.2}$$

To seek positivity, we can write

$$\begin{aligned} \frac{\mathrm{d}S(t)}{\mathrm{d}t} &\geq \varphi - \mu_0 S(t) \\ \Rightarrow \frac{\mathrm{d}S(t)}{\mathrm{d}t} &+ \mu_0 S(t) \geq \varphi \end{aligned} \tag{3.3}$$

The integrating factor of (3.3) is given by

$$\therefore I.F. = \mathrm{e}^{\int \mu_0 \mathrm{d}t} = \mathrm{e}^{\mu_0 t}$$

Multiplying $\mathrm{e}^{\mu_0 t}$ on both sides of (3.3), we get

$$\frac{\mathrm{d}}{\mathrm{d}t}\left(\mathrm{e}^{\mu_0 t} S(t)\right) \geq \varphi \mathrm{e}^{\mu_0 t} \tag{3.4}$$

Now, by integrating (3.4), we have

$$S(t) \geq \frac{\varphi}{\mu_0} + c\mathrm{e}^{-\mu_0 t} \tag{3.5}$$

where $c$ is an integrating constant.

Considering the initial value at $t = 0$, $S(t) \geq S(0)$.

From (3.5), we attain

$$S(0) \geq \frac{\varphi}{\mu_0} + c \Rightarrow S(0) - \frac{\varphi}{\mu_0} \geq c$$

Substituting the value of $c$ into (3.5), we obtain

$$S(t) \geq \frac{\varphi}{\mu_0} + \left(S(0) - \frac{\varphi}{\mu_0}\right)\mathrm{e}^{-\mu_0 t} \tag{3.6}$$

So, at $t = 0$ and $t \to \infty$, $S(t) > 0$. By repeating the above procedure, we can prove the positivity of all other state variables.

Consequently, it is clear that $\forall t \geq 0$.

$$S(t) > 0,\ A_s(t) \geq 0,\ I(t) \geq 0,\ D_r(t) \geq 0.$$

Thus, Lemma 1 is undoubtedly proven.

## Equilibrium Points

Let, $E_{dfe}(S^*, A_s^*, I^*, D_r^*, R^*)$ be the disease-free equilibrium point of the model (3.1). In order to find the disease-free equilibrium point, we need to solve $\frac{\mathrm{d}S^*}{\mathrm{d}t} = \frac{\mathrm{d}A_s^*}{\mathrm{d}t} = \frac{\mathrm{d}I^*}{\mathrm{d}t} = \frac{\mathrm{d}D_r^*}{\mathrm{d}t} = \frac{\mathrm{d}R^*}{\mathrm{d}t} = 0$ of the model (3.1). At the disease-free equilibrium point $E_{dfe}(S^*, A_s^*, I^*, D_r^*, R^*)$, the model (3.1) takes the following form:

$$\begin{aligned}
&\varphi - \mu_0 S^*(t) - rI^*(t)S^*(t) - \sigma I^*(t)S^*(t) = 0\\
&\sigma I^*(t)S^*(t) - \beta A_s^*(t)S^*(t) - \alpha A_s^*(t) = 0\\
&\beta A_s^*(t)S^*(t) - \theta I^*(t) - \delta I^*(t) + rI^*(t)S^*(t) + \gamma\xi D_r^*(t)I^*(t) = 0\\
&\mu + \rho R^*(t) - \gamma D_r^*(t)I^*(t) = 0\\
&\alpha A_s^*(t) + \delta I^*(t) - \rho R^*(t) + \gamma D_r^*(t)I^*(t)(\xi - 1) = 0
\end{aligned} \tag{3.7}$$

Since infection not found at the $E_{dfe}(S^*, A_s^*, I^*, D_r^*, R^*)$ (i.e., $A_s^*(t) = 0,\ I^*(t) = 0,\ R^*(t) = 0$), from (3.7), we have

$$S^*(t) = \frac{\varphi}{\mu_0},\ D_r^*(t) = 0.$$

Hence, $E_{dfe}(S^*, A_s^*, I^*, D_r^*, R^*) = \left(\frac{\varphi}{\mu_0}, 0, 0, 0, 0\right)$.

Similarly, we solve the system of equations (3.7) for finding $E_{ee}(S^*, A_s^*, I^*, D_r^*, R^*)$ where

$$S^* = \frac{\varphi\theta}{(\mu+\varphi)(r+\sigma)},\ A_s^* = \frac{\sigma\varphi^2 + \mu\sigma\varphi}{\alpha\mu r + \alpha\mu\sigma + \alpha\varphi r + \alpha\varphi\sigma + \beta\varphi\theta},\ I^* = \frac{\mu+\varphi}{\theta},$$

$$Dr^* = \frac{\begin{array}{c}\alpha\delta\mu^2 r + \alpha\delta\mu^2\sigma + \alpha\delta\varphi^2 r + \alpha\delta\varphi^2\sigma + \beta\delta\varphi^2\theta + \beta\mu\varphi\theta^2 + \alpha\mu^2 r\theta\\ +\alpha\mu^2\sigma\theta + \alpha\varphi^2\sigma\theta + 2\alpha\delta\mu\varphi r + 2\alpha\delta\mu\varphi\sigma + \beta\delta\mu\varphi\theta + \alpha\mu\varphi r\theta + 2\alpha\mu\varphi\sigma\theta\end{array}}{\xi(\gamma\mu + \gamma\varphi)(\alpha\mu r + \alpha\mu\sigma + \alpha\varphi r + \alpha\varphi\sigma + \beta\varphi\theta)},$$

$$R^* = \frac{\begin{array}{c}\alpha\delta\mu^2 r + \alpha\delta\mu^2\sigma + \alpha\delta\varphi^2 r + \alpha\delta\varphi^2\sigma + \beta\delta\varphi^2\theta + \beta\mu\varphi\theta^2 + \alpha\mu^2 r\theta + \alpha\mu^2\sigma\theta\\ +\alpha\varphi^2\sigma\theta + 2\alpha\delta\mu\varphi r + 2\alpha\delta\mu\varphi\sigma + \beta\delta\mu\varphi\theta + \alpha\mu\varphi r\theta + 2\alpha\mu\varphi\sigma\theta\\ -\beta\mu\varphi\theta^2\xi - \alpha\mu^2 r\theta\xi - \alpha\mu^2\sigma\theta\xi - \alpha\mu\varphi r\theta\xi - \alpha\mu\varphi\sigma\theta\xi\end{array}}{\rho\theta\xi(\alpha\mu r + \alpha\mu\sigma + \alpha\varphi r + \alpha\varphi\sigma + \beta\varphi\theta)}.$$

### *Basic Reproduction Ratio*

We have used the next-generation matrix method for finding the basic reproduction number of our model [38]. To employ this method, we have to seek the classes that are disease or infection related term. From the infection subsystem, we get transmission and transition matrix $F$ and $V$.

Here, the matrix for the transmission terms:

$$F = \begin{pmatrix} 0 & \sigma \\ r & 0 \end{pmatrix}$$

And, the matrix for the transition terms:

$$V = \begin{pmatrix} -\alpha - \beta & 0 \\ \beta & -\delta - \gamma - \theta \end{pmatrix}$$

Then, we need to estimate a matrix $G$, so that $G = FV^{-1}$.

$$G = FV^{-1} = \begin{pmatrix} \frac{\beta\sigma}{(\alpha+\beta)(\delta+\theta+\gamma)} & \frac{\sigma}{\delta+\theta+\gamma} \\ \frac{r}{\alpha+\beta} & 0 \end{pmatrix}$$

Now, the characteristic polynomial is given by setting $|G - \lambda I| = 0$.

$$\Rightarrow \alpha\lambda^2\theta - r\sigma + \alpha\delta\lambda^2 + \alpha\gamma\lambda^2 + \beta\delta\lambda^2 - \beta\sigma\lambda + \beta\theta\lambda^2 + \beta\gamma\lambda^2 = 0$$

Solving this equation, we get the following basic reproduction number.

$$R_0 = \frac{\beta\sigma + \sqrt{\sigma\left(\beta^2\sigma + 4\alpha\delta r + 4\beta\delta r + 4\alpha\gamma r + 4\beta\gamma r + 4\alpha r\theta + 4\beta r\theta\right)}}{2(\alpha\delta + \beta\delta + \alpha\gamma + \beta\gamma + \alpha\theta + \beta\theta)}.$$

$R_0$ signifies an important role in disease modeling that if $R_0 > 1$, the disease will persist, and if $R_0 < 1$, the disease will die out.

### *Stability Analysis at DFE* ($\mathbf{E_{dfe}}$)

Here, we have investigated the stability at $E_{dfe}$ by establishing the Theorem 1.

**Theorem 1** *The disease-free equilibrium (DEF) point of the model (3.1) is asymptotically stable if the eigenvalues of the Jacobian matrix are negative.*

***Proof*** The Jacobian of (3.1) is given by

$$J = \begin{bmatrix} -\mu_0 - Ir - I\sigma & 0 & -Sr - S\sigma & 0 & 0 \\ I\sigma - A_s\beta & -\alpha - S\beta & S\sigma & 0 & 0 \\ A_s\beta + Ir & S\beta & Sr - \theta - D_r\gamma - \delta + D_r\gamma\xi & I\gamma\xi - I\gamma & 0 \\ 0 & 0 & -D_r\gamma\xi & -I\gamma\xi & \rho \\ 0 & \alpha & \delta - D_r\gamma(\xi - 1) & -I\gamma(\xi - 1) & -\rho \end{bmatrix} \tag{3.8}$$

The characteristic polynomial can be attained as $|J - \lambda I| = 0$.

$$\begin{vmatrix} -\mu_0 - Ir - I\sigma - \lambda & 0 & -Sr - S\sigma & 0 & 0 \\ I\sigma - A_s\beta & -\alpha - S\beta - \lambda & S\sigma & 0 & 0 \\ A_s\beta + Ir & S\beta & Sr - \theta - D_r\gamma - \delta + D_r\gamma\xi - \lambda & I\gamma\xi - I\gamma & 0 \\ 0 & 0 & -D_r\gamma\xi & -I\gamma\xi - \lambda & \rho \\ 0 & \alpha & \delta - D_r\gamma(\xi - 1) & -I\gamma(\xi - 1) & -\rho - \lambda \end{vmatrix}$$
$$= 0$$

Substituting $S^* = \frac{\varphi}{\mu_0}$, we obtain

$$\begin{pmatrix} -\lambda - \mu_0 & 0 & -\frac{\varphi r}{\mu_0} - \frac{\varphi\sigma}{\mu_0} & 0 & 0 \\ 0 & -\alpha - \lambda - \frac{\beta\varphi}{\mu_0} & \frac{\varphi\sigma}{\mu_0} & 0 & 0 \\ 0 & \frac{\beta\varphi}{\mu_0} & \frac{\varphi r}{\mu_0} - \lambda - \theta - \delta & 0 & 0 \\ 0 & 0 & 0 & -\lambda & \rho \\ 0 & \alpha & \delta & 0 & -\lambda - \rho \end{pmatrix} = 0 \tag{3.9}$$

By taking determinant and solving it for λ.

$$\lambda_1 = 0, \lambda_2 = -\mu_0, \lambda_3 = -\rho, \lambda_3 = -\frac{\alpha\mu_0 + \delta\mu_0 + \beta\varphi + \mu_0\theta - \varphi r + \sqrt{\psi}}{2\theta},$$
$$\lambda_5 = -\frac{\alpha\mu_0 + \delta\mu_0 + \beta\varphi + \mu_0\theta - \varphi r - \sqrt{\psi}}{2\mu_0}.$$

where

$$\begin{aligned} \psi &= \alpha^2\mu_0^2 + 2\alpha\beta\mu_0\varphi - 2\alpha\delta\mu_0^2 - 2\alpha\mu_0^2\theta + 2\alpha\mu_0\varphi r \\ &+ \beta^2\varphi^2 - 2\beta\delta\mu_0\varphi - 2\beta\mu_0\varphi\theta + 2\beta\varphi^2 r + 4\sigma\beta\varphi^2 \\ &+ \delta^2\mu_0^2 + 2\delta\mu_0^2\theta - 2\delta\mu_0\varphi r + \mu_0^2\theta^2 - 2\mu_0\varphi r\theta + \varphi^2 r^2 \end{aligned}$$

Since $\lambda_1, \lambda_2, \lambda_3, \lambda_4$ and $\lambda_5$ are all negative, $E_{dfe}$ point is asymptotically stable. Hence, Theorem 1 is proved.

### *Stability Analysis at EE* ($\mathbf{E_{ee}}$)

In this section, we have investigated the stability at $E_{dfe}$ by proving the Theorem 2.

**Theorem 2** *The endemic equilibrium (DEF) point of the model (3.1) is asymptotically stable if the eigenvalues of the Jacobian matrix are negative.*

***Proof*** The Jacobian of (3.1) at $E_{ee}$ point is

$$J = \begin{bmatrix} -\mu_0 - rI^* - \sigma I^* & 0 & -rS^* - \sigma S^* & 0 & 0 \\ \sigma I^* - \beta A_s^* & -\beta S^* - \alpha & \sigma S^* & 0 & 0 \\ \beta A_s^* + rI^* & \beta S^* & -\theta - \gamma D_r^* - \delta + rS^* + \gamma\xi D_r^* & -\gamma I^* + \gamma\xi I^* & 0 \\ 0 & 0 & -\gamma\xi D_r^* & -\gamma\xi I^* & \rho \\ 0 & \alpha & \delta - \gamma D_r^* & -\gamma I^* & -\rho \end{bmatrix} \tag{3.10}$$

In Echelon form, the above matrix is written as

$$J = \begin{bmatrix} -\mu_0 - rI^* - \sigma I^* & 0 & -rS^* - \sigma S^* & 0 & 0 \\ 0 & -\beta S^* - \alpha & \frac{\beta A_s^* S^*}{I^*} & 0 & 0 \\ 0 & 0 & -a_{33} & -\gamma I^* + \gamma\xi I^* & 0 \\ 0 & 0 & 0 & -a_{44} & \rho \\ 0 & 0 & 0 & 0 & -a_{55} \end{bmatrix} \tag{3.11}$$

Equation (3.11) is a $5 \times 5$ matrix and the characteristic polynomial having eigenvalue $\lambda$ is given by $|J - \lambda I| = 0$.

$$\begin{vmatrix} -\mu_0 - rI^* - \sigma I^* - \lambda & 0 & -rS^* - \sigma S^* & 0 & 0 \\ 0 & -\beta S^* - \alpha - \lambda & \frac{\beta As^* S^*}{I^*} & 0 & 0 \\ 0 & 0 & -a_{33} - \lambda & -\gamma I^* + \gamma\xi I^* & 0 \\ 0 & 0 & 0 & -a_{44} - \lambda & \rho \\ 0 & 0 & 0 & 0 & -a_{55} - \lambda \end{vmatrix} = 0$$

$$\Rightarrow (-\mu_0 - rI^* - \sigma I^* - \lambda)(-\beta S^* - \alpha - \lambda)(-a_{33} - \lambda)(-a_{44} - \lambda)(-a_{55} - \lambda) = 0$$

Here the eigenvalues are given by

$$\lambda_1 = -(\mu_0 + rI^* + \sigma I^*),\ \lambda_2 = -(\beta S^* + \alpha),\ \lambda_3 = -a_{33},\ \lambda_4 = -a_{44},\ \lambda_5 = -a_{55}.$$

where

$$a_{33} = \frac{\alpha(\theta + \delta + \gamma D_r^*) + \beta S^*(\theta + \delta + \gamma D_r^*) - \gamma\xi D_r^*(\alpha + \beta S^*)}{\beta S^* + \alpha} + \frac{\alpha\beta S^* A_s^*}{(\beta S^* + \alpha) Q^*}$$

$$a_{44} = \frac{\gamma\xi(\delta + \theta) Q^{*2}\{\alpha + \beta S^*\} + \alpha\beta\gamma\xi S^* A_s^* Q^*}{(\theta + \delta)\{\alpha + \beta S^*\} Q^* + (1 - \xi)\{\gamma\alpha + \gamma\beta S^*\} Q^* D_r^* + \alpha\beta S^* A_s^*}$$

$$a_{55} = \frac{\rho\left[\beta\{\xi\theta + 2\delta + \theta\}Q^* S^* + \alpha\{\xi\theta + 2\delta + \theta\}Q^* + 2\alpha\beta S^* A_s^*\right]}{\xi\left[(\delta + \theta)\{\alpha + \beta S^*\}Q^* + \alpha\beta S^* A_s^*\right]}$$

The eigenvalues of the characteristic polynomial are $\lambda_1 = -(rI^* + \sigma I^*)$, $\lambda_2 = -(\beta S^* + \alpha)$, $\lambda_3 = -a_{33}$, $\lambda_4 = -a_{44}$, and $\lambda_5 = -a_{55}$, which are all real numbers. Since all the eigenvalues ($\lambda_1, \lambda_2, \lambda_3, \lambda_4$ and $\lambda_5$) have a negative real part, $E_{ee}$ is asymptotically stable. Hence, Theorem 2 is proved.

## Results and Discussion

Computer simulations of any biological phenomena provide a rapid, cost-effective, and illuminating assessment. For this reason, numerical simulations of the developed COVID-19 model (3.1) have been carried out by the Runge–Kutta-Fehlberg method using MATLAB programming language. We have taken initial population at $S = 10,000,000$, $A_s = 30,000$, $I = 50,000$, $D_r = 3000$, $R = 20,000$. Firstly, we have solved the model (3.1) for the tabulated values in Table 3.1, representing all the parametric values considered for our model and the figure obtained from numerical simulation is presented in Fig. 3.2. Also, we simulated the model for different parameter values of $\sigma$ and $\beta$ keeping all other values the same to show their effects on the model. In this case, the graphs are shown in Figs. 3.3, 3.4, 3.5, 3.6, 3.7 and 3.8. The parameter $\sigma$ represents disease transmission rate of asymptomatic individuals for contacting an infected and susceptible population. Thus, this parameter has great value in respect to the disease transmissions. So, we perform simulations for different values of $\sigma$ and the graphs are presented in Figs. 3.4, 3.5, 3.6, and 3.7. Furthermore, $\beta$ is the rate at which the infected populations get infectious for contacting COVID-19 patients, and this parameter has a significant role for disease transmission among the doctors. So, considering different values of this parameter, Figs. 3.8 and 3.9 have been drawn. Finally, we show the graphs of the basic reproductive ratio presented in Figs. 3.10, 3.11, 3.12 and 3.13. It helps us to predict whether the disease will persist or die out.

Figure 3.2 shows the dynamics of five compartments such as susceptible, asymptomatic, infected, doctors, and recovered populations. We have observed that the susceptible population decreases from the initial state and it reaches to zero steadily. The asymptomatic population decreases for the first week and quickly reaches to zero after that it gradually booms. The infected populations progressively soar from the initial state and reach to the peak level leading to decrease the number of doctors. As coronavirus disease is highly infectious, doctors are frequently getting infected at the time of performing their novel duty. So, the doctors are decreasing surprisingly from the initial state. Despite this, the recovered populations and death rate increase from the initial state. It shows that the infection rate of coronavirus disease increases leading to an enormous death every day and causing a massive number of infections among doctors. Thus, the whole medical service system is getting down over time.

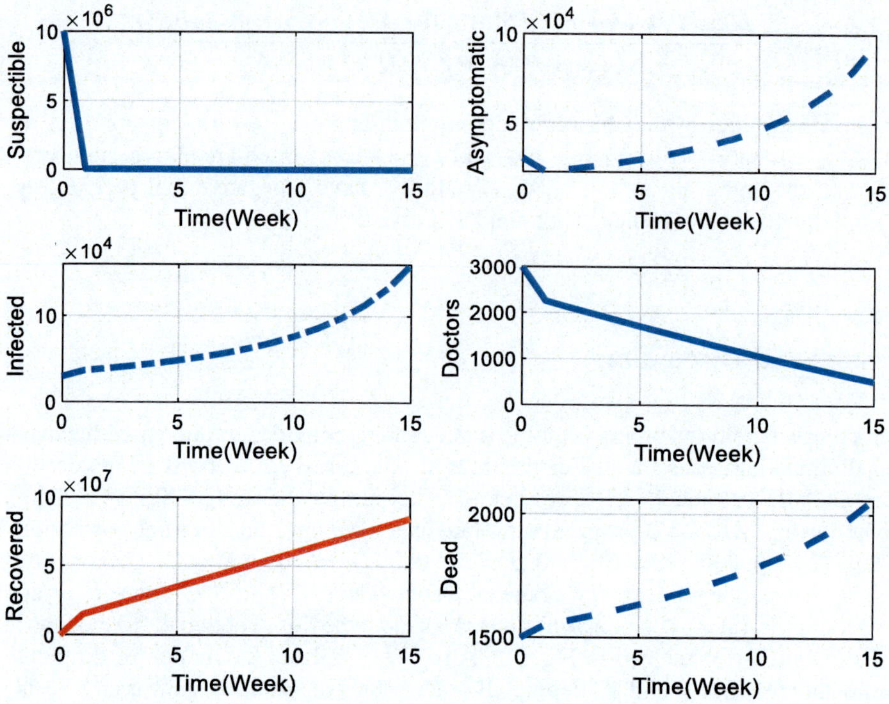


**Fig. 3.2** Trajectories of all the compartments of the epidemic model (3.1) (*Source* Own)

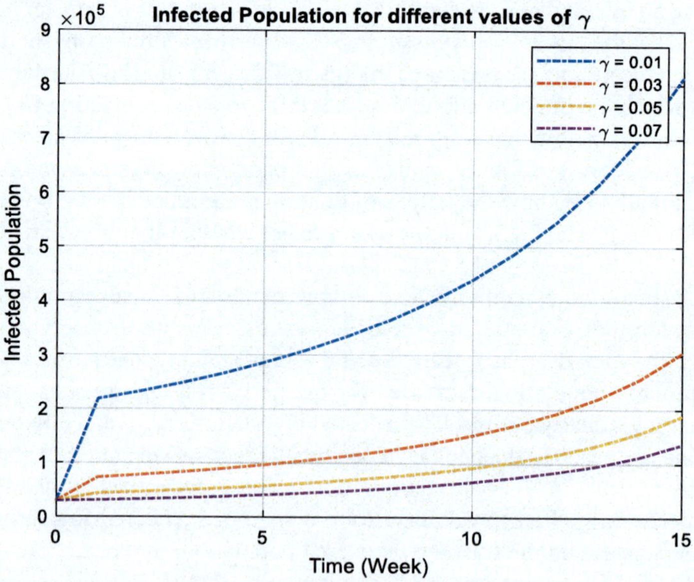


**Fig. 3.3** Change of infected population over 15 weeks for $\gamma = 0.01, 0.03, 0.05, 0.07$ (*Source* Own)

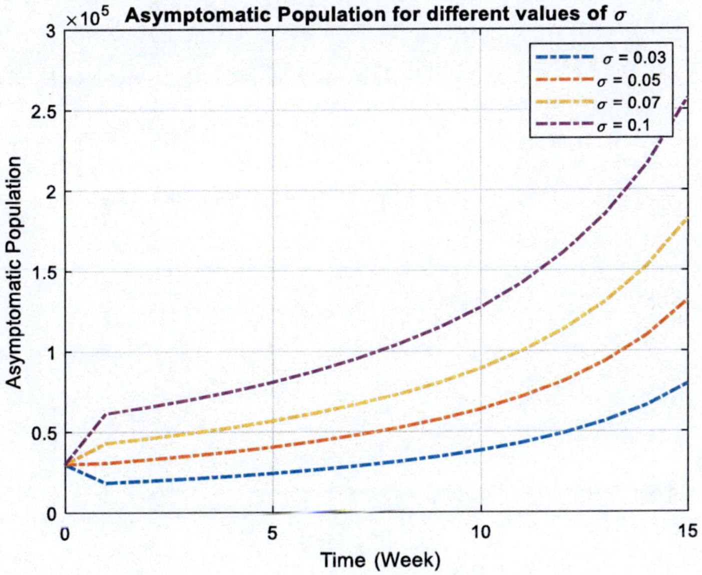


**Fig. 3.4** Change of asymptomatic population over 15 weeks for $\sigma = 0.03, 0.05, 0.07, 0.1$ (*Source* Own)

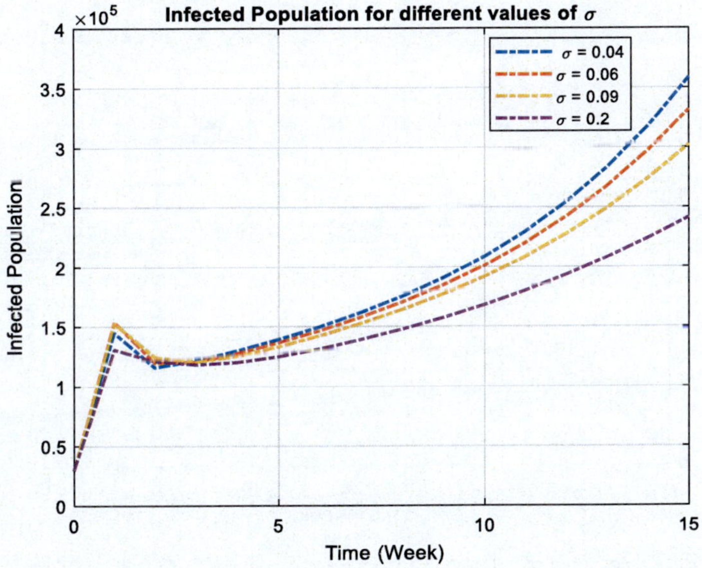


**Fig. 3.5** Change of infected population over 15 weeks for $\sigma = 0.04, 0.06, 0.09, 0.2$ (*Source* Own)

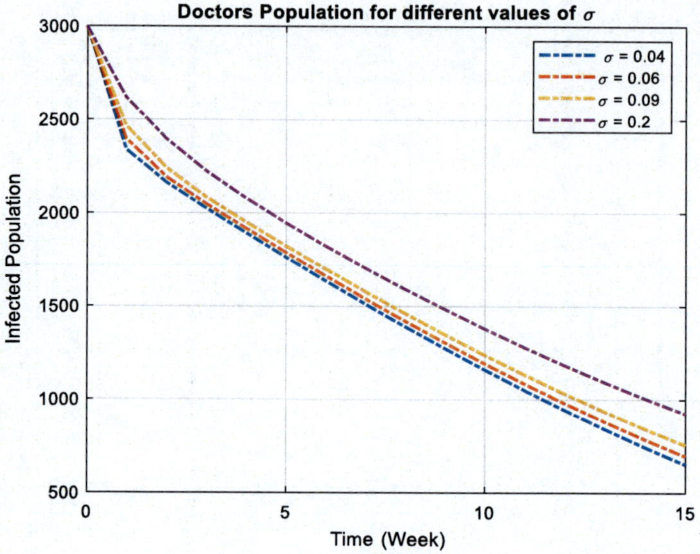


**Fig. 3.6** Change of doctor population over 15 weeks for $\sigma = 0.04, 0.06, 0.09, 0.2$ (*Source* Own)

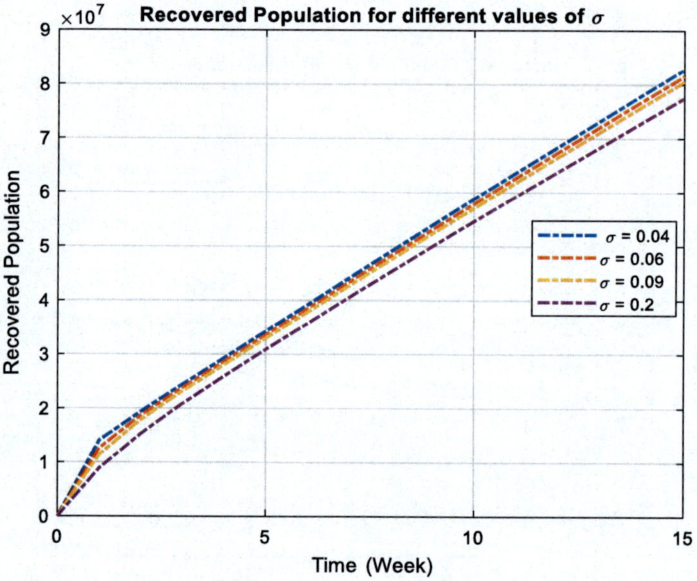


**Fig. 3.7** Change of recovered population (15 weeks) for $\sigma = 0.04, 0.06, 0.09, 0.2$ (*Source* Own)

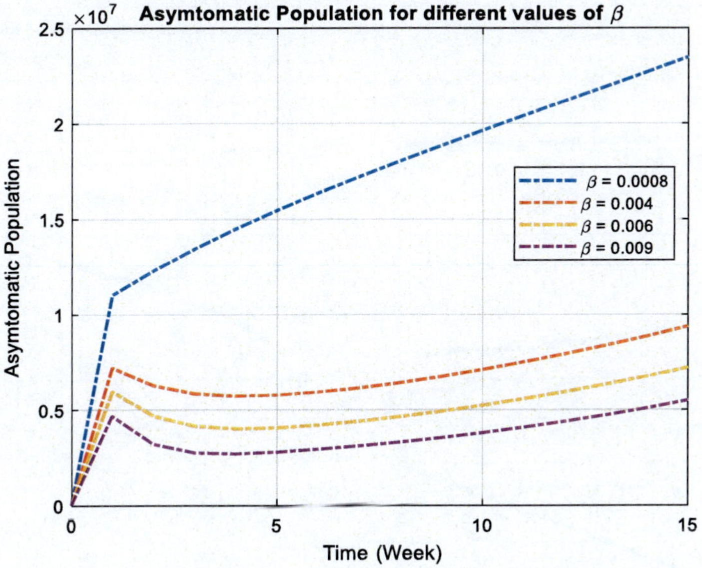


**Fig. 3.8** Change of asymptomatic population over (15 weeks) for $\beta = 0.0008, 0.004, 0.006, 0.009$ (*Source* Own)

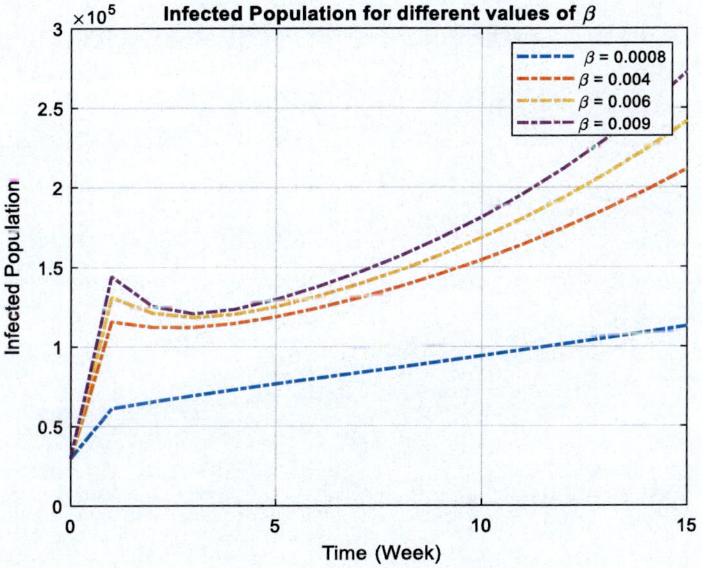


**Fig. 3.9** Change of asymptomatic population over (15 weeks) for $\beta = 0.0008, 0.004, 0.006, 0.009$ (*Source* Own)

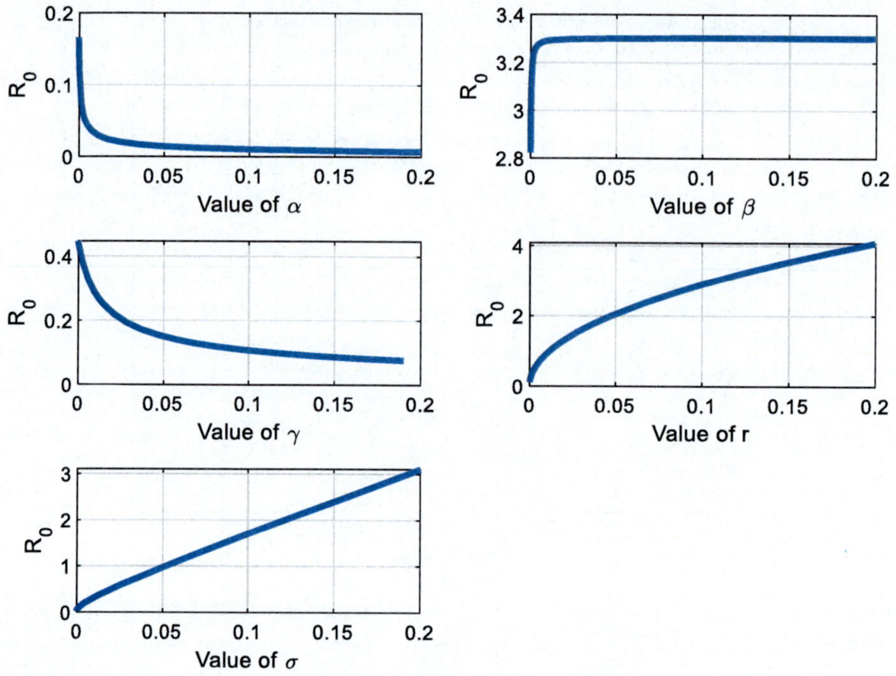


**Fig. 3.10** Graphical illustration of the basic reproductive ratio for different parameter value range (*Source* Own)

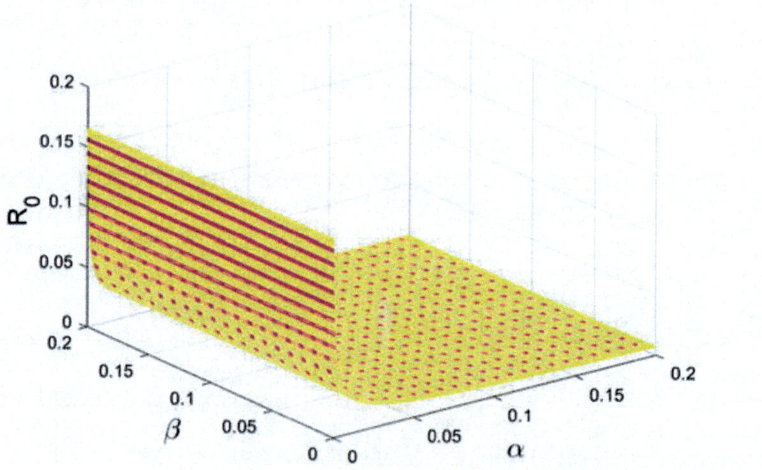


**Fig. 3.11** The graph of basic reproductive ratio ($R_0$) with respect to the infection rate ($\beta$) of infected populations and recovery rate ($\alpha$) of asymptomatic populations (*Source* Own)

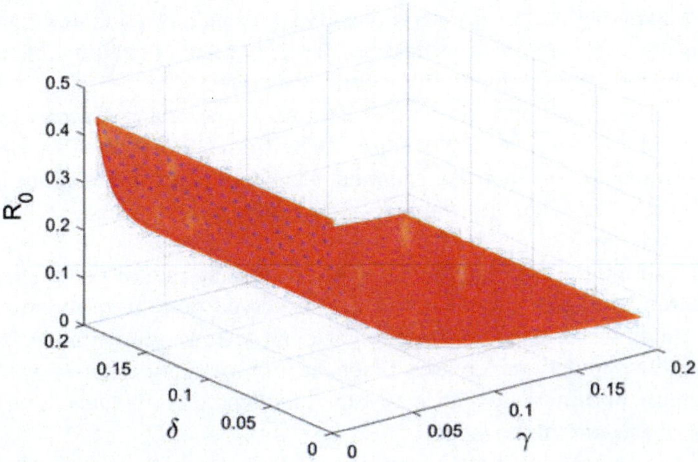


**Fig. 3.12** The graph of basic reproductive ratio ($R_0$) with respect to the recover rate ($\delta$) without consulting with doctors and recovered rate ($\gamma$) due to consulting with doctors (*Source* Own)

From Fig. 3.3, we can say that for a higher rate of $\gamma$, the infected population decrease with time because the majority of COVID-19 patients undertake treatment to relieve the disease. This leads to a less number of asymptomatic individuals. Hence, these reduced asymptomatic individuals do not infect the doctors massively.

Figure 3.4 indicates the influence of disease transmission rate ($\sigma$) on the asymptomatic populations for 15 weeks period. Asymptomatic populations are those who are infected by the coronavirus disease. The unique feature of these asymptomatic populations is that they don't reveal any signs or symptoms of COVID-19. For that reason, these asymptomatic populations spread the disease among people subconsciously. In Fig. 3.4, it has been noticed that the asymptomatic populations increase as the disease transmission rate (when $\sigma$ rises from 0.03 to 0.1) increases.

Figure 3.5 represents the effect of the disease transmission rate ($\sigma$) on the infected populations for 15 weeks period. Infected populations are those who are actually infected by the coronavirus disease, identified, and able to transmit it among people. From Fig. 3.5, we have observed that the infected populations decrease sequentially as the disease transmission rate $\sigma$ (when $\sigma$ rises from 0.04 to 0.2) of asymptomatic populations increases. This shows that the asymptomatic populations increase as the disease transmission rate $\sigma$ increases but the infected populations decrease as this parameter increases. This is because of the increasing number of asymptomatic populations lead to a decrease in the concentration of infected populations. As the more populations remain in asymptomatic condition, the more populations do not get identified as COVID-19 carriers. Thus, a massive population will remain unidentified and hence the number of identified infected populations will decrease.

Figure 3.6 shows the variation of doctor populations for different values of disease transmission rate ($\sigma$), the rate at which the susceptible populations becoming infectious by contacting COVID-19 patients. A remarkable number of doctors have been infected worldwide at the time of serving COVID-19 patients, and as a result, they have died by this novel coronavirus infection. Most of them are getting infected with coronavirus by the asymptomatic populations. This scenario is presented in Fig. 3.6, and from this figure, it has been observed that the doctor populations decrease quickly as the disease transmission rate ($\sigma$) (when $\sigma$ ranges from 0.04 to 0.2) of asymptomatic populations increases. The increasing number of asymptomatic populations leads the doctors to be infected. Thus, the number of doctors to serve patients is going down over time.

Figure 3.7 shows the effect of the disease transmission rate ($\sigma$) on the recovered populations for 15 weeks period. It has been observed that the recovered populations decrease as the disease transmission rate $\sigma$ (when $\sigma$ goes from 0.04 to 0.2) of asymptomatic population increases. Since the asymptomatic population increases with the increase of disease transmission rate $\sigma$, recovered individuals decrease with the increase of this parameter value.

Figure 3.8 presents the effect of infection rate ($\beta$) on the asymptomatic populations for 15 weeks period. From this figure, it has been noticed that the asymptomatic populations decrease as the infection rate $\beta$ (when $\beta$ goes from 0.0008 to 0.009) of the infected population increases. Since the infected population increases with the increase of infection rate $\beta$, the asymptomatic population decreases with the increase of this parameter value.

Figure 3.9 exhibits the effect of infection rate ($\beta$) on the infected populations for 15 weeks period. It has been observed that the infected populations increase swiftly for the first two weeks from the initial state but it slowly decreases to the next two weeks as the infection rate ($\beta$) (when $\beta$ goes from 0.0008 to 0.006) increases. After that, the infected populations gradually increase as the infection rate ($\beta$) increases. It means that when the infection rate ($\beta$) increases, the number of infected individuals also increases.

The simulated graphs presented in Fig. 3.10, shows the schematic view of the basic reproductive ratio of the model (3.1). We have performed the simulation of basic reproductive ratio ($R_0$) with respect to $\alpha$, $\beta$, $\gamma$, $r$, and $\sigma$. From Fig. 3.10, it has been noticed that the basic reproductive ratio $R_0 < 1$ for all values of $\alpha$ and $\gamma$ whereas $R_0 > 1$ for all values $\beta$, $r$, and $\sigma$. Hence, the disease-free equilibrium point is locally asymptotically stable if $R_0 < 1$ and unstable if $R_0 > 1$. Whereas, the endemic equilibrium point is locally asymptotically stable if $R_0 > 1$ and unstable if $R_0 < 1$ [37]. Figures 3.11, 3.12 and 3.13 present a 3-dimensional plot of basic reproductive ratio.

## Conclusions

People from the entire world are confined at the home due to life-threatening coronavirus disease (COVID-19). In spite of the deadliness of COVID-19, doctors are performing their novel duty to serve the coronavirus infected patients. Thus, they are acting as first-line soldiers keeping themselves vulnerable to coronavirus infections.

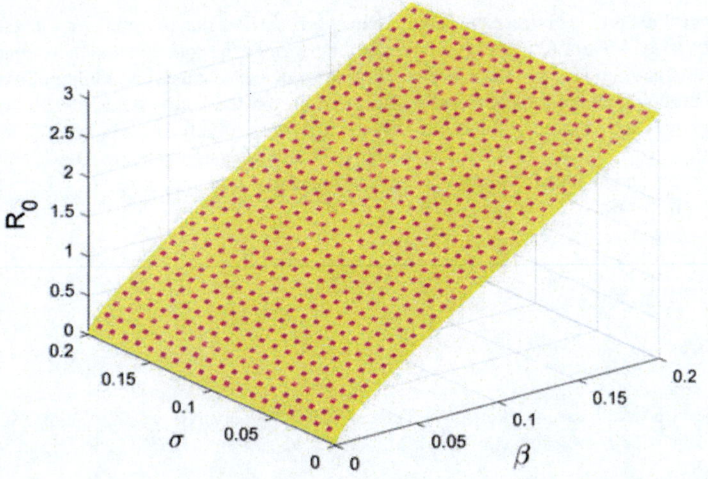


**Fig. 3.13** The graph of basic reproductive ratio ($R_0$) with respect to the disease transmission rate ($\sigma$) of asymptomatic and the infection rate ($\beta$) of infected populations (*Source* Own)

In this paper, we have developed a mathematical model of COVID-19 in terms of a set of non-linear ordinary differential equations showing that doctors are affecting more frequently at the time of serving coronavirus infected patients. So, the medical service system is going down over time. We have analyzed the model by determining of the basic reproductive ratio and related stability analysis at the disease-free and endemic equilibrium points. The graph of the basic reproductive ratio for different parameters has been carried out to show the disease behavior. Finally, numerical simulations have been performed to illustrate the analytic results. We have observed that the asymptomatic population increases as the disease (COVID-19) transmission rate increases and also the number of infected population increases when the infection rate increases. These increasing asymptomatic and infected populations lead the doctors to be infected by contact with them. Thus, the whole medical service system is getting down over time. So, it is time to save our supreme warriors (doctors) during this coronavirus outbreak by ensuring their proper safety like life-saving protective equipment.

**Acknowledgements** The authors greatly acknowledge the partial financial support provided by the Ministry of Science and Technology, Government of the People's Republic of Bangladesh under special allocation in 2019-2020 with the research grant Ref. No.-39.00.0000.009.06.024.19-12/410(EAS). Supports with Ref.: 17-392 RG/MATHS/AS_I–FR3240297753 funded by TWAS, Italy, and Ref. no.-6(74) UGC/ST/Physical-17/2017/3169 funded by the UGC, Bangladesh is also acknowledged.

**Data Availability** The data used to support the findings of this study are included in the article.

**Conflicts of Interest** The authors declare that there are no conflicts of interest regarding the publication of this manuscript.

**Authors' Contributions** This research is a group work carried out in collaboration among all authors. Authors MHAB and AKP designed the study, performed the conceptualization and methodological analysis and model formulation of the first draft of the manuscript. Authors SM and SA analyzed the model analytically and wrote some literature on the study. Author MSK wrote the programming codes and performed some part of the computational analysis. Author SAS contributed to literature searches and calculated the real data to estimate the parameters, MAI and MRK verified the parameters and checked the literature. All authors have read and agreed to publish the final version of the manuscript.

## References

1. Nearly 600 US health care workers have died of COVID-19. abc News (2020). https://abcnews.go.com/Health/600-us-health-care-workers-died-covid-19/story?id=71132667. Last accessed 2020/06/8.
2. Coronavirus: Doctors' mortality rate in Bangladesh 'highest in the world'. United News of Bangladesh (UNB). https://unb.com.bd/category/Special/coronavirus-doctors-mortality-rate-in-bangladesh-highest-in-the-world/53378. Last accessed 2020/06/21.
3. Global: Health workers silenced, exposed and attacked. Amnesty International. https://www.amnesty.org/en/latest/news/2020/07/health-workers-rights-covid-report/. Last accessed 2020/07/13.
4. Biswas, M. H. A., Khatun, M. S., Paul, A. K., Khatun, M. R., Islam, M. A., Samad, S. A., & Ghosh, U. (2020). Modeling the effective control strategy for the transmission dynamics of global pandemic COVID-19. Preprint (2020). https://doi.org/10.1101/2020.04.22.20076158.
5. Sheffield, S., York, A., Swartwood, N., Bilinski, A., Williamson, A., & Fitzpatrick, M. (2020). Strict physical distancing may be more efficient: A mathematical argument for making the lockdowns count. Preprint (2020). https://doi.org/10.1101/2020.05.19.20107045
6. Bairagi, A. K., Masud, M., Kim, D. H., Munir, M. S., Nahid, A. A., Abedin, S. F., et al. (2020). Controlling the outbreak of COVID-19: A noncooperative game perspective. Preprint (2020). arxiv:2007.13305.
7. Chen, X., & Yu, B. (2020). First two months of the 2019 coronavirus disease (COVID-19) epidemic in China: Real time surveillance and evaluation with a second derivative model. *Global Health Research and Policy, 5*(7), 1–9. https://doi.org/10.1186/s41256-020-00137-4.
8. Chen, N., Zhou, M., Dong, X., & Zhang, L. (2020). Epidemiological and clinical characteristics of 99 cases of 2019 novel coronavirus pneumonia in Wuhan, China: A descriptive study. *The Lancet, 395*(10223), 507–513.
9. Sookaromdee, P., & Wiwanitkit, V. (2020). Imported cases of 2019-novel coronavirus (2019-nCoV) infections in Thailand: Mathematical modelling of the outbreak. *Asian Pacific Journal of Tropical Medicine, 13*(3), 139–140.
10. Zhao, S., Lin, Q., Ran, J., Musa, S. S., Yang, G., Wang, W., et al. (2020). Preliminary estimation of the basic reproduction number of novel coronavirus (2019-nCoV) in China, from 2019 to 2020: A data-driven analysis in the early phase of the outbreak. *International Journal of Infectious Diseases, 92,* 214–217.
11. Ndairou, F., Area, I., Nieto, J. J., & Torres, D. F. M. (2020). Mathematical modeling of COVID-19 transmission dynamics with a case study of Wuhan. Preprint (2020). arXiv:2004.10885v1.
12. Aguilar, J. B., Faust, G. S., Westafer, M. L. M., & Gutierrez, J. B. (2020). Investigating the impact of asymptomatic carriers on COVID-19 transmission. Preprint (2020). https://doi.org/10.1101/2020.03.18.20037994.
13. Carolina Garcia-Vidal, C., et al. (2020). Incidence of co-infections and superinfections in hospitalized patients with COVID-19: A retrospective cohort study. *Clinical Microbiology and Infection.* https://doi.org/10.1016/j.cmi.2020.07.041.

14. Casas-Rojo, J. M., et al. (2020). Clinical characteristics of patients hospitalized with COVID-19 in Spain: Results from the SEMI-COVID-19 registry. *Revista Clinica Espanola*. https://doi.org/10.1016/j.rce.2020.07.003.
15. Chitnis, N., Hyman, J. M., & Cushing, J. M. (2020). Determining important parameters in the spread of malaria through the sensitivity analysis of a mathematical model. *Bulletin of Mathematical Biology, 70,* 1272–1296. https://doi.org/10.1007/s11538-008-9299-0.
16. Chen, T. M., Rui, J., Wang, Q., Zhao, Z., Cui, J., & Yin, L. (2020). A Mathematical model for simulating the phase-based transmissibility of a novel coronavirus. *Infectious Diseases of Poverty, 9*(24), 1–8. https://doi.org/10.1186/s40249-020-00640-3.
17. Huo, H. F., & Feng, L. X. (2012). Global stability of an epidemic model with incomplete treatment and vaccination. *Discrete Dynamics in Nature and Society, 2012,* 1–14. https://doi.org/10.1155/2012/530267.
18. Kabir, M. H., Gani, M. O., Mandal, S., & Biswas, M. H. A. (2020). Modeling the dispersal effect to reduce the infection of COVID-19 in Bangladesh. *Sensors International, 1,* 100043. https://doi.org/10.1016/j.sintl.2020.100043.
19. Queen Elizabeth Hospital Birmingham COVID-19 airway team, safety and 30-day outcomes of tracheostomy for COVID-19: A prospective observational cohort study. *British Journal of Anaesthesia*. https://doi.org/10.1016/j.bja.2020.08.023.
20. Magalhaesa, J. J. F. D., et al. (2020). Epidemiological and clinical characteristics of the first 557 successive patients with COVID-19 in Pernambuco state, Northeast Brazil. *Travel Medicine and Infectious Disease, 38,* 101884.
21. Melliani, S., Allaoui, A. E., & Chadli, L. S. (2020). A simple mathematical model for coronavirus (COVID-19). *Discrete Dynamics in Nature and Society*. Preprint (2020). https://doi.org/10.1101/2020.04.23.20076919.
22. Nadeem, S. (2020). Coronavirus Covid-19: Available free literature provided by various companies. *Journals and Organizations Around the World, 5*(1), 7–13.
23. Prompetchara, E., Ketloy, C., & Palaga, T. (2020). Immune responses in COVID-19 and potential vaccines: Lessons learned from SARS and MERS epidemic. *Asian Pacific Journal of Allergy and Immunology, 38*(1), 1–9. https://doi.org/10.12932/AP-200220-0772.
24. Safi, M. A., & Garba, S. M. (2012). Global stability analysis of SEIR model with Holling Type II incidence function. *Computational and Mathematical Methods in Medicine, 2012*(826052), 1–8. https://doi.org/10.1155/2012/826052.
25. Xu, X., Chen, P., Wang, J., Feng, J., Zhou, H., Li, X., et al. (2020). Evolution of the novel coronavirus from the ongoing wuhan outbreak and modeling of its spike protein for risk of human transmission. *Science China Life Sciences, 63,* 457–460.
26. Zhang, J., Weili, W., Zhao, X., & Zhang, W. (2020). Recommended psychological crisis intervention response to the 2019 novel coronavirus pneumonia outbreak in China: A model of West China hospital. *Precision Clinical Medicine, 3*(1), 3–8.
27. Zhang, S., Diao, M., Wenbo, Y., Pei, L., Lin, Z., & Chen, D. (2020). Estimation of the reproductive number of novel coronavirus (COVID-19) and the probable outbreak size on the diamond princess cruise ship: A data-driven analysis. *International Journal of Infectious Diseases, 93,* 201–214.
28. Biswas, M. H. A., Haque, M. M., & Mallick, U. K. (2019). Optimal control strategy for the immunotherapeutic treatment of HIV infection with state constraint. *Optimal Control, Applications and Methods, 40*(3), 1–12.
29. Biswas, M. H. A., Paiva, L. T., & de Pinho, M. D. R. (2014). A SEIR model for control of infectious diseases with constraints. *Mathematical Biosciences and Engineering, 11*(4), 761–784.
30. Biswas, M. H. A. (2014). On the immunotherapy of HIV infections via optimal control with constraint. In *Proceedings of the 18th International Mathematics Conference* (pp. 51–54), Dhaka, 20–22 March 2014.
31. Biswas, M. H. A. (2013). *Necessary conditions for optimal control problems with state constraints: Theory and applications* (Ph.D. thesis). Department of Electrical and Computer Engineering, Faculty of Engineering, University of Porto, Portugal.

32. Biswas, M. H. A. (2012). Model and control strategy of the deadly Nipah Virus (NiV) infections in Bangladesh. *Research & Reviews in BioSciences, 6*(12), 370–377.
33. Biswas, M. H. A. (2012). AIDS epidemic worldwide and the millennium development strategies: A light for lives. *HIV and AIDS Review, 11*(4), 87–94.
34. Biswas, M. H. A., Ara, M., Haque, M. N., & Rahman, M. A. (2011). Application of control theory in the efficient and sustainable forest management. *International Journal of Scientific & Engineering Research, 2*(3), 26–33.
35. Khatun, M. S., & Biswas, M. H. A. (2020). Mathematical analysis and optimal control applied to the treatment of leukemia. *Journal of Applied Mathematics and Computing, 64,* 331–354. https://doi.org/10.1007/s12190-020-01357-0.
36. Khatun, M. S., & Biswas, M. H. A. (2020). Optimal control strategies for preventing hepatitis B infection and reducing chronic liver cirrhosis incidence. *Infectious Disease Modelling, 5,* 91–110.
37. Khatun, M. S., & Biswas, M. H. A. (2019). Modeling the effect of adoptive T cell therapy for the treatment of leukemia. *Computational and Mathematical Method, 2,* e1069. https://doi.org/10.1002/cmm4.1069.
38. Driessche, P. V. D., & Watmough, J. (2002). Reproduction numbers and sub-threshold endemic equilibria for compartmental models of disease transmission. *Mathematical Biosciences, 180,* 29–48.

# Chapter 4
# Transmission Dynamics of Covid-19 from Environment with Red Zone, Orange Zone, Green Zone Using Mathematical Modelling

**Bijal M. Yeolekar and Nita H. Shah**

**Abstract** The novel corona virus or Covid-19 spread had its inception in November of 2019, and in March 2020, it was declared as a pandemic. Since its initial stage, it has now already infected over 5 million people, leading to the lockdown of countries around the world, and a halt on global as well as national travel across the globe. Based on this, the research proposes a mathematical Covid-19 model to study the outcome of these classified zones under different control strategies. In the nonlinear mathematical model, the total population has been divided into seven compartments, namely Susceptible, Exposed, Red zone, Orange zone, Green zone, Hospitalized, and Recovered. The spectral radius is calculated to analyze dynamics of the Covid-19. To control the spread of the virus, the parameters of controls are Medical Intervention, Partial Lockdown, and Strict Lockdown. This model has been validated with numerical data. The conclusion validates the implementation of lockdown in curbing Covid-19 cases.

**Keywords** Covid-19 · Red zone · Orange zone · Green zone · Basic reproduction number · Optimal control

**MSC Code** 37C75 · 37N25

## Introduction

World were faced Flu pandemic in nineteenth century, Spanish and Asian flu pandemics in 20th Century, Swine flu pandemic in 2009–10 in the first decade of twenty-first Century and nowadays, world is facing Covid-19 pandemic disease.

---

B. M. Yeolekar
Department of Mathematics and Humanities, Nirma University, Ahmedabad, Gujarat 382481, India
e-mail: bijalyeolekar28@gmail.com

N. H. Shah (✉)
Department of Mathematics, Gujarat University, Ahmedabad, Gujarat 380009, India
e-mail: nitahshah@gmail.com



N. H. Shah and M. Mittal (eds.), *Mathematical Analysis for Transmission of COVID-19*, Mathematical Engineering, https://doi.org/10.1007/978-981-33-6264-2_4

Covid-19 is a sort of infection that causes irresistible maladies in well-evolved creatures and feathered creatures. Normally, the infections cause respiratory contaminations among individuals and the primary distinguishing proof are during the 1960s [1]. The primary transmission of Covid-19 resembles different infections: through sniffling, hacking, coming into contact with the tainted individuals, or contacting every day utilized things [2]. The principle perceived novel Covid-19 cases in China were represented as a dark etiologic lung infection in Wuhan dated December 26, 2019. Affirmations featuring the individual-to-individual transmission in crisis facilities and families are found in audit analyses [3–7]. It takes a few days to energize people's thought and Chinese Center for Control and Prevention of the disease detached the principal strain of the relevant contamination (Covid-19) adequately on January 7, 2020. With Chinese New Year development, the gigantic scourges occur in China and spread to various countries rapidly. The World Health Organization (WHO) declares that the pneumonia scene realized by Covid-19 as a general prosperity emergency of overall concern on January 31, 2020 [1]. So, the Govt. of India has reported 21 days across the nation lockdown from March 25, 2020 to April 14, 2020, to forestall stage-III spreading of the infection transmission. The most significant methods of transmission of Covid-19 are respiratory beads and contact transmission (individual-to-individual transmission) [8–11]. The probability of Covid-19 infected individuals by air transport reach to India in different places [12] is of most risky environment for country.

From the affirmed instances of coronavirus, the side effects extend from fever, cough, or even a runny nose, dry hack weakness breathing issue and lung invasion to seriously sick and death [1]. Assemble the tremendous data related to Covid-19 and separate the trademark linkage are exceptionally huge for the accompanying stage control technique. Plague elements and populace environment are the key techniques to contemplate irresistible infections. Numerical demonstrating assumes a crucial job to more readily comprehend the illness elements and planning strategies to oversee rapidly spreading irresistible infections in absence of powerful immunization or explicit antivirals [13].

Different mathematical *SEIR-models* discussed by number of researchers [2, 7, 14–24]. Based on this, Nowadays, so many researchers discussed mathematical system to recognize the complicated dynamics of Covid-19 models [2, 8, 25–33] and add more outbreaks [34–40]. Imai et al. established a Covid-19 model find the basic reproduction number [27] in China which mainly focused on human-to-human transmission by add in symptomatic ill class to obtain the patients' epidemiological status discussed by Tang et al. [41]. Nadim et al. [42] contemplated a scientific to research the Covid-19 ailment, where they performed dependability investigation and they approved their model with the information from Hubei, the city of China. Their model is initially settled by Gumel et al. [34] for Severe Acute Respiratory Condition (SARS) episode. Covid-19 model by considering four classes S, E, I, R to study the transmission dynamics of an individuals based on the data from December 31, 2019, to January 28 by Wu et al. [10]. Jiang et al. [4] assessed that passing pace of this infection is close about 4.5% however for the age bunch 70–79 it has gone up to 8%. This illness is increasingly basic for senior individuals who have different illnesses like diabetes, asthma, cardiovascular infection [43]. Researchers are trying their best to discover vaccines and treatment, huge news coverage, educational campaigns and rapid information flow can create great psychological effects on the population and thus remarkably change the publics' behavior and affect the implementation of individuals' intervention and control strategies [34]. It is mainly the awareness program due to media which make the individuals enlighten regarding the disease to take safeguards such as vaccination, wearing protective masks, social distancing etc., to minimize their risks of being infected [44]. Several mathematical models have been incorporated with the assumption that the media related awareness has an impact on reducing the contact rate of susceptible and infected population [40, 45–48].

**Summary of Covid-19 Disease in India**
The main instance of Covid-19 was accounted for on January 30, 2020, trailed by two comparable cases on February second and third in India and afterward sharp increment in numbers at that point followed. To control this pandemic, On March 22nd, Hon. PM Narendra Modi started the lockdown procedure with a 14-h 'Janta Curfew', trailed by lockdowns. An across the country India announced its first Lockdown from March 24, 2020 for 21 days, which was later stretched out to its subsequent lockdown stage from fifteenth of April till the third of May. In its third phase of lockdown from 4th May, various cities were classified under Red, Orange, and Green zones as per the cases. From 18th May, the fourth phase of the lockdown started with varying levels of restrictions as per different zones to control the spread as well as to start the economy. Fig. 4.1 shows that a list of all the Indian districts in each zone.

On the base of this, we formulated nonlinear mathematical Covid-19. In Sect. Notations, the Notation with parametric values are given. Mathematical model is formulated in Sect. Mathematical Model and spectral radius of Covid-model is calculated using next generation matrix method in Sect. Introduction. Stability analysis for Covid-19 model is calculated in Sect. Stability Analysis. Optimal control theory in terms of following strict lock down, partial lock down and minimize treatment cost for Covid-19 in Sect. Optimal Control Problem. Covid-19 model is validated by numerical data in Sect. Numerical Simulation and ends with conclusion.

## Notations

In this section, we outline scientific model for Covid-19, the accompanying images and parametric qualities in Table 4.1 are thought about.

Now, using the above parameter, we construct a diagram for Covid-19 system in Fig. 4.2.

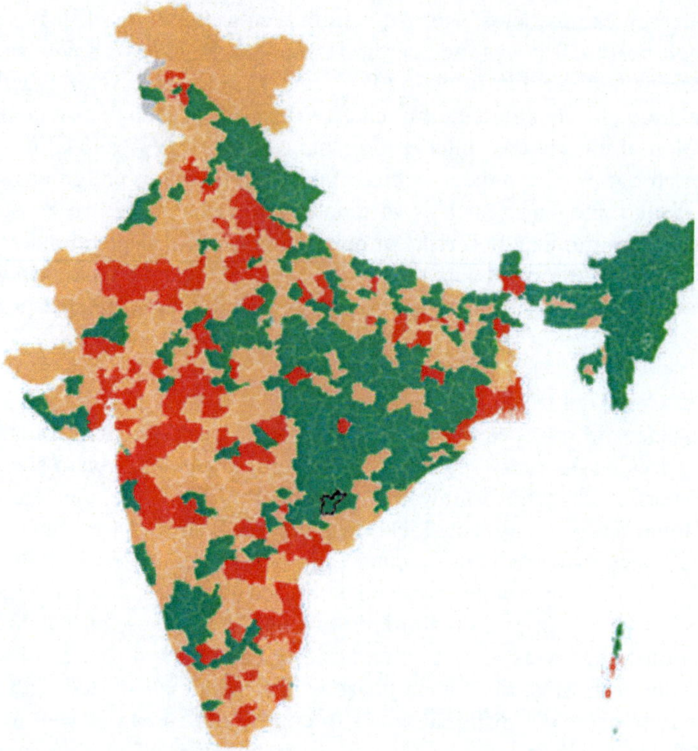

**Fig. 4.1** Red, orange and green zone in India. https://www.thehindu.com/data/article31485463.ece/inline/. *Note* District boundaries as of 2019. Data collated by Srinivasan Ramani and Naresh Singaravelu. Map based on shapefiles provided by Guneet Narula

## Mathematical Model

The prevalent model of Covid-19 focuses on Red, Orange, and Green zone with optimal preventions by complete lockdown, partial lockdown and social distances. The total infected population at $t$ is divided into seven mutually exclusive compartments. Susceptible individual for Covid-19 denoted by $S(t)$, Exposed to Covid-19 $E(t)$, number of individuals who are in Red zone $R(t)$, number of individuals in Orange zone $O(t)$, number of individuals in Green zone $G(t)$, hospitalized individuals $H(t)$, recovered individual $R_c(t)$. The nonlinear mathematical model for Covid -19 under these zones are

$$\frac{dS}{dt} = B - \beta SE - \mu S$$

$$\frac{dE}{dt} = \beta SE - \gamma_1 E - \gamma_2 E - \gamma_3 E - \mu E$$

**Table 4.1** Description of notation and parametric values

| Rates | Notation | Parametric values |
|---|---|---|
| $B$ | Natural birth rate | 0.8 (Calculated) |
| $\beta$ | Transmission rate | 0.55 (Calculated) |
| $\mu$ | Mortality rate | 0.01 (Assumed) |
| $\gamma_1$ | Rate from exposed to Red zone | 0.33 [20] |
| $\gamma_2$ | Rate from exposed to Orange zone | 0.43 [20] |
| $\gamma_3$ | Rate from exposed to Green zone | 0.24 [20] |
| $\delta_1$ | The rate at which individuals of Green zone moving in Orange zone | 0.2 (Assumed) |
| $\delta_2$ | The rate at which individuals of Orange zone moving into Red zone | 0.4 (Assumed) |
| $\delta_3$ | The rate at which individuals of Red zone moving into Orange zone | 0.3 (Assumed) |
| $\delta_4$ | The rate at which individuals of Orange zone moving into Green zone | 0.5 (Assumed) |
| $u_1$ | Control rate on investment for counseling persons to go for medical intervention as complete lockdown | [0, 1] |
| $u_2$ | Control rate on investment for counseling persons to go for medical intervention as partial lockdown | [0, 1] |
| $u_3$ | Control rate in terms of maintaining social distances | [0, 1] |

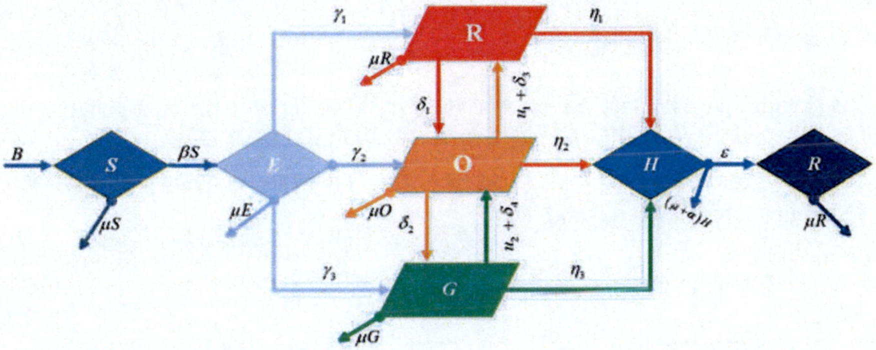


**Fig. 4.2** Diagram for Covid-19 model

$$\frac{dR}{dt} = \gamma_1 E + (\delta_3 + u_1)O - \delta_1 R - \eta_1 R - (\mu + \alpha)R$$
$$\frac{dO}{dt} = \delta_1 R - \delta_3 O + (\delta_4 + u_2)G - \delta_2 O - \gamma_2 E - \eta_2 O - \mu O$$
$$\frac{dG}{dt} = \delta_2 O + \gamma_3 E - \delta_4 G + u_3 R - \eta_3 G - \mu G$$
$$\frac{dH}{dt} = \eta_1 R + \eta_2 O + \eta_3 G - (\mu + \alpha)H - \varepsilon H$$

$$\frac{dR_c}{dt} = \varepsilon H - \mu R_c \tag{4.1}$$

where $S, E, R, O, G, H$ and $R_c$ are non-negative real numbers and $S + E + R + O + G + H + R_c = N$.

Let us sum of all the above differential equations of Covid-system and we get,

$$\frac{d}{dt}(S + E + R + O + G + H + R_c) = B - \mu N$$

This gives,

$$\lim_{t\to\infty} \sup(S + E + R + O + G + H + R_c) \le \frac{B}{\mu}$$

Therefore, the feasible region for (4.1) is

$$\Lambda = \left\{S, E, R, O, G, H, R_e \in R^7_+ / S + E + R + O + G + H + R_c \le \frac{B}{\mu}\right\}$$

## *Spectral Radius Ro*

In this section, we calculate the spectral radius [39] using with the next-generation matrix for Covid-free equilibrium $E = R = O = G = H = R_c = 0$.

Hence, let $X_0 = \left(\frac{B}{\mu}, 0, 0, 0, 0, 0, 0\right)$ be Covid-free equilibrium point of system. Using next generation method, let

$$\therefore X' = \frac{dX}{dt} = F_1(x) - v_1(x)$$

where $F_1 = \begin{bmatrix} \beta SE \\ 0 \\ 0 \\ 0 \\ 0 \\ 0 \\ 0 \end{bmatrix}$ and

$$v_1 = \begin{bmatrix} -B + \mu S \\ \gamma_1 E + \gamma_2 E + \gamma_3 E + \mu E \\ -\gamma_1 E - (\delta_3 + u_1)O + \delta_1 R + \eta_1 R + (\mu + \alpha)R \\ -\delta_1 R + \delta_3 O - (\delta_4 + u_2)G + \delta_2 O + \gamma_2 E + \eta_2 O + \mu O \\ -\delta_2 O - \gamma_3 E + \delta_4 G - u_3 R + \eta_3 G + \mu G \\ -\eta_1 R - \eta_2 O - \eta_3 G + (\mu + \alpha)H + \varepsilon H \\ -\varepsilon H + \mu R_c \end{bmatrix}.$$

Using $F = \left[\frac{\partial F_{1i}(X_0)}{\partial X_j}\right]$ and $V = \left[\frac{\partial v_{1i}(X_0)}{\partial X_j}\right]$ for $i, j = 1, 2, 3, \ldots, 7$.
Therefore, we have

$$F = \begin{bmatrix} \beta E & \beta S & 0 & 0 & 0 & 0 & 0 \\ 0 & 0 & 0 & 0 & 0 & 0 & 0 \\ 0 & 0 & 0 & 0 & 0 & 0 & 0 \\ 0 & 0 & 0 & 0 & 0 & 0 & 0 \\ 0 & 0 & 0 & 0 & 0 & 0 & 0 \\ 0 & 0 & 0 & 0 & 0 & 0 & 0 \\ 0 & 0 & 0 & 0 & 0 & 0 & 0 \end{bmatrix} \text{ and}$$

$$V = \begin{bmatrix} \mu & 0 & 0 & 0 & 0 & 0 & 0 \\ 0 & \mu + \gamma_1 + \gamma_2 + \gamma_3 & 0 & 0 & 0 & 0 & 0 \\ 0 & -\gamma_1 & \delta_3 + \eta_1 + \mu & -\delta_2 & 0 & 0 & 0 \\ 0 & -\gamma_2 & -\delta_3 & \delta_2 + \eta_2 + \mu & -\delta_1 & 0 & 0 \\ 0 & -\gamma_3 & 0 & -\gamma & \delta_1 + \eta_3 + \mu & 0 & 0 \\ 0 & 0 & -\eta_1 & 0 & 0 & \varepsilon + \mu & 0 \\ 0 & 0 & 0 & 0 & -\eta_3 & -\varepsilon & \mu \end{bmatrix}$$

$R_0$ is the spectral radius of matrix $FV^{-1}$ of the system (4.1).

$$R_0 = \frac{\beta B}{\mu(\gamma_1 + \gamma_2 + \mu + \gamma_3)} = 0.4726 < 1.$$

## Stability Analysis

In this section, we discussed the local stability of the system (4.1) at Covid-free equilibrium point.

### *Local Stability*

**Theorem 1** *The Covid-free equilibrium point $X_0$ is locally asymptotic stable if $R_0 < 1$. If $R_0 = 1$, $X_0$ is locally stable and if $R_0 > 1$ then $X_0$ is unstable.*

***Proof*** The Jacobian matrix of the system (4.1) at equilibrium point $X_0$ is

$$J = \begin{bmatrix} -\mu & -\frac{B\beta}{\mu} & 0 & 0 & 0 & 0 & 0 \\ 0 & \frac{B\beta}{\mu} - (\mu + \gamma_1 + \gamma_2 + \gamma_3) & 0 & 0 & 0 & 0 & 0 \\ 0 & \gamma_1 & -\delta_1 - \eta_1 - \mu & \delta_3 & 0 & 0 & 0 \\ 0 & \gamma_2 & \delta_1 & -\delta_2 - \delta_3 - \eta_2 - \mu & -\delta_4 & 0 & 0 \\ 0 & \gamma_3 & 0 & \delta_2 & -\delta_4 - \eta_3 - \mu & 0 & 0 \\ 0 & 0 & \eta_1 & \eta_2 & \eta_3 & \varepsilon - \mu & 0 \\ 0 & 0 & 0 & 0 & 0 & -\varepsilon & -\mu \end{bmatrix}$$

The roots of the Jacobian matrix are

$$J_1 = -\mu < 0,\ J_2 = -\mu < 0,\ j_3 = -\varepsilon - \mu < 0,\ J_4 = -\mu - \delta_4 - \delta_2 - \eta_3,$$
$$J_{5,6} = -\mu \pm \sqrt{\delta_1 - \delta_3 - \eta_2},\ J_7 = \frac{B\beta}{\mu} - \mu - \gamma_1 - \gamma_2 - \gamma_3 < 0 \text{ if } \mathrm{R}_0 < 1$$

Thus, $X_0$ is locally asymptotically stable if $R_0 < 1$. For $R_0 = 1$, $j_i < 0, i = 1, 2, \ldots, 6$ and $j_7 = 0$, $X_0$ is locally stable. If $R_0 > 1$, then the characteristic equation has a real positive eigenvalue. So, $X_0$ is unstable. This Covid-system is not globally stable as there is no effective vaccine found in the globe.

## Optimal Control Problem

Given the Covid-19 model (4.1), we want to design the optimal control problem with the control rates $u_1, u_2, u_3$ to improvise effect of lockdown and social distances on the number of an individual. Obviously, this task can always be accomplished by imposing extremely high recommending for complete lockdown, partial lockdown and medical intervention. However, maximize strict lockdown in Red zone area and partial lockdown in Orange zone area and minimize social gathering which increase the prevention of Covid-19. We consider the following objective cost-functional

$$J(X_i, u_i) = \int_0^T \left( A_1 S^2(t) + A_2 E^2(t) + A_3 R^2(t) + A_4 O^2(t) + A_5 G^2(t) + A_6 H^2(t) + A_7 R_c^2(t) + \sum_{i=1}^{3} w_i u_i^2(t) \right) dt$$

The weighted parameters $w_i$ and $u_i$ are used for the state variables *S*, *E*, *R*, *O*, *G*, *H* and $R_c$ and the control variables $u = (u_1, u_2, u_3)$, respectively. Since we are mainly interested in maximizing the lockdown, limit the expense of treatment given in quadratic structure in the cost work. The cost of counseling might come from cost of medical intervention; any cost related with awareness campaigns. So, our aim is to maximize the lockdown among the individuals in red zone and maximize partial lockdown in Orange zone and to minimize the expenses of treatment in the society.

We seek for an optimal control pair $\left(u_1^*, u_2^*, u_3^*\right)$ such that

$$J\left(u_1^*, u_2^*, u_3^*\right) = \max \min\{(u_1, u_2, u_3)/(u_1, u_2, u_3) \in \Lambda\}$$

subject to system (4.1), where the control parameters are piecewise continuous on $[0, T]$ which is defined as

$$\Lambda = \{(u_1(t), u_2(t), u_3(t))/a_i \le u_i \le b_i, i = 1, 2, 3\} \tag{4.2}$$

Here, $a_i$ and $b_i$ are constants in [0, 1]. The existence of an optimal control pair can be proved by using results from Fleming and Rishel [49]. We use the results on existence from Fleming and Rishel [49] in the optimal control for a free terminal point problem, where the initial time and state along with the final time are fixed, and there are no conditions on the final state.

The integrand $L(X, U)$ of the objective function j is convex on $\Lambda$ and satisfies the inequality

$$L(S, E, R, O, G, H, R_c, u_1, u_2, u_3) \ge c_1\left(|u_1|^2 + |u_2|^2\right)^{\alpha/2} - c_2,$$

where $c_i$ are positive constants and $\alpha > 1$.

Here, we used theorem for the solution of the state equations are bounded and state equations are Lipchitz and continuous in state variable by Lukes [50], then exists a unique solution corresponding to every control set in $\Lambda$. Using this fact for all $X = (S, E, R, O, G, H, R_c) \in \Lambda$, all the state variables are bounded, then the solutions of the state equations are bounded.

The Lagrangian function is

$$\begin{aligned} L(X, U) = {} & A_1 S^2 + A_2 E^2 + A_3 R^2 + A_4 O^2 + A_5 G^2 + A_6 H^2 \\ & + A_7 R_c^2 + w_1 u_1^2 + w_2 u_2^2 + w_3 u_3^2 \end{aligned}$$

Here, we depict the optimal controls which give the optimal levels for the various states and the corresponding controls. From Lewis and Syrmos [36], we present the optimality system. Using the Pontryagin's maximum principle (1986), the necessary conditions for the optimal controls are obtained. The terminal conditions of the adjoint variables are obtained based on the transversal condition [51].

Hamiltonian function is formulated as

$$\begin{aligned} H(X, U, \lambda) = {} & L + \lambda_1(B - \beta SE - \mu S) + \lambda_2(\beta SE - \gamma_1 E - \gamma_2 E - \gamma_3 E - \mu E) \\ & + \lambda_3(\gamma_1 E + \delta_2 O - (\delta_3 + u_2)R - (\mu + \alpha)R - \eta_1 R) \\ & + \lambda_4(\delta_3 R - \delta_2 O - (\delta_4 + u_1)O + \delta_1 G + \gamma_2 E - \eta_2 O - \mu O) \end{aligned}$$

$$+\lambda_5(\delta_4 O + \gamma_3 E - \delta_1 G - \eta_3 G - \mu G)$$
$$+\lambda_6(\eta_1 R + \eta_2 O + \eta_3 G - \varepsilon H - (\mu + \alpha)H) + \lambda_7(\varepsilon H - \mu R_c)$$

We develop adjoint equations by using Pontryagin's maximum principle from Pontriagin et al. [6],

$$\dot{\lambda}_1 = -\frac{\partial H_1}{\partial S} = -2A_1 S + (\lambda_1 - \lambda_2)\beta E + \lambda_1\mu$$
$$\dot{\lambda}_2 = -\frac{\partial H_1}{\partial E} = -2A_2 E + (\lambda_1 - \lambda_2)\beta S + (\lambda_2 - \lambda_3)\gamma_1 + (\lambda_2 - \lambda_4)\gamma_2 + (\lambda_2 - \lambda_5)\gamma_3 + \lambda_2\mu$$
$$\dot{\lambda}_3 = -\frac{\partial H_1}{\partial R} = -2A_3 R + (\lambda_3 - \lambda_4)(\delta_3 + u_1) + (\lambda_3 - \lambda_6)\eta_1 + \lambda_3(\mu + \alpha)$$
$$\dot{\lambda}_4 = -\frac{\partial H_1}{\partial O} = -2A_4 O + (\lambda_4 - \lambda_3)\delta_1 + (\lambda_4 - \lambda_5)(\delta_4 + u_2) + (\lambda_4 - \lambda_6)\eta_2 + \lambda_4\mu$$
$$\dot{\lambda}_5 = -\frac{\partial H_1}{\partial G} = -2A_5 G + (\lambda_5 - \lambda_4)\delta_2 + (\lambda_5 - \lambda_6)\eta_3 + (\lambda_3 - \lambda_5)u_3 + \lambda_5\mu$$
$$\dot{\lambda}_6 = -\frac{\partial H_1}{\partial H} = -2A_6 + (\lambda_7 - \lambda_6)\varepsilon + \lambda_6\mu$$
$$\dot{\lambda}_7 = -\frac{\partial H_1}{\partial R_c} = -2A_7 R_c + \lambda_7\mu$$

The transversal condition gives the form as in equation, since all the state equations are free at terminal time. $H(X, U, \lambda)$ is maximize with respect to the all control pairs, to obtain the optimality condition by differentiating $H$ with respect to $u_1$, $u_2$ and $u_3$, respectively
$\frac{\partial H}{\partial u_i} = 0$ for $i = 1, 2, 3,$
Thus, we have

$$\frac{\partial H}{\partial u_1} = 2w_1 u_1 - (\lambda_3 - \lambda_4)R = 0, \quad \frac{\partial H}{\partial u_2} = 2w_2 u_2 - (\lambda_4 - \lambda_5)O = 0 \text{ and}$$
$$\frac{\partial H}{\partial u_3} = 2w_3 u_3 - (\lambda_3 - \lambda_5)G = 0$$

By substituting $u_1 = u_1^*$, $u_2 = u_2^*$ and $u_3 = u_3^*$ and solving for the $\left(u_1^*, u_2^*, u_3^*\right)$ as

$$u_1^* = \frac{(\lambda_3 - \lambda_4)R}{2w_1}, \quad u_2^* = \frac{(\lambda_4 - \lambda_5)O}{2w_2}, \quad u_3^* = \frac{(\lambda_3 - \lambda_5)G}{2w_3} \text{ and}$$

Now, we execute the bounds $0 \le u_1 \le u_{1\max}$, $0 \le u_2 \le u_{2\max}$ and $0 \le u_3 \le u_{3\max}$ on the control pairs to get

$$u_1^* = \max\left\{\min\left(0, \frac{(\lambda_3 - \lambda_4)R}{2w_1}\right), u_{1\max}\right\},$$
$$u_2^* = \max\left\{\min\left(0, \frac{(\lambda_4 - \lambda_5)O}{2w_2}\right), u_{2\max}\right\} \text{ and}$$

$$u_3^* = \max\left\{\min\left(0, \frac{(\lambda_3 - \lambda_5)G}{2w_3}\right), u_{3\,\min}\right\}$$

Now solve the optimal Covid-19 model which comprises of the state system with its initial conditions and the adjoint system with its transversal conditions joined with the control pairs characterization (10).

## Numerical Simulation

In this section, we have authenticated our numerical data in Table 4.1 with MATLAB. The simulation process is done iteratively until result converges.

Figure 4.3 shows that rates of convergence from exposed to Green, Orange and Red zone are zero that means there are no increments or decrement in the size of zone.

Green zone is rapidly converge to Orange zone with high rate of convergence (Fig. 4.4a) while Fig. 4.4b shows that if Orange zone is more in size, then it converges to Red zone and if Red zone will be large, then it converges to Orange zone. This can be happened due to natural recovery of individual or death of individual and

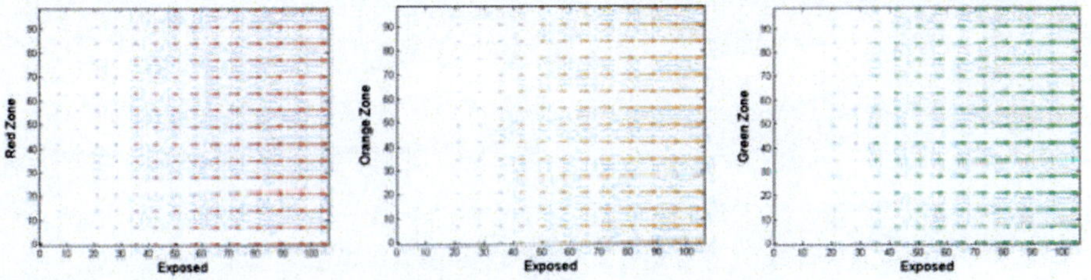


**Fig. 4.3** Directed graph of exposed versus Red zone, Orange zone and Green zone

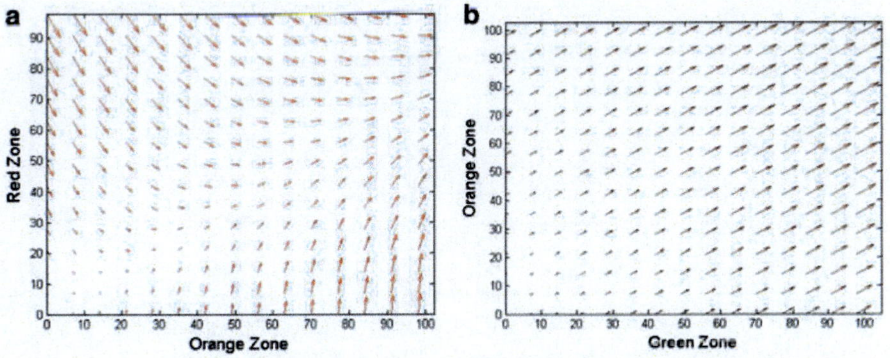


**Fig. 4.4** **a** Orange zone versus Red zone. **b** Green zone versus Orange zone

this process will not be controlled then after short time, we can found high mortality ratio. So, the implementation of control is necessary as soon as possible to reduce the mortality ratio.

As we know that once single individual infected by Covid-19, its spread in the society is very fast and so it became pandemic for the world. As per data announced by Indian government, the impact of Covid-19 in India is shown in the above figures with classification of area in Green, Orange and Red zone with basic reproduction number $R_0 > 1$ (Fig. 4.5a) with $R_0 < 1$ (Fig. 4.5b).

Figure 4.6a, b shows that the effect of control parameters in zones. If the individuals in red zone are followed strict lockdown means unnecessary do not go out then infected individual reduces by 20–30. In Orange zone, if society follows partial lockdown, then the ratio of infected individual decreases up to 70–80 and if proper treatment at proper time provided with lower cost than individuals in Green zone are increase.

All the three types of controls like strict lockdown, partial lockdown and medical treatment are applied on the different zones as per their class. Figure 4.7a–c shows

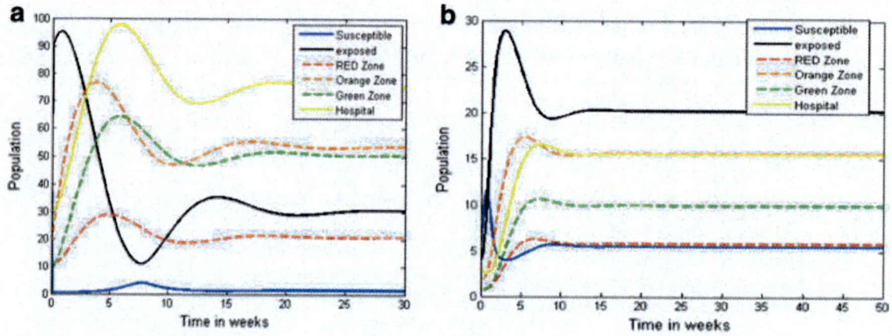


**Fig. 4.5** **a** Compartments with $R_0 > 1$. **b** Compartments with $R_0 < 1$

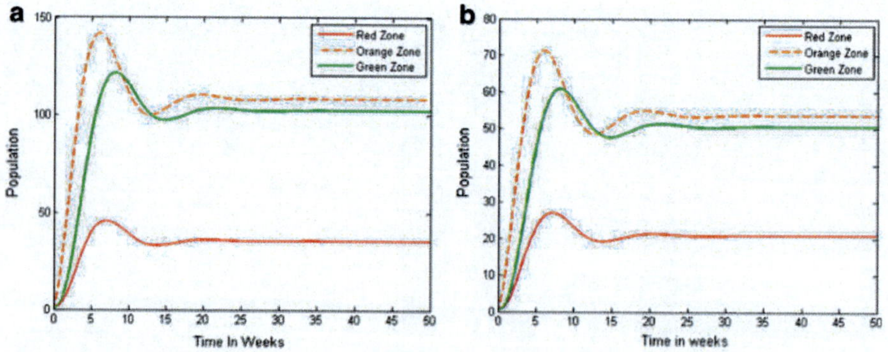


**Fig. 4.6** **a** Red, Orange, Green zones without all control. **b** Red, Orange, Green zones with all control

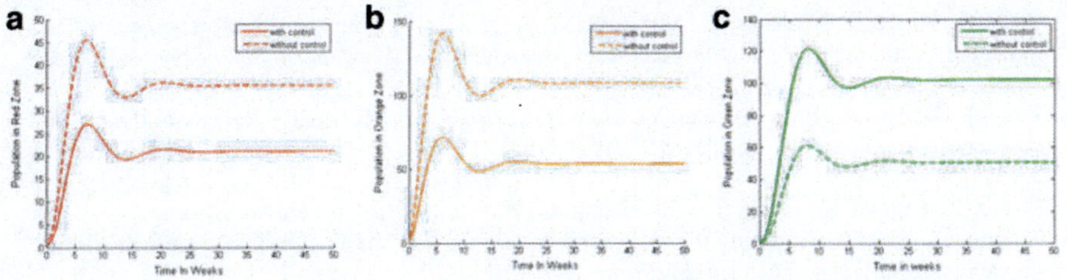


**Fig. 4.7** **a** Red zone with and without control. **b** Orange zone with and without control. **c** Green zone with and without control

that after implementation of controls, the size of Green zone increases while the Orange and Red zone shrink and made it flat after 20–25 weeks.

## Conclusion

Dynamical model for Covid-19 with various zones is anticipated and nonlinear scientific investigation of the arrangement of differential condition is done to get knowledge into the subjective elements in nearness of control in opportunity of Red zone, Orange zone and Green zone with various limitations according to government rules. The epidemiological and numerical consequences of the proposed Covid-model are talked about in this exploration. The Covid-model is locally–asymptotically stable at Covid-free harmony at whatever point the related viable limit is not as much as solidarity. In this research, we studied optimal control with strict lockdown, partial lockdown and minimize treatment cost. This study suggested that in this pandemic situation lockdown in the society is necessary. One needs to find vaccine for the Covid-19 Disease. In the future these study extended with added other parameters like liberty in Red zone, proper sanitization, reduction in the cost of necessary parameters like mask, sanitizer, medical treatment and also one important parameter is migration of labors in the society.

**Acknowledgements** The second author thanks DST-FIST file # MS1-097 for support to the department of Mathematics.

## References

1. World Health Organization Website. https://www.who.int/.
2. Castillo-Chavez, C., Blower, S., Driessche, P., Kirschner, D., & Yakubu, A. A. (Eds.). (2002). *Mathematical approaches for emerging and reemerging infectious diseases: Models, methods, and theory*. New York: Springer.

3. Funk, S., Gilad, E., Watkins, C., & Jansen, V. A. A. (2009). The spread of awareness and its impact on epidemic outbreaks. *Proceedings of the National Academy of Sciences of the United States of America, 106,* 6872–6877.
4. Jiang, F., Deng, L., Zhang, L., Cai, Y., Cheung, C. W., & Xia, Z. (2020). Review of the clinical characteristics of coronavirus disease 2019 (COVID-19). *Journal of General Internal Medicine, 35*(5), 1545–1549.
5. Muniz-Rodriguez, K., Chowell, G., Cheung, C.-H., Jia, D., Po-Kumar, A., Srivastava, P. K., & Takeuchi, Y. (2017). Modeling the role of information and limited optimal treatment on disease prevalence. *Journal of Theoretical Biology, 414,* 103–119.
6. Pontriagin, L. S., Boltyanskii, V. G., Gamkrelidze, R. V., & Mishchenko, E. F. (1986). *The mathematical theory of optimal process* (pp. 4–5). New York: Gordon and Breach Science Publishers.
7. Safi, M. A., & Gumel, A. B. (2013). Dynamics of a model with quarantine-adjusted incidence and quarantine of susceptible individuals. *Journal of Mathematical Analysis and Applications, 399*(2), 565–575.
8. Fuk-Woo, C. J., et al. (2020). A familial cluster of pneumonia associated with the 2019 novel coronavirus indicating person-to-person transmission: A study of a family cluster. *Lancet, 395,* 514–523.
9. Chan, J. F.-W., Yuan, S., Kok, K.-H., To, K. K.-W., Chu, H., Yang, J., et al. (2020). A familial cluster of pneumonia associated with the 2019 novel coronavirus indicating person-to-person transmission: A study of a family cluster. *Lancet, 395,* 514–523.
10. Wu, J. T., Leung, K., & Leung, G. M. (2020). Nowcasting and forecasting the potential domestic and international spread of the 2019-nCoV outbreak originating in Wuhan, China: A modelling study. *Lancet, 395,* 689–697.
11. Rothe, C., Schunk, M., Sothmann, P., Bretzel, G., Froeschl, G., Wallrauch, C., et al. (2020). Transmission of 2019-nCoV infection from an asymptomatic contact in Germany. *New England Journal of Medicine.* https://doi.org/10.1056/NEJMc2001468.
12. Event Horizon—COVID-19. (2020). Coronavirus COVID-19 Global Risk Assessment. https://rocs.hu-berlin.de/corona/relative-import-risk.
13. Egger, M., Johnson, L., Althaus, C., Schni, A., Salanti, G., Low, N. & Norris, S. L. (2017). Developing WHO guidelines: Time to formally include evidence from mathematical modelling studies. *F1000Research, 6,* 1584.
14. Driessche, P., & Watmough, J. (2002). Reproduction numbers and sub-threshold endemic equilibria for compartmental models of disease transmission. *Mathematical Biosciences, 180,* 29–48.
15. Freedman, H. I., Ruan, S., & Tang, M. (1994). Uniform persistence and flows near a closed positively invariant set. *Journal of Dynamics and Differential Equations, 6*(4), 583–600.
16. Heffernan, L., Smith, R., & Wahl, L. (2005). Perspectives on the basic reproduction ratio. *Journal of the Royal Society Interface, 2,* 281–293; Sanche, S., Lin, Y. T., Xu, C., Romero-Severson, E., Hengartner, N. W., & Ke, R. (2020). The novel coronavirus, 2019-nCoV, is highly contagious and more infectious than initially estimated.
17. Hethcote, H. W. (2000). The mathematics of infectious diseases. *Society for Industrial and Applied Mathematics Review, 42*(4), 599–653.
18. Sahu, G. P., & Dhar, J. (2012). Analysis of an SVEIS epidemic model with partial temporary immunity and saturation incidence rate. *Applied Mathematical Modelling, 36*(3), 908–923.
19. Shah, N. H., Patel, Z. A., & Yeolekar, B. M. (2016). Stability analysis of vertical transmission of HIV with delays in treatment and Pre-AIDS. *Journal of Basic and Applied Research International*, 88–102.
20. Shah, N. H., Satia, M. H., & Yeolekar, B. M. (2017). Optimum control for spread of pollutants through forest resources. *Applied Mathematics, 8*(5), 607–620.
21. Shah, N. H., Thakkar, F. A., & Yeolekar, B. M. (2017). Simulation of dengue disease with control. *International Journal of Scientific and Innovative Mathematical Research, 5*(7), 14–28.
22. Shah, N. H., Yeolekar, B. M., & Patel, Z. A. (2020). Mathematical model to analyze effect of demonetization. In *Mathematical models of infectious diseases and social issues* (pp. 270–286). Pennsylvania: IGI Global.

23. Shah, N. H., Yeolekar, B. M., & Shukla, N. J. (2015). Liquor habit transmission model. *Applied Mathematics, 6*(08), 1208.
24. Shah, N. H., Yeolekar, B. M., & Patel, Z. A. (2016). Stability analysis of vertical transmission of HIV with delays in treatment and pre-AIDS. *Journal of Basic and Applies Research International, 17*(2), 88–102.
25. Chen, T., Rui, J., Wang, Q., Zhao, Z., Cui, J., & Yin, L. (2020). A mathematical model for simulating the phase-based transmissibility of a novel coronavirus. *Infectious Diseases Poverty, 9,* 24.
26. (2020) Coronavirus in Vietnam. *New England Journal of Medicine, 382*, 872–874.
27. Imai, N., et al. (2020). Estimating the potential total number of novel coronavirus cases in Wuhan City, China. https://www.preventionweb.net/news/view/70092.
28. Mandal, S., Bhatnagar, T., Arinaminpathy, N., Agarwal, A., Chowdhury, A., Murhekar, M., et al. (2020). Prudent public health intervention strategies to control the coronavirus disease 2019 transmission in India: A mathematical model-based approach. *Indian Journal of Medical Research.* https://doi.org/10.4103/ijmr.IJMR-504-20. medRxiv preprint.
29. Phan Lan, T., Nguyen Thuong, V., Luong Quang, C., et al. (2020). Importation and human-to-human transmission of a novel coronavirus in Vietnam. *New England Journal of Medicine, 382,* 872–874.
30. Van den Driessche, P., & Watmough, J. (2002). Reproduction numbers and sub-threshold endemic equilibria for compartmental models of disease transmission. *Mathematical Biosciences, 180*(1), 29–48.
31. Victor, A. (2020). Mathematical predictions for COVID-19 as a global pandemic. Available at SSRN 3555879.
32. Yang, Y., Lu, Q., Liu, M., Wang, Y., Zhang, A., Jalali, N, & Zhang, X. (2020). Epidemiological and clinical features of the 2019 novel coronavirus outbreak in China. MedRxiv.
33. Xiao, X., Tang, S., & Wu, J. (2015). Media impact switching surface during an infectious disease outbreak. *Scientific Reports, 5,* 7838.
34. Gumel, A. B., Ruan, S., Day, T., Watmough, J., Brauer, F., Van den Driessche, P., et al. (2004). Modelling strategies for controlling SARS outbreaks. *Proceedings of the Royal Society of London, Series B: Biological Sciences, 271,* 2223–2232.
35. Nishiura, H., Linton, N. M., & Akhmetzhanov, A. R. (2020). Serial interval of novel coronavirus (2019-ncov) infections. medRxiv.
36. Lewis, F. L., & Syrmos, V. L. (1995). *Optimal control*. Hoboken: Wiley.
37. Simonsen, & Fung, I. C.-H. (2020). Epidemic doubling time of the 2019 novel spread of the 2019-nCoV outbreak originating in Wuhan, China: a modeling study. *The Lancet, 395,* 689–697.
38. Sanche, S., Lin, Y. T., Xu, C., Romero-Severson, E., NickSun, C., Yang, W., et al. (2011). Effect of media-induced social distancing on disease transmission in a two patch setting. *Mathematical Biosciences, 230,* 87–95.
39. Guan, W., Ni, Z., Hu, Y., et al. Clinical characteristics of 2019 novel coronavirus infection in China. MedRxiv.
40. Zhu, N., Zhang, D., Wang, W., Li, X., Yang, B., Song, J. ... Tan, W. (2020). A novel coronavirus from patients with pneumonia in China, 2019. *New England Journal of Medicine.*
41. Tang, B., Wang, X., Li, Q., Bragazzi, N. L., Tang, S., Xiao, Y., & Wu, J. (2020). Estimation of the transmission risk of the 2019-ncov and its implication for public health interventions. *Journal of Clinical Medicine, 9*(2), 462.
42. Nadim, S. K., Ghosh, I., & Chattopadhyay, J. (2020). Short-term predictions and prevention strategies for COVID-2019: A model based study. arXiv:2003.08150.
43. https://www.who.int/emergencies/diseases/novel-coronavirus-2019.
44. d'Onofrio, A., & Manfredi, P. (2009). Information-related changes in contact patterns may trigger oscillations in the endemic prevalence of infectious diseases. *Journal of Theoretical Biology, 256,* 473–478.
45. Buonomo, B., d'Onofrio, A., & Lacitignola, D. (2008). Global stability of an SIR epidemic model with information dependent vaccination. *Mathematical Biosciences, 216,* 9–16.

46. Coronavirus: common symptoms, preventive measures, & how to diagnose it. Caringly Yours, January 28, 2020. https://www.caringlyyours.com/coronavirus/. Retrieved January 28, 2020.
47. Das, D. K., Khajanchi, S., & Kar, T. K. (2020). The impact of the media awareness and optimal strategy on the prevalence of tuberculosis. *Applied Mathematics and Computation, 366,* 124732.
48. Sun, C., Yang, W., Arino, J., & Khan, K. (2011). Effect of media-induced social distancing on disease transmission in a two patch setting. *Mathematical Biosciences, 230,* 87–95.
49. Fleming, W. H., & Rishel, R. W. (2012). *Deterministic and stochastic optimal control* (Vol. 1). Berlin: Springer Science and Business Media.
50. Lukes, D. L. (1982). *Differential equations: Classical to controlled* (Vol. 162). Cambridge: Academic Press.
51. Hartl, R. F., & Sethi, S. P. (1983). A note on the free terminal time transversality condition. *Zeitschrift für Operations Research, 27*(1), 203–208.
52. https://www.thehindu.com/news/national/coronavirus-india-lists-red-zones-as-it-extends-lockdown-till-may-17/article31478592.ece
53. Liu, Z., Magal, P., Seydi, O., & Webb, G. (2020). Predicting the cumulative number of cases for the COVID-19 epidemic in China from early data. arXiv:2002.12298.
54. Wikipedia Website. https://en.wikipedia.org/wiki/Coronavirus/.

# Chapter 5
# A Comparative Study of COVID-19 Pandemic in Rajasthan, India

**Mahesh Kumar Jayaswal, Navneet Kumar Lamba, Rita Yadav, and Mandeep Mittal**

**Abstract** The treatment of corona virus disease is not possible without any vaccine. However, spreading of the deadly virus can be controlled by various measures being imposed by Government like lockdown, quarantine, isolation, contact tracing, social distancing and putting face mask on mandatory basis. As per information from the Department of Medical Health and Family Welfare of Rajasthan on 19 September 2020, corona virus COVID-19 severely affected the state of Rajasthan, resulting in cumulative positive cases 113,124, cumulative recovered 93,805 and cumulative deaths 1322. Without any appropriate treatment, it may further spread globally as it is highly communicable and because potentially affecting the human body respiratory system, which could be fatal to mankind. Therefore, to reduce the spread of infection, authors are motivated to construct a predictive mathematical model with sustainable conditions as per the ongoing scenario in the state of Rajasthan. Mathematica software has been used for numerical evaluation and graphical representation for variation of infection, recovery, exposed, susceptibles and mortality versus time. Moreover, comparative analysis of results obtained by predictive mathematical model has been done with the exact data plotting by curve fitting as obtained from Rajasthan government website. As a part of analysis and result, it is noted that due to the variation of transmission rate from person to person corresponding rate of infection goes on increasing monthly and mortality rate found high as shown and discussed numerically. Further, we can predict that the situation will become worse in the winter months especially in month of December due to unavailability of proper

M. K. Jayaswal · R. Yadav
Department of Mathematics and Statistics, Banasthali Vidyapith, Banasthali, Rajasthan, India
e-mail: maheshjayaswal17@gmail.com

R. Yadav
e-mail: ritayadav76@gmail.com

N. K. Lamba
Department of Mathematics, S.L.P.M., Mandhal, R.T.M. Nagpur University, Nagpur, India
e-mail: navneetkumarlamba@gmail.com

M. Mittal (✉)
Amity Institute of Applied Sciences, Amity University Uttar Pradesh, Noida, U.P., India
e-mail: mittal_mandeep@yahoo.com

 
N. H. Shah and M. Mittal (eds.), *Mathematical Analysis for Transmission of COVID-19*,
Mathematical Engineering, https://doi.org/10.1007/978-981-33-6264-2_5

vaccine. This model may become more efficient when the researchers, experts from medical sciences and technologist work together.

**Keywords** Covid-19 · Corona virus · Curve fitting · Modelling

**MSC Code** 00A71 · 93A30 · 03C99

## Introduction

Firstly, this pandemic COVID-19 has been identified in Wuhan city situated in China in the month of December 2019 and is an infectious disease. Yang and Wang [1] developed a mathematical model for the novel coronavirus epidemic in Wuhan city, China. Day by day, this pandemic spread due to the high transmission rate because no vaccine at present available. Some renowned researchers have dedicated nice work based on COVID-19 disease. COVID-19 disease came in the state of Rajasthan on 2nd March and identified in the SMS hospital which is situated in Jaipur city. Initially, the pandemic was slow increasing due to slow transmission rate. Nowadays, this disease is rapidly distributing to all over Rajasthan. The government organizations have taken some nice primary decision and tried to control it with the implementation of lockdown as well as manage the social distancing. Guckenheimer and Holmes [2] proposed a mathematical model along with nonlinear oscillation, dynamical systems and bifurcations of vector field. Driessche and Watmough [3] presented a mathematical model with the help of reproduction factor and endemic equilibriums for disease transmission. Mizumoto and Chowell [4] have explained a mathematical model based on potential of novel corona virus. Rothan and Byrareddy [5] have presented a mathematical model concern with the epidemiology and pathogenesis coronavirus disease (COVID-19). Sohrabi et al. [6] developed a mathematical model based on COVID-19 disease using data of World Health Organization. They reviewed the latest situation. World Health Organization declared the highly infectious COVID-19 as a global pandemic in March 2020. Thevarajan et al. [7] explained the symptoms of COVID-19 which includes high fever, cough and difficulties in inhalation. Riou and Althaus [8] studied a mathematical model in which they discussed that people may get infected while breathing in an infected environment or touching any infected surface or by coming in contact of an infected person. Some authors like Khot and Nadkar [9], Wang [10], have contributed their work wherein they studied that aged people, children and adults with low persons immunity are prone to catch infection and may be severely affected from this disease, even lead to death. Some experts like Sahin et al. [11], Cheng and Shan [12] considered the risk factor in their study as people are unaware of reason behind the spreading of virus.

All mathematical models discussed above are predictive model and have not given an exact solution. The above said references motivate us to develop a mathematical model under different strategies to discuss COVID-19 pandemic situation in Rajasthan. The present mathematical model explains the transmission of COVID-19 disease in the cold season and sensitivity analysis has been studied with the help of modelling graph under different cases. Moreover, a comparative analysis of results obtained by predictive mathematical model has been developed with the exact data plotting by curve fitting as obtained from Rajasthan government website. We have taken some input data from the site of Rajasthan, India (as shown in Annexures 5.1, 5.2, 5.3 and 5.4).

The spread of COVID-19 is found to be pandemic in nature for the state of Rajasthan. From the graphical plotting of the last four months date (June, July, August and September), it is noted that there is drastic change in variation of infection, recovery and mortality observed. Here, the input data used for the analysis is taken from the Rajasthan government website and as reflected in Annexures 5.1, 5.2, 5.3 and 5.4.

From the graphical plotting as shown in Figs. 5.1, 5.2 and 5.3 for the month of June 2020 in the state of Rajasthan, it is clearly noted that there is rapid growth in

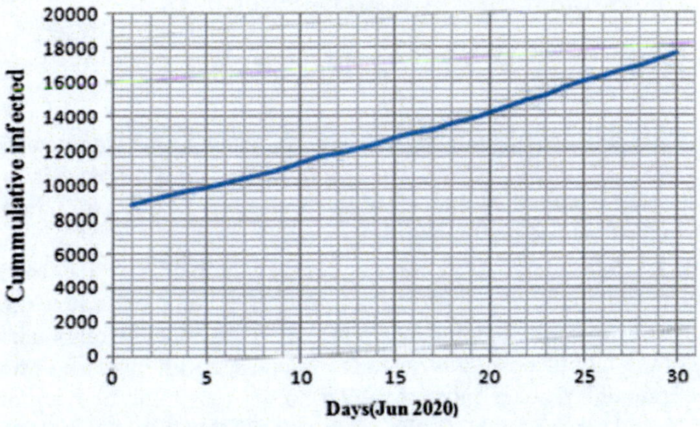


**Fig. 5.1** Variation of infection versus time for the month of June 2020. *Source* own

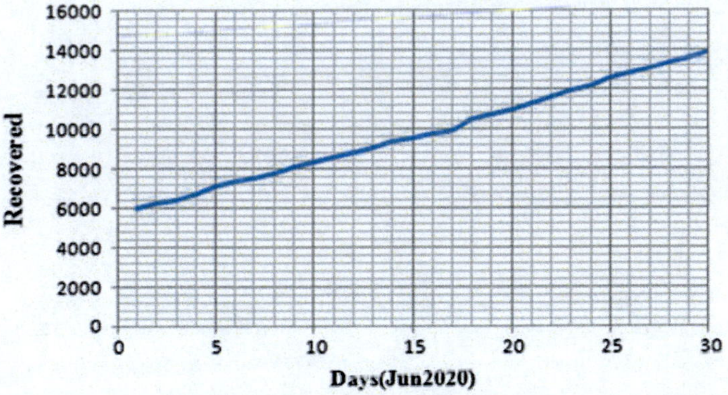


**Fig. 5.2** Variation of recovery versus time for the month of June 2020. *Source* own

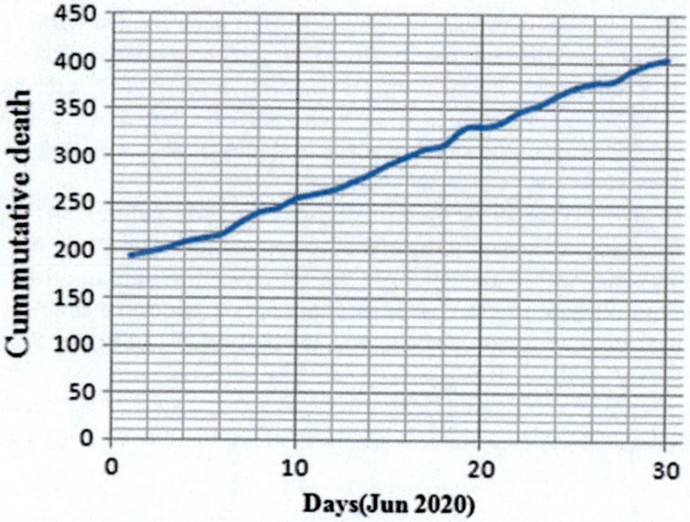


**Fig. 5.3** Variation of death versus time for the month of June 2020. *Source* own

the rate of infection but corresponding recovery rate also found significant but rate of deaths increases as time increases. The commutative infection reaches to 18,000 approximately whereas total recovery observed close to 14,000 and commutative deaths touch 400 till end of the June month.

From Figs. 5.4, 5.5 and 5.6, it is observed, for the month of July 2020 for the state of Rajasthan, growth in the rate of infection is double as compare to June month but there is small decay in corresponding recovery rate. Also, infection outbreak is noted near the end of the month as well as death rate increases with increase in time. The commutative infection reaches more than 40,000 whereas total recovery observed close to 29,000 and commutative deaths touch 680 till end of the July month.

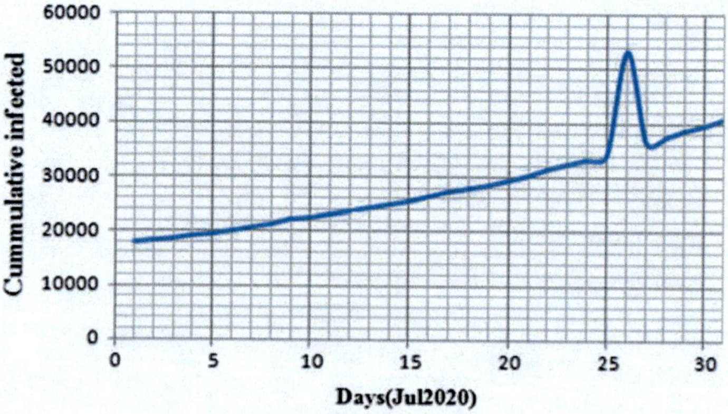


**Fig. 5.4** Variation of infection versus time for the month of July 2020. *Source* own

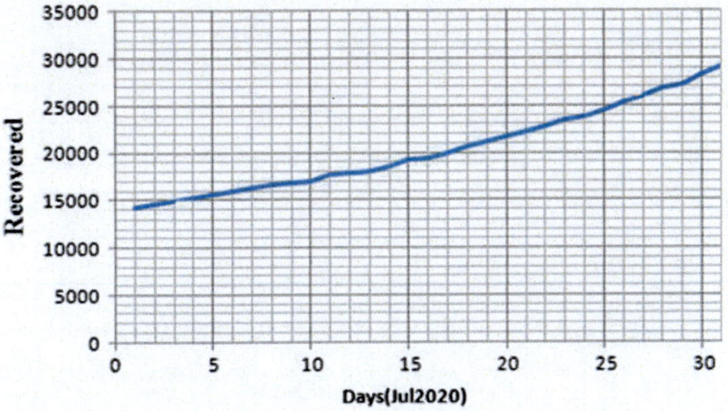


**Fig. 5.5** Variation of recovery versus time for the month of July 2020. *Source* own

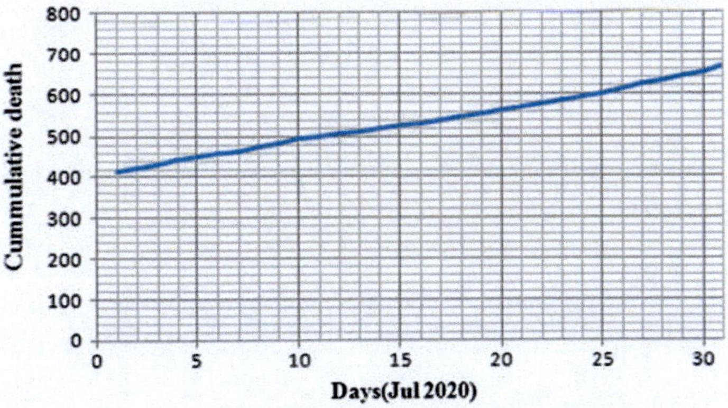


**Fig. 5.6** Variation of death versus time for the month of July 2020. *Source* own

From Figs. 5.7, 5.8, 5.9, 5.10, 5.11 and 5.12, a similar behaviour is noted for the overall variation of infection, recovery and mortality in the month of August and September, respectively. Infection touches 80,000 in August, which is double as happened in July month. Close 110,000 infected people noted in September, but corresponding recovery is not significantly increasing, and death rates become high.

## Mathematical Modelling of COVID-19

By looking the overall scenario, here we construct a mathematical modelling of COVID-19 for the state of Rajasthan by making use of the parameters described in Table 5.1. We divide the total population of Rajasthan into mainly five compartments, the susceptible (denoted by $S$), the exposed (denoted by $E$), the infected (denoted by

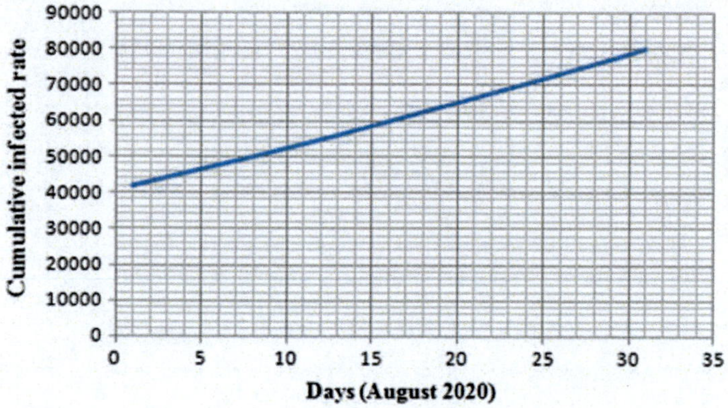


**Fig. 5.7** Variation of infection versus time for the month of August 2020. *Source* own

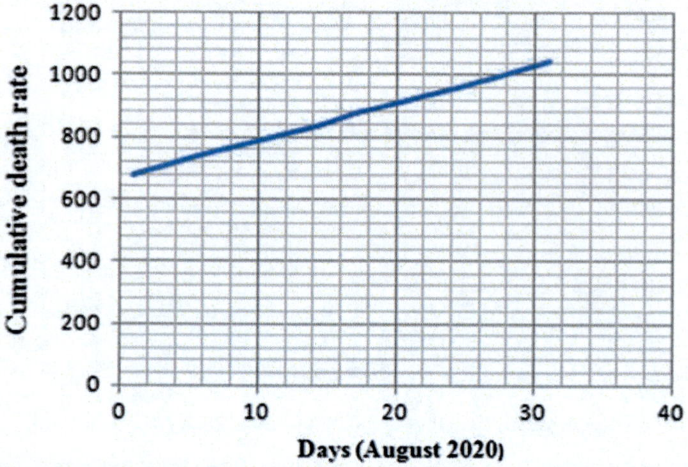


**Fig. 5.8** Variation of death versus time for the month of August 2020. *Source* own

$I$), and the recovered (denoted by $R$) and mortality (denoted by $M$). In Fig. 5.13, we describe the flowchart presentation of model.

Here it is assumed that infected individual class has fully developed disease symptoms and has the capacity to infect others. Also, exposed class individuals are those who are in incubation period and they do not show symptoms but still capable to pass infection to others. In this model, $E$ and $I$ compartment can be interpreted as they contain asymptomatic and symptomatic infected individuals.

The mathematical model suits the current COVID-19 situation to describe the transmission dynamics represented as:

$$\frac{\mathrm{d}S}{\mathrm{d}t} = \Lambda - C_1 SE - C_2 SI \tag{5.1}$$

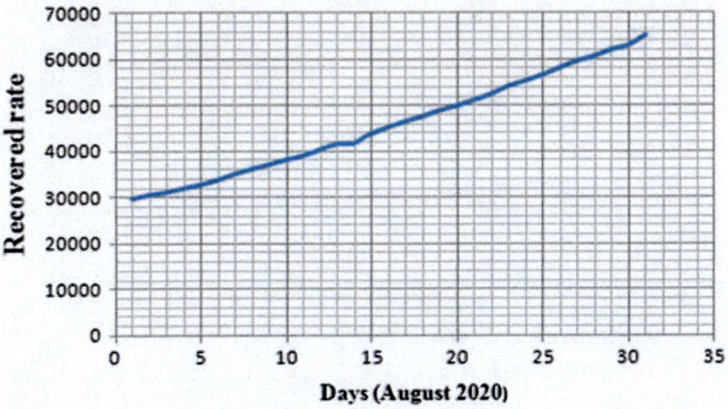


**Fig. 5.9** Variation of recovered versus time for the month of August 2020. *Source* own

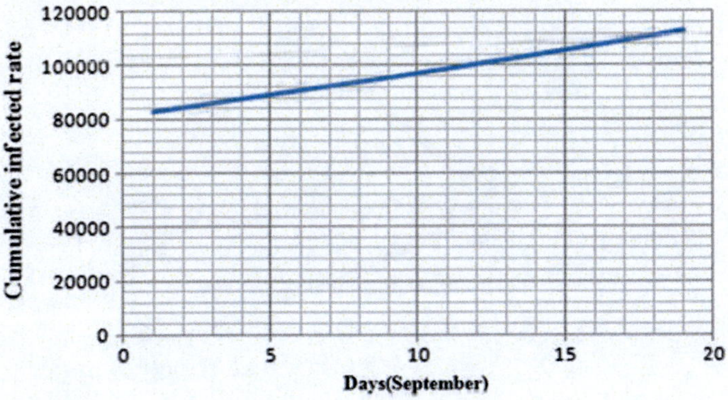


**Fig. 5.10** Variation of infection versus time for the month of September (up to 19th September 2020). *Source* own

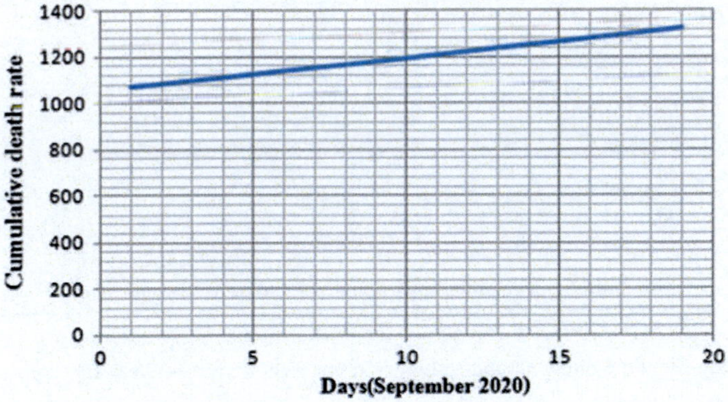


**Fig. 5.11** Variation of death versus time for the month of September (up to 19 September 2020). *Source* own

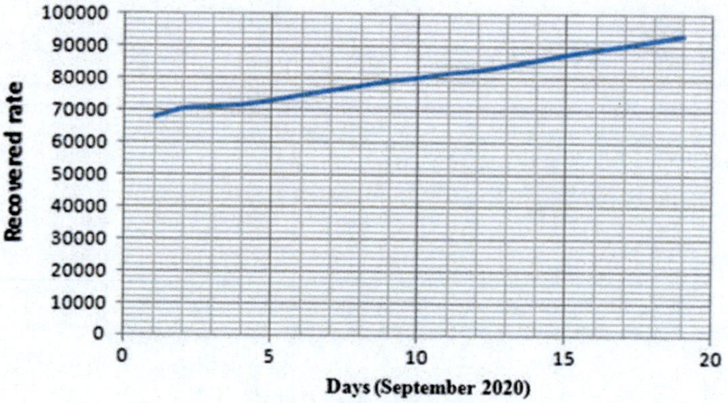


**Fig. 5.12** Variation of recovered versus time for the month of September (up to 19 September 2020). *Source* own

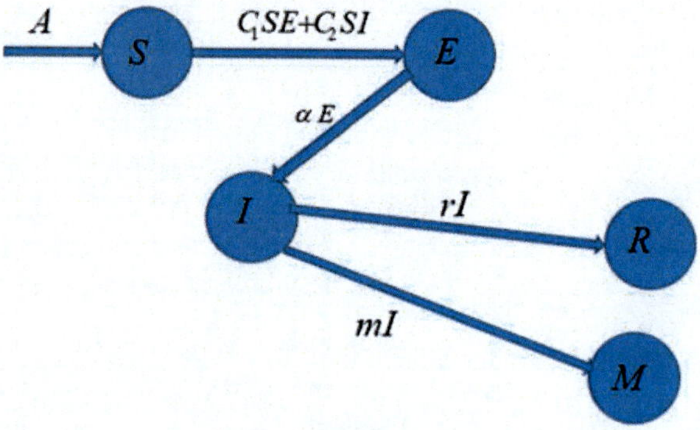


**Fig. 5.13** Mathematical representation of model. *Source* own

**Table 5.1** Notation and its description

| Parameter | Description |
|---|---|
| $\Lambda$ | Population parameter |
| $S$ | Susceptible |
| $E$ | Exposed |
| $I$ | Infected |
| $R$ | Recovery |
| $M$ | Mortality rate |
| $\frac{1}{\alpha}$ | Incubation period between the infection and the onset of symptoms |
| $r$ | Recovery rate |
| $C_1$ | Represent the direct human-to-human transmission rates between the exposed and susceptible individuals |
| $C_2$ | Represent the direct human-to-human transmission rates between the infected and susceptible individuals |

$$\frac{dE}{dt} = C_1 SE + C_2 SI - \alpha E \tag{5.2}$$

$$\frac{dI}{dt} = \alpha E - (m + r)I \tag{5.3}$$

$$\frac{dR}{dt} = rI \tag{5.4}$$

$$\frac{dM}{dt} = mI \tag{5.5}$$

Here, $C_1$ and $C_2$ are assumed as non-increasing functions, given that higher values of $E$ and $I$ would motivate stronger control measures that could reduce the transmission rates.

All the initial conditions of the system are assumed non negative as $S(0) \geq 0$, $E(0) \geq 0$, $I(0) \geq 0$, $R \geq 0$ and $M(0) \geq 0$.

## Numerical Analysis

The basic input parameters of this model are physical in nature and taken by looking the current situation as per the data shown in annexure.

Figures 5.14, 5.15 and 5.16 indicate the variation of infection versus time for the month of June, July and August, respectively. From the figures, it is observed that infection goes on increasing with increase in time. Also, growth of infection

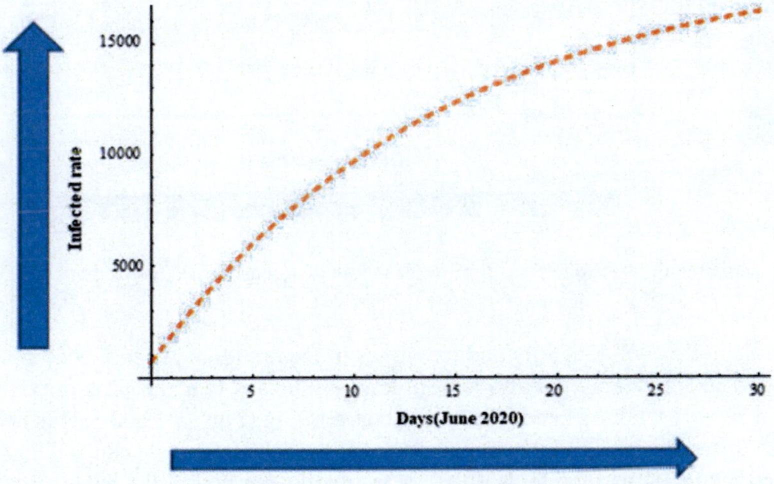


**Fig. 5.14** Variation of infection versus time in the month of June 2020. *Source* own

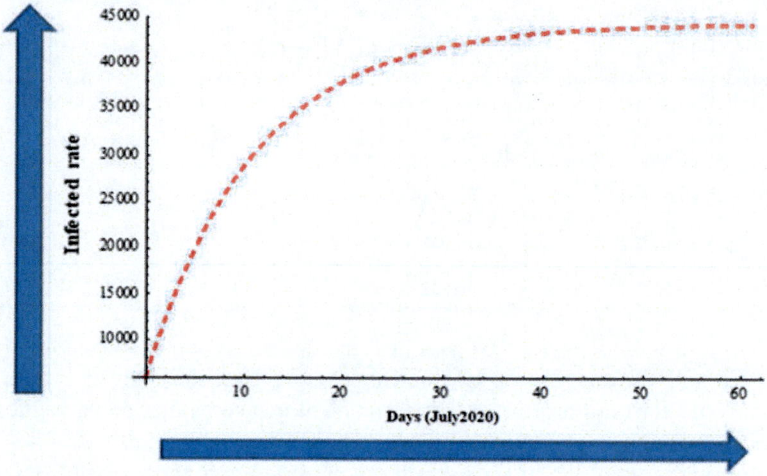


Fig. 5.15 Variation of infection versus time in the month of July 2020. *Source* own

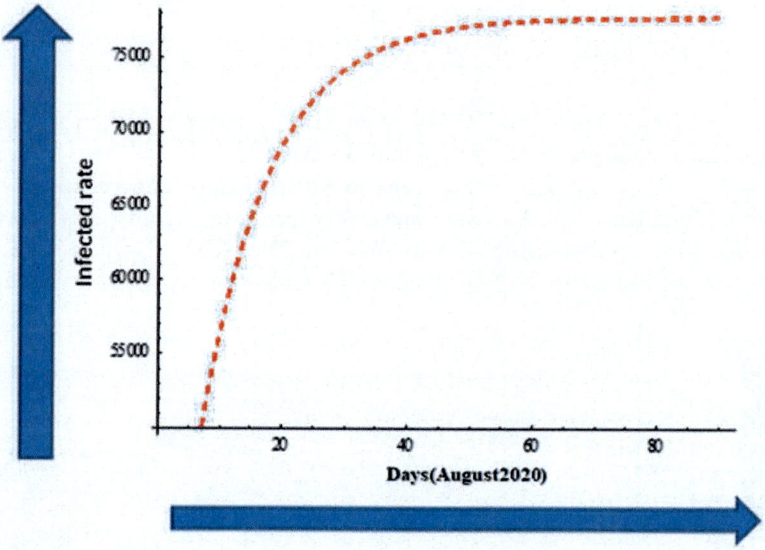


Fig. 5.16 Variation of infection versus time in the month of August 2020. *Source* own

found triple in the month of July and five times in the month of August as compared to infection in June. This variation of infection indicates that the virus growth as discussed above by taking exact data using curve fitting (Figs. 5.1 and 5.4) is nearly similar to the analysis done by our mathematical model. Hence, we can say that our developed mathematical model is efficient and works parallel to the exact situation of spread of infection as recorded in Rajasthan state.

Figures 5.17, 5.18 and 5.19 show the variation of recovery versus time for the month of June, July and August, respectively. It is observed that recovery goes on increasing with increase in time and found significant in the month of June but there is small decay noted in the month of July and August as compared to corresponding rate of infection. The same behaviour of recovery rate was observed by curve fitting as shown in Figs. 5.2 and 5.5 for the exact data, hence our mathematical model work exactly parallel in case of rate of recovery also and gives efficient results.

From Figs. 5.20, 5.21 and 5.22, it is clear that rate of mortality goes on increasing with increase in time but at a steady rate, also as compare to growth in infection the rate of mortality is very low. A similar characteristic of mortality was found during curve fitting for the exact recorded data.

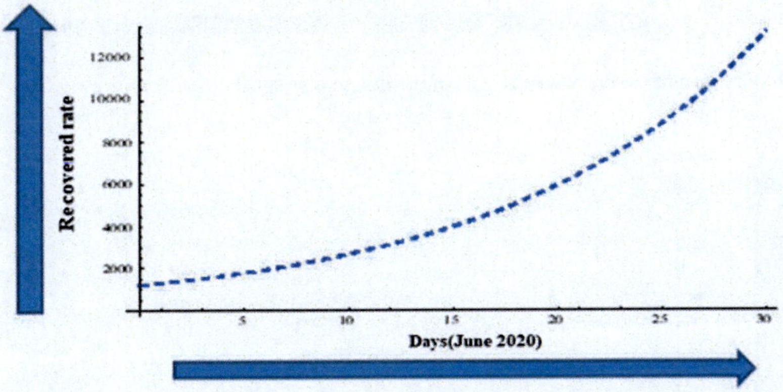


**Fig. 5.17** Variation of recovery versus time in the month of June 2020. *Source* own

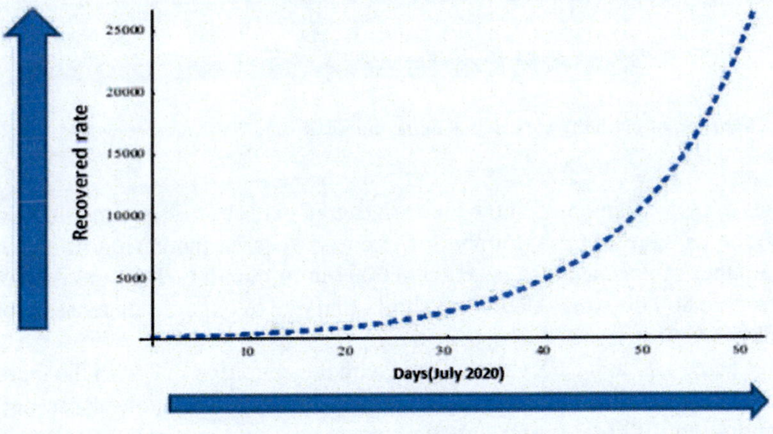


**Fig. 5.18** Variation of recovery versus time in the month of July 2020. *Source* own

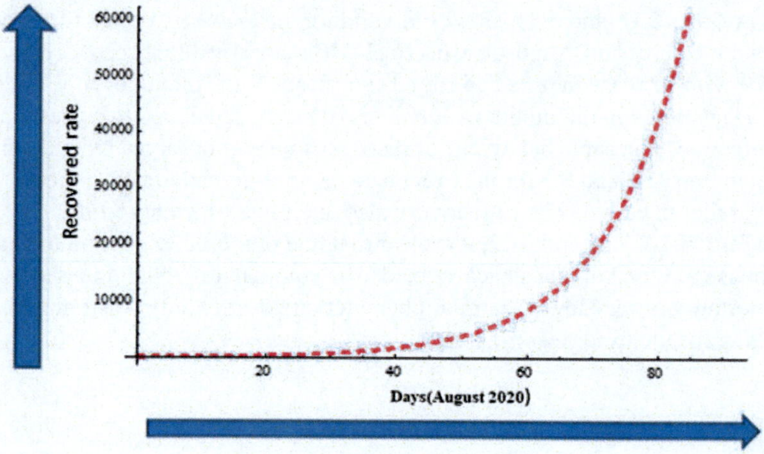


**Fig. 5.19** Variation of recovery versus time in the month of August 2020. *Source* own

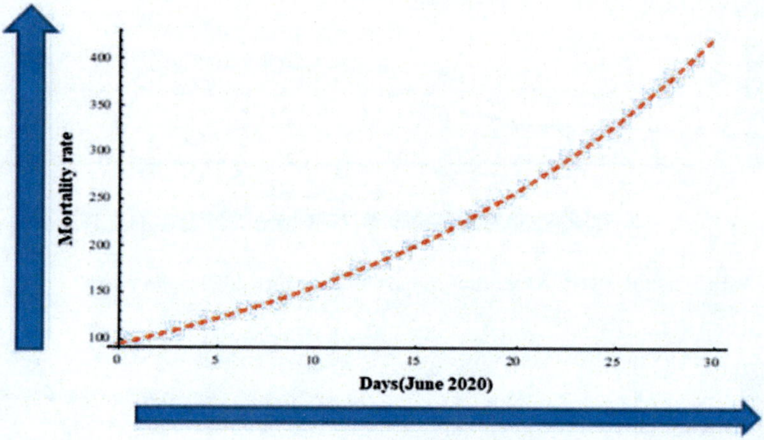


**Fig. 5.20** Variation of mortality versus time in the month of June 2020. *Source* own

Figures 5.23, 5.24 and 5.25 show the variation of exposed versus time. It is noted that with the passage of time, number of exposed goes on increasing. In the June month, number of exposed was nearly 500,000 but in month of July and August, it increases from 800,000 to 1,000,000. It shows that rate of exposed increases rapidly day by day.

Figures 5.26, 5.27 and 5.28 clearly agree with the condition that with the increase in infection corresponding decay is susceptible observed. Variation of susceptible corresponding to different months shown versus time and completely agrees with current scenario of COVID-19 in Rajasthan state.

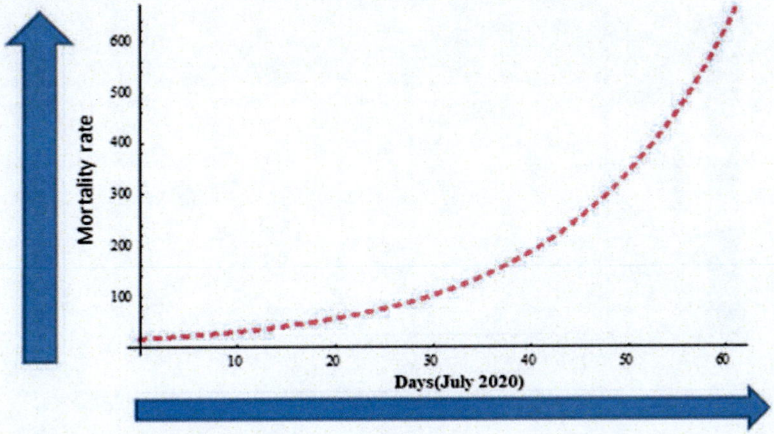


**Fig. 5.21** Variation of mortality versus time in the month of July 2020. *Source* own

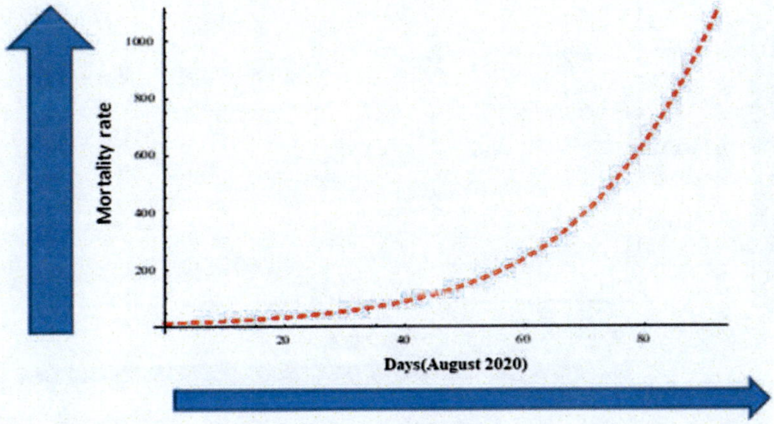


**Fig. 5.22** Variation of mortality versus time in the month of August 2020. *Source* own

Figures 5.29, 5.30, 5.31, 5.32 and 5.33 show the overall variation of infection, mortality, recovery, exposed and susceptibles versus for the June to December month. After observing overall plotting, it is clearly visible that the impact of corona virus will increase significantly in the winter months especially in December 2020. It is forecast that till the end of December month infected population will touches 178,000 and becomes steady, mortality may nearly 3500; recovery may approximately 120,000 (which is insignificant variation as compared to infection), number of exposed may increases 200,000 (double as in August month). The impact can be reduced only if the guideline provided by the Government of India should be strictly followed by entire population with the arrangement of proper vaccination.

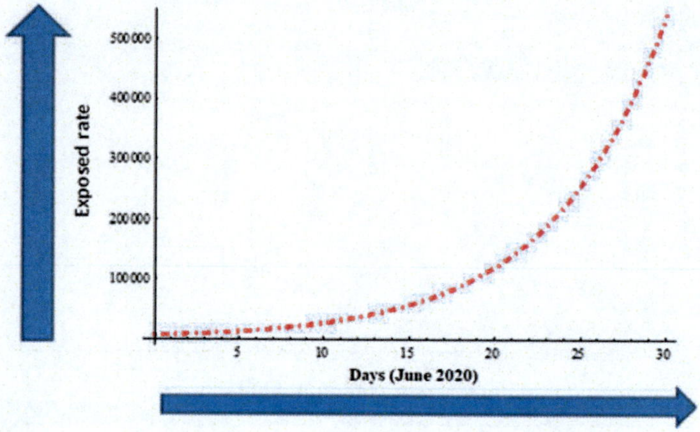


Fig. 5.23 Exposed versus time in the month of June 2020. *Source* own

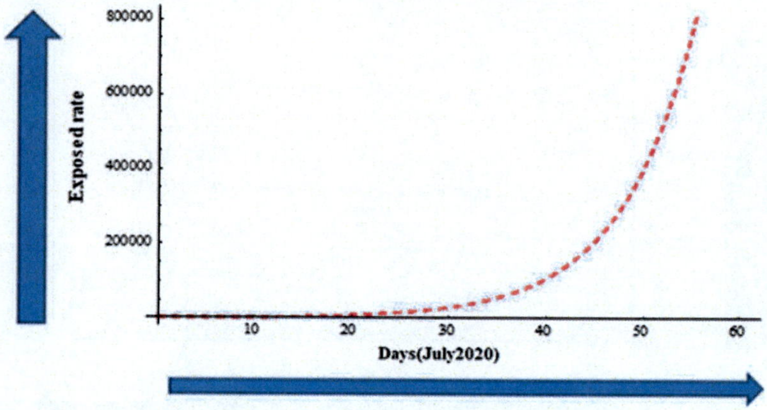


Fig. 5.24 Exposed versus time in the month of July 2020. *Source* own

## Conclusion

In the present study, a comparison model has been constructed to examine the transmission of dynamic COVID-19 for the state of Rajasthan, India. Numerical results obtained by the mathematical modelling which have been compared with exact curve fitting method, where data is taken from government website of Rajasthan state. Numerical plotting in mathematical and curve fitting shows same characteristic in the case of infection, recovery, mortality, susceptible and exposed. The presented mathematical model is hypothetical in nature but for some particular functions and parameters, it is feasible for the real-world problems. Finally, on the basis of the presented model, we can predict that the situation will become worse in the month of December due to unavailability of proper vaccine. Hence, to control the spreading of

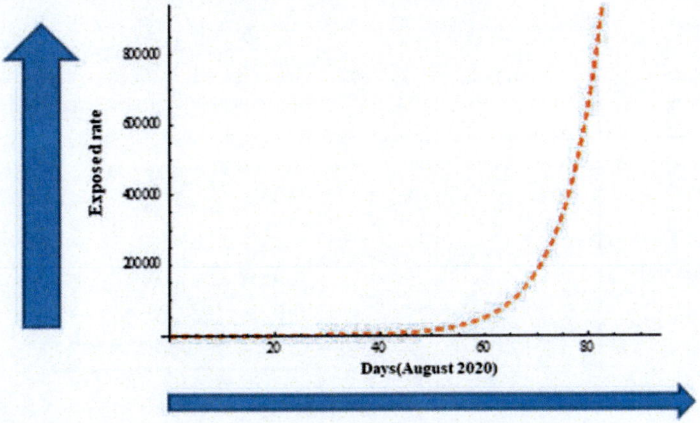


**Fig. 5.25** Exposed versus time in the month of August 2020. *Source* own

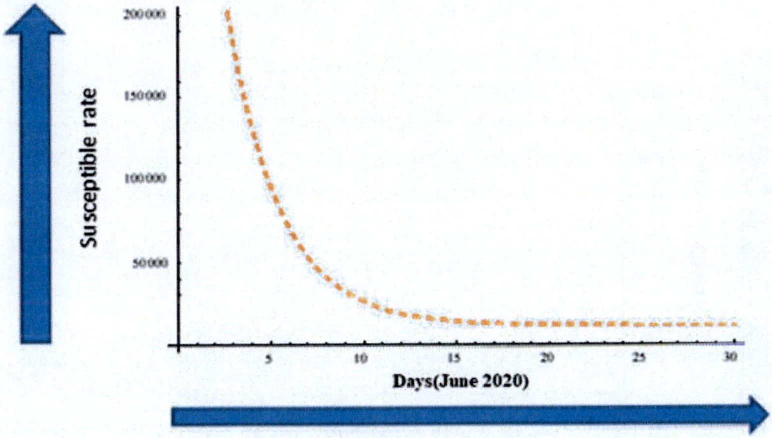


**Fig. 5.26** Susceptible versus time in the month of June 2020. *Source* own

virus infection, it is very much essential to follow all the restrictions like lockdown, social distancing, contact tracing, mask cover, etc., as laid down by Government of Rajasthan and India.

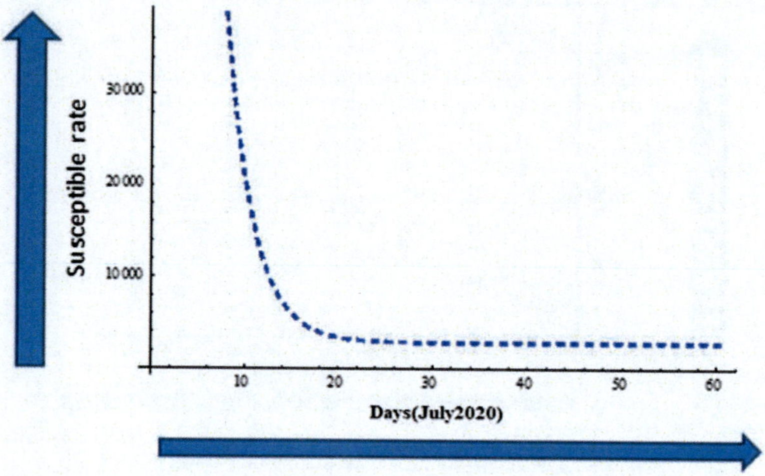


Fig. 5.27 Susceptible versus time in the month of July 2020. *Source* own

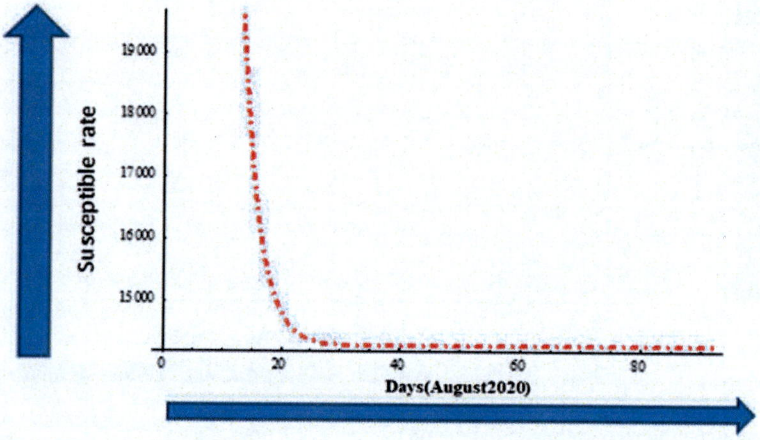


Fig. 5.28 Susceptible versus time in the month of August 2020. *Source* own

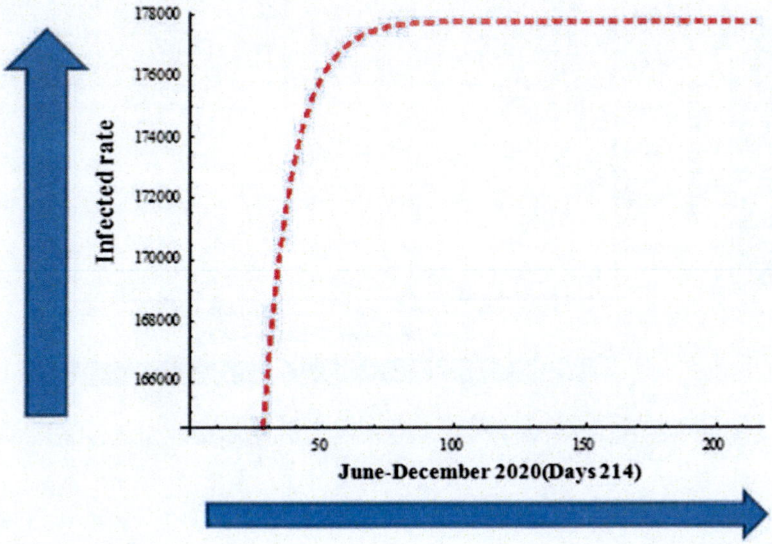


Fig. 5.29 Overall variation of infection versus time from June to December 2020. *Source* own

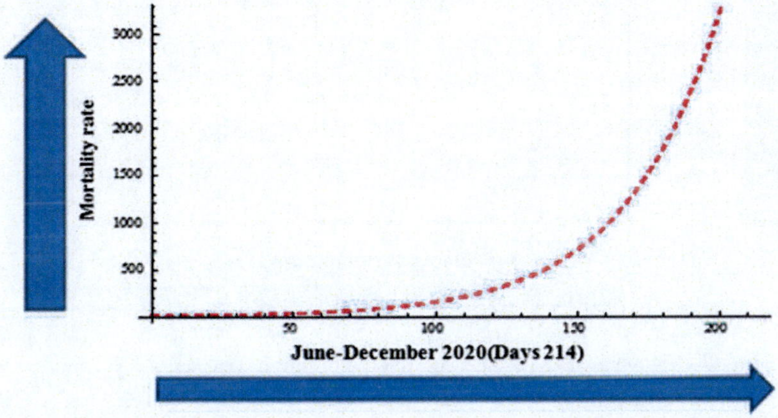


Fig. 5.30 Overall variation of mortality versus time from June to December 2020. *Source* own

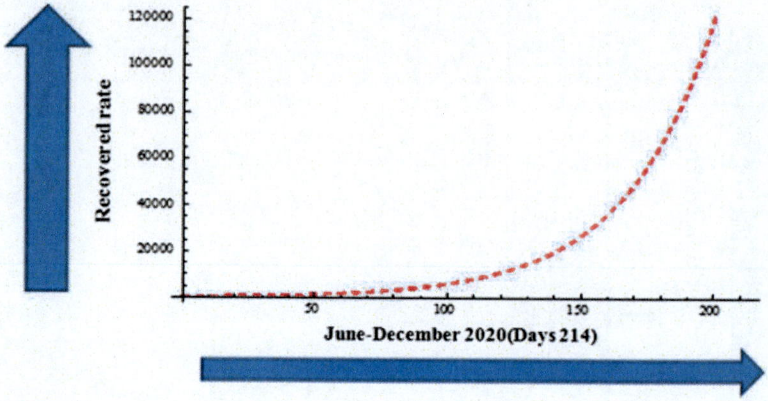


Fig. 5.31 Overall variation of recovery versus time from June to December 2020. *Source* own

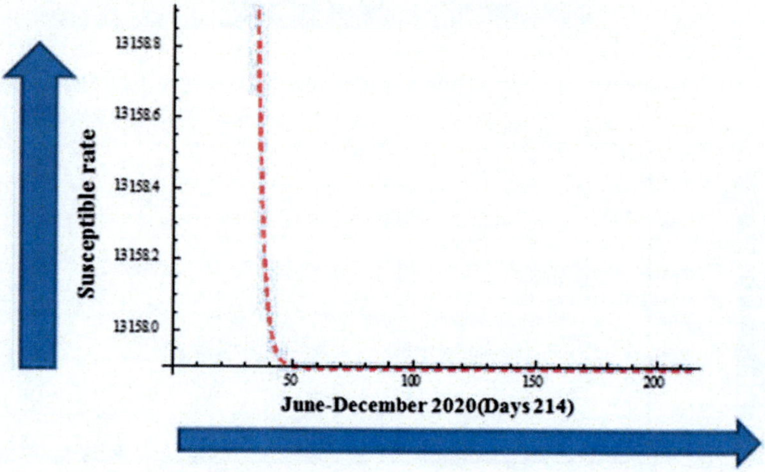


Fig. 5.32 Overall variation of susceptible versus time from June to December 2020. *Source* own

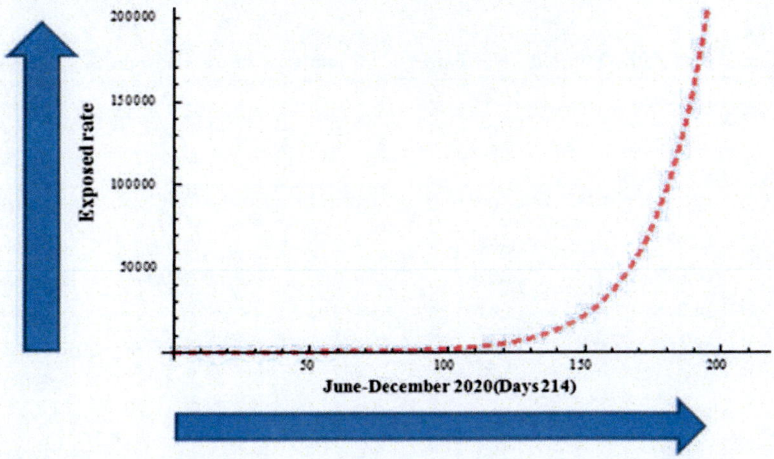


**Fig. 5.33** Overall variation of Exposed versus time from June to December 2020. *Source* own

## Annexure 1

The status of COVID-19 disease in the month of June

| Date (June) | Cumulative infected | New infected | Cumulative death | New death | Recovered |
|---|---|---|---|---|---|
| 1 | 8831 | 214 | 194 | 1 | 6032 |
| 2 | 9100 | 269 | 199 | 5 | 6213 |
| 3 | 9373 | 273 | 203 | 4 | 6435 |
| 4 | 9652 | 279 | 209 | 6 | 6744 |
| 5 | 9862 | 210 | 213 | 4 | 7104 |
| 6 | 10,084 | 222 | 218 | 5 | 7359 |
| 7 | 10,337 | 253 | 231 | 13 | 7501 |
| 8 | 10,599 | 262 | 240 | 9 | 7754 |
| 9 | 10,876 | 277 | 246 | 6 | 8117 |
| 10 | 11,245 | 369 | 255 | 9 | 8328 |
| 11 | 11,600 | 355 | 259 | 4 | 8569 |
| 12 | 11,838 | 238 | 265 | 6 | 8775 |
| 13 | 12,068 | 230 | 272 | 7 | 9011 |
| 14 | 12,401 | 333 | 282 | 10 | 9337 |
| 15 | 12,694 | 293 | 292 | 10 | 9566 |
| 16 | 12,981 | 287 | 301 | 9 | 9785 |
| 17 | 13,216 | 235 | 308 | 7 | 9962 |
| 18 | 13,542 | 326 | 313 | 5 | 10,467 |
| 19 | 13,857 | 315 | 330 | 17 | 10,742 |

(continued)

(continued)

| Date (June) | Cumulative infected | New infected | Cumulative death | New death | Recovered |
|---|---|---|---|---|---|
| 20 | 14,156 | 299 | 333 | 3 | 10,997 |
| 21 | 14,555 | 399 | 337 | 4 | 11,274 |
| 22 | 14,930 | 393 | 349 | 12 | 11,597 |
| 23 | 15,232 | 302 | 356 | 7 | 11,910 |
| 24 | 15,627 | 395 | 365 | 9 | 12,213 |
| 25 | 16,009 | 382 | 375 | 10 | 12,611 |
| 26 | 16,296 | 287 | 379 | 4 | 12,840 |
| 27 | 16,660 | 364 | 380 | 1 | 13,062 |
| 28 | 16,944 | 284 | 391 | 11 | 13,367 |
| 29 | 17,271 | 327 | 399 | 8 | 13,611 |
| 30 | 17,660 | 389 | 405 | 6 | 13,921 |

*Source* [13]

## Annexure 2

The status of COVID-19 disease in the Month of July

| Date (July) | Cumulative infected | New infected | Cumulative death | New death | Recovered |
|---|---|---|---|---|---|
| 1 | 18,014 | 354 | 413 | 8 | 14,220 |
| 2 | 18,312 | 298 | 421 | 8 | 14,574 |
| 3 | 18,662 | 350 | 430 | 9 | 14,948 |
| 4 | 19,052 | 390 | 440 | 10 | 15,281 |
| 5 | 19,532 | 480 | 447 | 7 | 15,640 |
| 6 | 20,164 | 632 | 456 | 9 | 15,928 |
| 7 | 20,688 | 524 | 461 | 5 | 16,278 |
| 8 | 21,404 | 716 | 472 | 11 | 16,575 |
| 9 | 22,063 | 659 | 482 | 10 | 16,866 |
| 10 | 22,563 | 500 | 491 | 9 | 17,070 |
| 11 | 23,174 | 611 | 497 | 6 | 17,620 |
| 12 | 23,748 | 574 | 503 | 6 | 17,869 |
| 13 | 24,392 | 644 | 510 | 7 | 18,103 |
| 14 | 24,936 | 544 | 518 | 8 | 18,630 |
| 15 | 25,571 | 635 | 524 | 6 | 19,169 |
| 16 | 26,437 | 866 | 530 | 6 | 19,502 |
| 17 | 27,174 | 737 | 538 | 8 | 19,970 |

(continued)

(continued)

| Date (July) | Cumulative infected | New infected | Cumulative death | New death | Recovered |
|---|---|---|---|---|---|
| 18 | 27,789 | 615 | 546 | 8 | 20,626 |
| 19 | 28,500 | 711 | 553 | 7 | 21,144 |
| 20 | 29,434 | 934 | 559 | 6 | 21,730 |
| 21 | 30,390 | 956 | 568 | 9 | 22,195 |
| 22 | 31,373 | 983 | 577 | 9 | 22,744 |
| 23 | 32,334 | 961 | 583 | 6 | 23,364 |
| 24 | 33,220 | 886 | 594 | 11 | 23,815 |
| 25 | 34,178 | 958 | 602 | 8 | 24,547 |
| 26 | 53,298 | 1120 | 613 | 11 | 25,306 |
| 27 | 36,430 | 1132 | 624 | 11 | 25,954 |
| 28 | 37,564 | 1134 | 633 | 9 | 26,834 |
| 29 | 38,636 | 1072 | 644 | 11 | 27,317 |
| 30 | 39,780 | 1144 | 654 | 10 | 28,309 |
| 31 | 40,936 | 1156 | 667 | 13 | 29,231 |

*Source* [13]

# Annexure 3

The status of COVID-19 disease in the month of August

| Date (August) | Cumulative infected | New infected | Cumulative death | New death | Recovered |
|---|---|---|---|---|---|
| 1 | 42,083 | 1147 | 680 | 13 | 29,845 |
| 2 | 43,243 | 1160 | 694 | 14 | 30,668 |
| 3 | 44,410 | 1167 | 706 | 12 | 31,216 |
| 4 | 45,555 | 1145 | 719 | 13 | 32,051 |
| 5 | 46,679 | 1124 | 732 | 13 | 32,832 |
| 6 | 47,845 | 1166 | 745 | 13 | 33,849 |
| 7 | 48,996 | 1151 | 757 | 12 | 35,131 |
| 8 | 50,157 | 1161 | 767 | 10 | 36,195 |
| 9 | 51,328 | 1171 | 778 | 11 | 37,163 |
| 10 | 52,497 | 1169 | 789 | 11 | 38,235 |
| 11 | 53,670 | 1173 | 800 | 11 | 39,060 |
| 12 | 54,887 | 1217 | 811 | 11 | 40,399 |
| 13 | 56,100 | 1213 | 822 | 11 | 41,648 |

(continued)

(continued)

| Date (August) | Cumulative infected | New infected | Cumulative death | New death | Recovered |
|---|---|---|---|---|---|
| 14 | 57,414 | 1264 | 833 | 11 | 41,819 |
| 15 | 58,692 | 1278 | 846 | 13 | 43,897 |
| 17 | 61,296 | 1317 | 876 | 14 | 46,604 |
| 18 | 62,630 | 1334 | 887 | 11 | 47,654 |
| 19 | 63,977 | 1347 | 898 | 11 | 48,960 |
| 20 | 65,289 | 1312 | 910 | 12 | 49,963 |
| 21 | 66,619 | 1330 | 921 | 11 | 51,190 |
| 22 | 67,954 | 1335 | 933 | 12 | 52,496 |
| 23 | 69,264 | 1310 | 944 | 11 | 54,144 |
| 24 | 70,609 | 1345 | 955 | 11 | 55,324 |
| 25 | 71,955 | 1346 | 967 | 12 | 56,600 |
| 26 | 73,325 | 1370 | 980 | 13 | 58,126 |
| 27 | 74,670 | 1345 | 992 | 12 | 59,579 |
| 28 | 76,015 | 1345 | 1005 | 13 | 60,585 |
| 29 | 77,370 | 1355 | 1017 | 12 | 62,033 |
| 30 | 78,777 | 1407 | 1030 | 13 | 62,971 |
| 31 | 80,227 | 1450 | 1043 | 13 | 65,093 |

*Source* [13]

# Annexure 4

The status of COVID-19 disease up to 19 September 2020

| Date (September) | Cumulative infected | New infected | Cumulative death | New death | Recovered |
|---|---|---|---|---|---|
| 1 | 83,163 | 1470 | 1069 | 13 | 68,124 |
| 2 | 84,674 | 1511 | 1081 | 12 | 70,674 |
| 3 | 86,227 | 1553 | 1095 | 14 | 71,220 |
| 4 | 87,797 | 1570 | 1108 | 13 | 71,899 |
| 5 | 89,363 | 1566 | 1122 | 14 | 73,245 |
| 6 | 90,956 | 1593 | 1137 | 15 | 74,861 |
| 7 | 92,536 | 1580 | 1151 | 14 | 76,427 |
| 8 | 94,126 | 1590 | 1164 | 13 | 77,872 |
| 9 | 95,736 | 1610 | 1178 | 14 | 79,450 |

(continued)

(continued)

| Date (September) | Cumulative infected | New infected | Cumulative death | New death | Recovered |
|---|---|---|---|---|---|
| 10 | 97,376 | 1640 | 1192 | 14 | 80,482 |
| 11 | 99,036 | 1660 | 1207 | 15 | 81,970 |
| 12 | 100,705 | 1669 | 1221 | 14 | 82,902 |
| 13 | 102,408 | 1703 | 1236 | 15 | 84,518 |
| 14 | 104,138 | 1730 | 1250 | 14 | 86,162 |
| 15 | 105,898 | 1760 | 1264 | 14 | 87,873 |
| 16 | 107,680 | 1782 | 1279 | 15 | 89,352 |
| 17 | 109,473 | 1793 | 1293 | 14 | 90,685 |
| 18 | 111,290 | 1817 | 1308 | 15 | 92,265 |
| 19 | 113,124 | 1834 | 1322 | 14 | 93,805 |

*Source* [14]

# References


1. Yang, C., & Wang, J. (2020). A mathematical model for the novel coronavirus epidemic in Wuhan, China. *Mathematical Biosciences and Engineering, 17,* 2708–2724.
2. Guckenheimer, J., & Holmes, P. (2002). *Nonlinear oscillations, dynamical systems, and bifurcations of vector fields.* Mathematics (462 p). New York: Springer.
3. Driessche, P., & Watmough, J. (2002). Reproduction numbers and sub-threshold endemic equilibria for compartmental models of disease transmission. *Mathematical Biosciences, 180,* 29–48. https://doi.org/10.1016/S0025-5564(02)00108-6.
4. Mizumoto, K., & Chowell, G. (2020). Transmission potential of the novel coronavirus (COVID-19) onboard the diamond Princess Cruises Ship. *Infectious Disease Modeling, 5,* 264–270.
5. Rothan, H. A., & Byrareddy, S. N. (2020). The epidemiology and pathogenesis of coronavirus disease (COVID-19) outbreak. *Journal of Autoimmunity, 109,* Article ID 102433.
6. Sohrabi, C., Alsafi, Z., O'Neill, N., Khan, M., Kerwan, A., Al-Jabir, A., et al. (2020). World Health Organization declares global emergency: A review of the 2019 novel coronavirus (COVID-19). *International Journal of Surgery, 76,* 71–76.
7. Thevarajan, I., Nguyen, T. H. O., Koutsakos, M., Druce, J., Caly, L., van de Sandt, C. E., et al. (2020). Breadth of concomitant immune responses prior to patient recovery: A case report of non-severe COVID-19. *Nature Medicine,* 453–455.
8. Riou, J., & Althaus, C. L. (2020). Pattern of early human-to-human transmission of Wuhan 2019 novel coronavirus (2019-nCoV), December 2019 to January 2020. *Eurosurveillance, 25.*
9. Khot, W. Y., & Nadkar, M. Y. (2020). The 2019 novel coronavirus outbreak—A global threat. *The Journal of the Association of Physicians of India, 68,* 67–71.
10. Wang. (2020) Coronavirus disease 2019 (COVID-19): situation report. WHO, Geneva, Switzerland.

11. Sahin, A. R., Erdogan, A., Agaoglu, P. M., Dineri, Y., Cakirci, A. Y., Senel, M. E., et al. (2020). Novel coronavirus (COVID-19) outbreak: A review of the current literature. *EJMO, 4*, 1–7.
12. Cheng, Z. J., & Shan, J. (2020). Novel coronavirus: Where we are and what we know. *Infection, 48,* 153–155.
13. https://en.wikipedia.org/wiki/COVID-19_pandemic_in_Rajasthan.
14. www.covid19india.org.

# Chapter 6
# A Mathematical Model for COVID-19 in Italy with Possible Control Strategies

**Sumit Kumar, Sandeep Sharma, Fateh Singh, PS Bhatnagar, and Nitu Kumari**

**Abstract** Italy faced the COVID-19 crisis in the early stages of the pandemic. In the present study, a SEIR compartment mathematical model has been proposed. The model considers four stages of infection: susceptible($S$), exposed ($E$), infected ($I$) and recovered ($R$). Basic reproduction number $R_0$ which estimates the transmission potential of a disease has been calculated by the next-generation matrix technique. We have estimated the model parameters using real data for the Coronavirus transmission. To get a dipper insight into the transmission dynamics, we have also studied four of the most pandemic affected regions of Italy. Basic reproduction number stood differently for different regions of Italy i.e. Lombardia (2.1382), Veneto (1.7512), Emilia Romagna (1.6331), Piemonte (1.9099) and for Italy at 2.0683. The sensitivity of $R_0$ corresponding to various disease transmission parameters has also been demonstrated via numerical simulations. Besides, it has been demonstrated with the help of simulations that earlier lockdown and rapid isolation of infective individuals would have been helpful in a dual way; by substantially reducing the number of susceptible people on one hand and preponing the end of the pandemic on the other. This paper also includes complete theoretical analysis of the proposed model including the epidemic feasibility of the model and existence of endemic equilibrium point. We have also derived the conditions under which the disease became endemic. Since the existence of an endemic equilibrium point refers to the possibility of backward bifurcation, we have given a detailed analysis regarding the same. All the theoretical analysis is supported by detailed numerical simulations to understand the transmis-

S. Kumar · N. Kumari (✉)
School of Basic Sciences, Indian Institute of Technology Mandi,
Mandi, Himachal Pradesh 175001, India
e-mail: nitu@iitmandi.ac.in

S. Kumar
e-mail: sumitgodara08.95@gmail.com

S. Sharma · F. Singh
Department of Mathematics, DIT University, Dehradun, Uttarakhand 248009, India

P. Bhatnagar
Zoological Survey of India, Ministry of Environment,
Forest and Climate Change, Pune 411044, India

 
N. H. Shah and M. Mittal (eds.), *Mathematical Analysis for Transmission of COVID-19*,
Mathematical Engineering, https://doi.org/10.1007/978-981-33-6264-2_6

sion dynamics of COVID-19 While analyzing different regions of Italy it was found that Lombardia was the hardest hit and had the highest number of infectives. We have also forecasted the future scenario of the pandemic in Italy. The model predicts that the COVID-19 epidemic shall die out from the worst affected Lombardia region by approximately by November 2020.

**Keywords** SEIR Model · Basic reproduction number · Stability analysis · Backward bifurcation · Parameter estimation

**Mathematics Subject Classification** Primary: 92B05 · 93A30 · Secondary: 34C23

## Introduction

Epidemics and pandemics have shaped human history since long. It has been proposed that a majority of human Coronovirus have been derived from bat reservoir [1]. At the initial stage it was believed that transmission of disease took place through animal to human mode but later it has been established that direct transmission of the disease is also possible and is the primary reason for transmission to various countries [2–4]. Hospital related transmission is also suspected in 41 percent of the patients [3].

Along with high transmission efficiency, global travel has also contributed to SARS-CoV-2 spread across the globe [5]. On 30 January 2020, the WHO initially declared COVID-19 a public health emergency of international concern and later a pandemic on 11 March 2020 [6]. Interestingly African region has lowest reported cases of COVID-19 which is otherwise host to many infectious diseases. Italy recorded its first case of COVID-19 on February 20, 2020, at Lodi40 (Lombardy) [7]. In the next 24 h, the infected cases increased to 36 [7].

In the begning, Italian data followed similar trend observed in Hubei Province, China. Till date, Italy is one of the countries that have faced grievous consequences of COVID-19. Till 20th May 2020, Italy has 2,27,364 recorded cases and 32,330 deaths due to COVID-19. In terms of reported cases, it is the sixth highest while it is third when it comes to the number of deaths across the globe. 42.2% of the patients who died were 80–89 years old, 32.4% were 70–79 years, 8.448% were 60–69 years, and 2.8% were 50–59 years old. The male to female ratio is 80–20% with older median age for women (83.4 years for women vs 79.9 years for men) [8]. Moreover, the estimated mean age of those who lost their lives in Italy was 81 years [8]. The COVID-19 outbreak completely disturbed the economic condition of Italy. Several family owned small sectors are suering [9]. The condition of Italy surprised the research fraternity because Italy stands in the top ve countries in terms of medical facilities. The ongoing dismal scenario in Italy has forced the government to admit that they do not have any control over the spread of the disease as well as do not know when the ongoing web of COVID-19 will stop.

Besides, the situation is so grim that Italian College of Anaesthesia, Analgesia, Resuscitation and Intensive Care has published guidelines that said resources that could be every scarce are reserved for those who have the greatest chances of survival first and secondly to those who have more years of life to be saved, with a view to maximizing the benefits for the most number of people. Due to the catastrophic impacts of the COVID-19 outbreak, efforts have been made to analyze the trend of the disease and predict the future of the epidemic [8, 10]. The work carried out in [8] predicts that in the absence of timely implementation of available medical resources, the authorities will not be able to control the outbreak of the disease. The study further concludes that together with the medical facilities people's movement and social activities should be restricted immediately in order to curtail the burden of the COVID-19. The work carried out in [8], collected the day to day data and measured the possible similarity between Italy and Hubei Province (China). Further, the study also shows that the number of deaths increased almost five times as the available treatment facilities reached the limit. As of now, there is no vaccination available for COVID-19.

There are many models for infectious diseases and in the class of compartmental models, they may range from very classical SIR to more complex ones. In the current work, we have developed a compartmental model to examine the case of Italy. The proposed model incorporates four different compartments namely—Susceptible, Exposed, Infected and Recovered population. The present COVID-19 has an latency period which ranges up to 14 days. Therefore, to make our model realistic, we include the exposed population along with the infected population, which certainly results in improved prediction. We collected the data of Italy for COVID-19 available on the Worldometers website [11]. Further we trained the proposed mathematical model using the data available till 20th May 2020. Using this enhanced model we analyze the effect of lock down on the spread of COVID-19 in Italy. Also, we predict the possible end of the current outbreak of COVID-19 in Italy. Moreover, to make our predictions more realistic, we have trained and validated our model with COVID-19 data of some the highly affected regions of Italy.

## The Mathematical Model

The proposed model describes the transmission mechanism of COVID-19. In the modelling process, we have divided the human population into four mutually exclusive compartments, namely, susceptible ($S$), exposed ($E$) infected but asymptomatic, infected ($I$) symptomatic and infectious and recovered ($R$). The flow chart of the model is presented in Fig. 6.1.

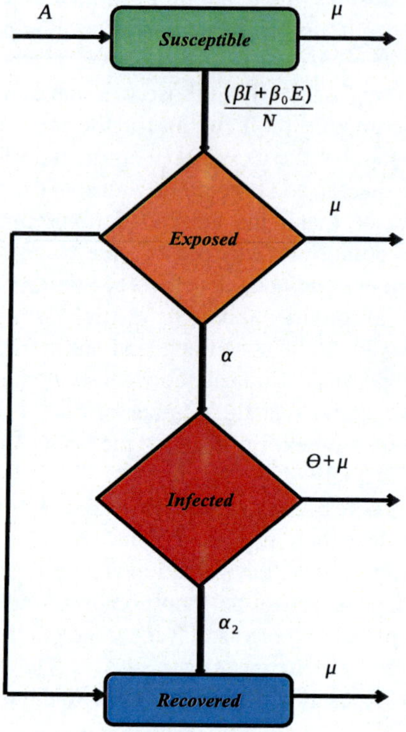


**Fig. 6.1** Flow chart for the COVID-19 Italy model

Based on the above assumptions, the model is governed by the following system of equations:-

$$\begin{aligned}\frac{dS}{dt} &= A - \beta S\frac{I}{N} - \beta_0 S\frac{E}{N} - \mu S\\ \frac{dE}{dt} &= \beta S\frac{I}{N} + \beta_0 S\frac{E}{N} - \alpha E - \alpha_1 E - \mu E\\ \frac{dI}{dt} &= \alpha E - \theta I - \alpha_2 I - \mu I\\ \frac{dR}{dt} &= \alpha_1 E + \alpha_2 I - \mu R\end{aligned} \tag{6.1}$$

In the above model, $N$ represents the total population. We assume that all the new recruiters joined the susceptible class at a constant rate $A$. $\beta$ is the disease transmission rate from the infected individuals to susceptible individuals. We further assume that susceptible individuals once come into the contact of infected individual will not directly join the infected class. They first join the exposed class ($E$) and after certain period of time shows visible symptoms of the disease and enters into the infected class ($I$). Exposed class individuals are assumed to be less infectious as compared to the infected class individuals. Therefore, $\beta_0$ represents the disease transmission rate for exposed individuals. Clearly, $\beta_0 \leq \beta$. Here, $\alpha$ is the rate by which exposed population moves to infected compartment. $\alpha_1$ is the recovery rate of exposed individuals and $\alpha_2$ is the recovery rate of infected individuals. $\theta$ is the disease induced death rate. $\mu$ is the natural death rate.

Our proposed model system involves human population. Hence for the initial state, all the compartmental values are assumed to be non-negative. The model will be studied under the following initial conditions:

$$S(0) \geq 0, E(0) \geq 0, I(0) \geq 0, R(0) \geq 0 \tag{6.2}$$

## Basic Properties

In this section, we check the mathematical feasibility of the proposed model. For this purpose, we check whether all the solutions of the proposed model will remain positive and bounded or not.

### *Non-negativity of the Solution*

To show the epidemiological feasibility of the proposed model system (6.1), it is required that all the solutions remain non-negative. Hence, in the following theorem we verify that all the solutions with non-negative initial condition will remain non-negative.

**Theorem 1** *The solution* $(S(t), E(t), I(t), R(t))$ *of the proposed model system is non-negative for all* $t \geq 0$ *with non-negative initial condition (6.2).*

***Proof*** From the first equation of system (6.1), we have

$$\frac{\mathrm{d}S}{\mathrm{d}t} = A - \left(\frac{\beta I}{N} + \frac{\beta_0 E}{N} + \mu\right) S$$

From this equation, we can deduce that

$$\frac{\mathrm{d}}{\mathrm{d}t}\left[S(t)\exp\left\{\int_0^t\left(\frac{\beta I(v)}{N} + \frac{\beta_0 E(v)}{N}\right).\mathrm{d}v + \mu t\right\}\right] = A\exp\left\{\int_0^t\left(\frac{\beta I(v)}{N} + \frac{\beta_0 E(v)}{N}\right).\mathrm{d}v + \mu t\right\}$$

Now, integrating the equation on both sides

$$S(t_1)\exp\left\{\int_0^{t_1}\left(\frac{\beta I(v)}{N} + \frac{\beta_0 E(v)}{N}\right).\mathrm{d}v + \mu t_1\right\} - S(0) = \int_0^{t_1}\frac{A}{2}\exp\left\{\int_0^{u}\left(\frac{\beta I(v)}{N} + \frac{\beta_0 E(v)}{N}\right).\mathrm{d}v + \mu t\right\}.\mathrm{d}u$$

On simplification we have,

$$S(t_1) = S(0)\exp\left\{-\left(\int_0^{t_1}\left(\frac{\beta I(v)}{N} + \frac{\beta_0 E(v)}{N}\right).dv + \mu t_1\right)\right\}$$
$$+\exp\left\{-\left(\int_0^{t_1}\left(\frac{\beta I(v)}{N} + \frac{\beta_0 E(v)}{N}\right).dv + \mu t_1\right)\right\} \times$$
$$\int_0^{t_1}\frac{A}{2}\exp\left\{\int_0^{u}\left(\frac{\beta I(v)}{N} + \frac{\beta_0 E(v)}{N}\right).dv + \mu t\right\}.du$$

This gives

$$S(t_1) \geq 0$$

where $t_1 \geq 0$ is arbitrary. Similarly we can show the non-negativity for compartments $E$, $I$ and $R$.

Hence, the solution $(S(t), E(t), I(t), R(t))$ will remain positive for non-negative initial condition.

## *Boundedness of the Solution*

In order to accurately predict the epidemic, the solutions of the mathematical model should be bounded.

**Theorem 2** *All solutions of the proposed model are bounded.*

***Proof*** We need to show that $(S(t), E(t), I(t), R(t))$ is bounded for each value of $t \geq 0$. From our model system (6.1) we obtain:

$$(S + E + I + R)' = A - \mu(S + E + I + R) - \theta I \leq A - \mu(S + E + I + R)$$

which gives us

$$\lim_{t\to\infty} \text{Sup}\,(S + E + I + R) \leq \frac{A}{\mu}$$

which implies that each individual component is also bounded.

## Disease Free Equilibrium and Basic Reproduction Number

The basic reproduction number is one of the critical parameters to examine the long term behaviour of an epidemic. It can be defined as the expected number of cases directly generated by one case in a population where all individuals are susceptible to infection. We have used next-generation matrix technique explained in [12], to obtain the expression of reproduction number $R_0$.

In order to reduce the number of parameter in proposed model system (6.1), we normalize the model by considering $X_1 = \frac{S}{N}, X_2 = \frac{E}{N}, X_3 = \frac{I}{N}, X_4 = \frac{R}{N}$ and $A = \mu N$. For convenience, we represent the variables $X_1$, $X_2$, $X_3$ and $X_4$ by the same variables as $S, E, I$ and $R$. The proposed model takes the following form:

$$
\begin{aligned}
\frac{dS}{dt} &= \mu - \beta SI - \beta_0 SE - \mu S \\
\frac{dE}{dt} &= \beta SI + \beta_0 SE - \alpha E - \alpha_1 E - \mu E \\
\frac{dI}{dt} &= \alpha E - \theta I - \alpha_2 I - \mu I \\
\frac{dR}{dt} &= \alpha_1 E + \alpha_2 I - \mu R
\end{aligned}
\tag{6.3}
$$

The disease fee equilibrium (DFE)of the model system (6.3) can be given as:

$$E^0 = \{1, 0, 0, 0\}$$

The infection states of the model are $E$ and $I$. The progression from $E$ to $I$ is not considered as new infection but rather a progression of infection through compartment. Therefore,

$$F = \begin{pmatrix} \beta_0 SE + \beta SI \\ 0 \end{pmatrix} \text{ and } V = \begin{pmatrix} -\alpha E - \alpha_1 E - \mu E \\ \alpha E - \alpha_2 I + \theta I + \mu I \end{pmatrix}$$

At DFE $E^0$ the transmission matrix $\mathcal{F}$ and the transition matrix $\mathcal{V}$ are given as:

$$\mathcal{F} = \begin{pmatrix} \beta_0 & \beta \\ 0 & 0 \end{pmatrix} \text{ and } \mathcal{V} = \begin{pmatrix} \alpha + \alpha_1 + \mu & 0 \\ -\alpha & \alpha_2 + \theta + \mu \end{pmatrix}$$

Which gives,

$$\mathcal{F}\mathcal{V}^{-1} = \frac{1}{(\alpha+\alpha_1+\mu)(\alpha_2+\theta=\mu)} \begin{pmatrix} (\beta_0 (\alpha_2 + \theta + \mu) + \beta\alpha) & \beta (\alpha + \alpha_1 + \mu) \\ 0 & 0 \end{pmatrix}$$

Hence, the $R_0$ takes the following expression

$$R_0 = \rho\left(\mathcal{F}\mathcal{V}^{-1}\right) = \frac{\beta_0 (\alpha_2 + \theta + \mu) + \beta\alpha}{(\alpha + \alpha_1 + \mu)(\alpha_2 + \theta + \mu)}$$

We will analyse the variation in $R_0$ for different values of the parameters involved in the model system. Figure 6.2 illustrates the simultaneous variation in the basic reproduction number for different values of corresponding parameters. The parameter values used are given in Tables 6.1 and 6.2.

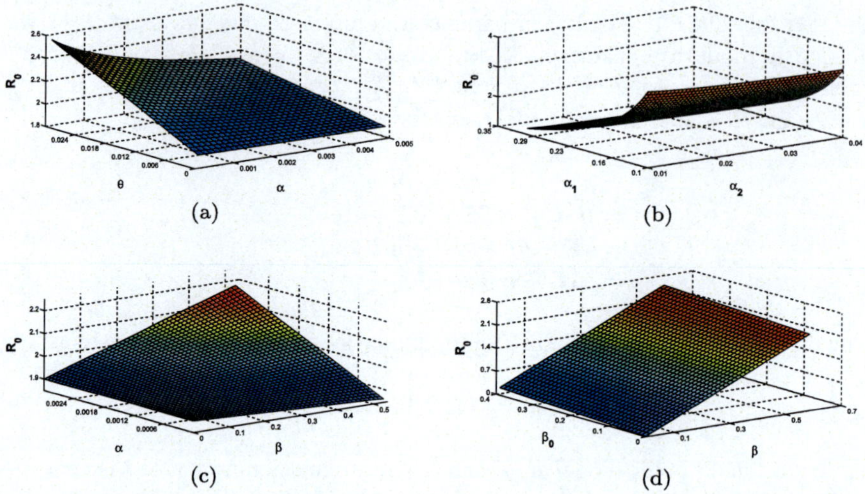


**Fig. 6.2** Variation in the basic reproduction number $R_0$ for different values of sensitive parameters. **a** Effect of $\alpha$ and $\theta$ on $R_0$. **b** Effect of $\alpha_1$ and $\alpha_2$ on $R_0$. **c** Effect of $\beta$ and $\alpha$ on $R_0$. **d** Effect of $\beta$ and $\beta_0$ on $R_0$

**Table 6.1** Parameter values used for the model

| Parameters | Value | Source |
|---|---|---|
| $\alpha_2$ | 0.02 | Estimated |
| $\theta$ | 0.01 | [11] |
| $\mu$ | 3.3009e−05 | [16] |

**Table 6.2** Estimated parameter values of the model for Lombardia, Veneto, Emilia Romagna, Piemonte and Italy

| Parameters | Lombardia | Veneto | Emilia Romagna | Piemonte | Italy |
|---|---|---|---|---|---|
| $A$ | 100.4365 | 45.4183 | 110.0589 | 160.4357 | 675.0309 |
| $\beta$ | 0.6290 | 0.7191 | 0.5006 | 0.5040 | 0.5036 |
| $\beta_0$ | 0.2241 | 0.4595 | 0.4689 | 0.2630 | 0.3159 |
| $\alpha$ | 0.0023 | 0.0017 | 0.0025 | 0.0017 | 0.0019 |
| $\alpha_1$ | 0.1250 | 0.2839 | 0.3101 | 0.1509 | 0.1662 |

## Existence of Endemic Equilibrium

The endemic equilibrium state is the state when the disease can not be completely eradicated but remains in the population. For the disease to persist in the population; the exposed class and infected class must not be zero at the equilibrium state.

Let $E^* = (S^*, E^*, I^*, R^*)$ be the endemic equilibria. Now, $E^*$ can be achieved by solving the equations below:

$$\begin{aligned} &\mu - \beta S^* I^* - \beta_0 S^* E^* - \mu S^* = 0 \\ &\beta S^* I^* + \beta_0 S^* E^* - (\alpha + \alpha_1 + \mu) E^* = 0 \\ &\alpha E^* - (\theta + \alpha_2 + \mu) I^* = 0 \\ &\alpha_1 E^* + \alpha_2 I^* - \mu R^* = 0 \end{aligned} \tag{6.4}$$

From the third equation, we have

$$E^* = \frac{\theta + \alpha_2 + \mu}{\alpha} I^*$$

Also, from fourth equation, we have

$$R^* = \frac{\alpha_1 E^* + \alpha_2 I}{\mu} = \left( \frac{\alpha_1 (\theta + \alpha_2 + \mu) + \alpha \alpha_2}{\alpha \mu} \right) I^*$$

Now, using the values of $E^*$ and $R^*$ in second equation we will have

$$S^* = \frac{(\alpha + \alpha_1 + \mu) E^*}{\beta I^* + \beta_0 E^*} = \frac{(\alpha + \alpha_1 + \mu)(\theta + \alpha_2 + \mu)}{\alpha\beta + \beta_0 (\theta + \alpha_2 + \mu)}$$

Now using the values of $E^*$, $R^*$ and $S^*$, first equation implies

$$I^* = \frac{\mu - \mu S^*}{\left( \beta + \frac{\beta_0(\theta + \alpha_2 + \mu)}{\alpha} \right) S^*}$$

On simplification, we have

$$\begin{aligned} S^* &= \frac{\Lambda \Upsilon}{\psi} \\ E^* &= \frac{\mu[\alpha\beta + \Upsilon(\beta_0 - \alpha - \alpha_1 - \mu)]}{\Lambda \psi} \\ I^* &= \frac{\alpha\mu[\alpha\beta + \Upsilon(\beta_0 - \alpha - \alpha_1 - \mu)]}{\psi \Lambda \Upsilon} \\ R^* &= \frac{[\alpha\beta + \Upsilon(\beta_0 - \alpha - \alpha_1 - \mu)](\alpha_1 \Upsilon + \alpha\alpha_2)}{\psi \Lambda \Upsilon} \end{aligned} \tag{6.5}$$

where,

$$\begin{aligned} \psi &= \alpha\beta + \beta_0 (\theta + \alpha_2 + \mu) \\ \Lambda &= \alpha + \alpha_1 + \mu \\ \Upsilon &= \theta + \alpha_2 + \mu \end{aligned} \tag{6.6}$$

Hence, there exist a unique endemic equilibrium point for our proposed model system 6.3.

## Backward Bifurcation

The analysis conducted in the previous section on the occurrence of endemic equilibrium $E^*$ suggests the probability of backward bifurcation. We have used the resuts based on the center manifold theorem (theorem 4.1) given in [13], to check the occurrence of backward bifurcation. From the expression of $R_0$ it is clear that $R_0$ is directly related to $\beta$ as well as $\beta_0$. Therefore we will consider two separate cases one for each $\beta$ and $\beta_0$.

- ***Case 1:*** If we select $\beta$ as the bifurcation parameter. Moreover, $R_0 = 1$ implies

$$\beta^* = \frac{(\alpha_2 + \theta + \mu)\,(\alpha + \alpha_1 + \mu - \beta_0)}{\alpha}$$

Now,

$$J_0(E_0, \beta^*) = D_x f(E_0, \beta^*) = \begin{pmatrix} -\mu & -\beta_0 & -\beta^* & 0 \\ 0 & \beta_0 - (\alpha + \alpha_1 + \mu) & \beta^* & 0 \\ 0 & \alpha & -(\theta + \alpha_2 + \mu) & 0 \\ 0 & \alpha_1 & \alpha_2 & -\mu \end{pmatrix}$$

It is clear that 0 is a simple eigenvalue of $J_0$. Let $w = (w_1, w_2, w_3, w_4)$ be the associated right eigenvector, then

$$\begin{aligned} &-\mu w_1 - \beta_0 w_2 - \beta^* w_3 = 0 \\ &(\beta_0 - (\alpha + \alpha_1 + \mu))\, w_2 + \beta^* w_3 = 0 \\ &\alpha w_2 - (\theta + \alpha_2 + \mu)\, w_3 = 0 \\ &\alpha_1 w_2 + \alpha_2 w_3 - \mu w_4 = 0 \end{aligned}$$

On solving the above system of equation and substituting the value of $\beta^*$, we obtain

$$(w_1, w_2, w_3, w_4) = \left(-\frac{(\theta+\alpha_2+\mu)(\alpha+\alpha_1+\mu)}{\alpha\mu}, \frac{\theta+\alpha_2+\mu}{\alpha}, 1, \frac{\alpha_1(\theta+\alpha_2+\mu)+\alpha\alpha_2}{\alpha\mu}\right)$$

Similarly, Let $v = (v_1, v_2, v_3, v_4)$ be the corresponding left eigenvector satisfying $w.v = 1$. After evaluation, we have

$$(v_1, v_2, v_3, v_4) = \left(0, \frac{\alpha}{(\theta+\alpha_2+\mu)+(\alpha+\alpha_1+\mu-\beta_0)}, \frac{\alpha+\alpha_1+\mu-\beta_0}{(\theta+\alpha_2+\mu)+(\alpha+\alpha_1+\mu-\beta_0)}, 0\right)$$

As discussed in Theorem 4.1 [13], the coefficients 'a'and 'b'can be computed as:

$$a = \sum_{k,i,j=1}^{4} v_k w_i w_j \frac{d^2 f_k}{dx_i dx_j} (E_0, \beta^*)$$

$$b = \sum_{k,i=1}^{4} v_k w_i \frac{d^2 f_k}{dx_i d\beta}(E_0, \beta^*)$$

Algebraic calculations shows that

$$\frac{d^2 f_2}{dx_1 dx_2} = \beta_0 = \frac{d^2 f_2}{dx_2 dx_1}, \qquad \frac{d^2 f_2}{dx_1 dx_3} = \beta^* = \frac{d^2 f_2}{dx_3 dx_1}$$

$$\frac{d2 f_2}{dx_3 d\beta} = S^0 = 1$$

The rest of the second derivatives appearing in the formula for ‘a’ and ‘b’ are all zero. Hence,

$$a = v_2\left(w_1 w_2 \beta_0 + w_1 w_3 \beta^* + w_2 w_1 \beta_0 + w_3 w_1 \beta^*\right)$$

After simplifying and substituting the value of $\beta^*$, we have

$$a = -\frac{2(\alpha_2 + \theta + \mu)^2 (\alpha + \alpha_1 + \mu)^2}{\alpha[(\alpha + \alpha_1 + \mu - \beta_0) + (\theta + \alpha_2 + \mu)]}$$

Similarly, we have

$$b = \frac{\alpha}{(\theta + \alpha_2 + \mu) + (\alpha + \alpha_1 + \mu - \beta_0)}$$

Now, from the above two expressions it is clear that:

$$\begin{aligned} a > 0 \quad &if \quad \beta_0 > \alpha + \alpha_1 + \alpha_2 + \theta + 2\mu \\ b > 0 \quad &if \quad \beta_0 < \alpha + \alpha_1 + \alpha_2 + \theta + 2\mu \end{aligned}$$

Now, from the above two conditions it is clear that the coefficients ‘$a$’ and ‘$b$’ can not be positive simultaneously. Hence, backward bifurcation is not possible at $R_0 = 1$ for this particular case.

- **Case 2:** In this case, we will consider ‘$\beta_0$’ as our the bifurcation parameter. Now $R_0 = 1$ implies:

$$\beta_0^* = \frac{(\alpha + \alpha_1 + \mu)(\alpha_2 + \theta + \mu) - \beta\alpha}{(\alpha_2 + \theta + \mu)}$$

On following the same procedure as in “case 1", the associated right eigenvector $w = (w_1, w_2, w_3, w_4)$ can be given as:

$$(w_1, w_2, w_3, w_4) = \left(-\frac{(\alpha+\alpha_1+\mu)(\alpha_2+\theta+\mu)}{\alpha\mu}, \frac{\theta+\alpha_2+\mu}{\alpha}, 1, \frac{\alpha_1(\theta+\alpha_2+\mu)+\alpha\alpha_2}{\alpha\mu}\right)$$

and the corresponding left eigenvector $v = (v_1, v_2, v_3, v_4)$ can be given as:

$$(v_1, v_2, v_3, v_4) = \left(0, \frac{\theta + \alpha_2 + \mu}{\theta + \alpha_2 + \mu + \beta}, \frac{\beta}{\theta + \alpha_2 + \mu + \beta}, 0\right)$$

Further calculations gives as the values of the coefficients '$a$' and '$b$, as:

$$a = \frac{-2}{\alpha^2 \mu} \frac{(\theta + \alpha_2 + \mu)^3 (\alpha + \alpha_1 + \mu)^2}{(\theta + \alpha_2 + \mu + \beta)}$$

$$b = \frac{\theta + \alpha_2 + \mu}{\theta \alpha_2 + \mu + \beta}$$

Hence it is clear that '$a$' is always negative and '$b$' is always positive. Hence, for this case also there does not exist backward bifurcation at $R_0 = 1$.

Therefore, from case 1 and case 2, we can conclude that there does not exist backward bifurcation at $R_0 = 1$. Only bifurcation which occur at $R_0 = 1$ will be forward in nature.

## Stability Analysis

### *Local Stability of Disease Free Equilibrium*

**Theorem 3** *The disease free equilibrium of the proposed model system is locally asymptotically stable when $R_0 < 1$ and unstable for $R_0 > 1$.*

***Proof*** The Jacobian matrix corresponding to the disease free equilibrium is

$$M = \begin{pmatrix} -\mu & -\beta_0 & -\beta & 0 \\ 0 & (\beta_0 - \alpha - \alpha_1 - \mu) & \beta & 0 \\ 0 & \alpha & -(\theta + \alpha_2 + \mu) & 0 \\ 0 & \alpha_1 & \alpha_2 & -\mu \end{pmatrix}$$

The characteristic equation of $M$ is

$$(-\mu - \Sigma)(-\mu - \Sigma)\left(-\left(\beta_0 - \alpha - \alpha_1 - \mu - \Sigma\right)\left(\theta + \alpha_2 + \mu + \Sigma\right) - \beta\alpha\right) = 0$$

The first two eigenvalues can be given as $\Sigma = -\mu, -\mu$.

In order to find the other two eigenvalues we will simplify

$$\left(-\left(\beta_0 - \alpha - \alpha_1 - \mu - \Sigma\right)\left(\theta + \alpha_2 + \mu + \Sigma\right) - \beta\alpha\right) = 0$$

The above equation can be written as

$$(\beta_0 - D_1 - \Sigma)(D_2 + \Sigma) + \alpha\beta = 0$$

Here, $D_1 = \alpha + \alpha_1 + \mu$ and $D_2 = \theta + \alpha_2 + \mu$. On simplification, the above equation will be reduced to

$$\Sigma^2 + (D_1 + D_2 - \beta_0)\,\Sigma + D_1 D_2 - \beta_0 D_2 - \alpha\beta = 0 \tag{6.7}$$

The basic reproduction number in terms of $D_1$ and $D_2$ is

$$R_0 = \frac{\beta_0 D_2 + \beta\alpha}{D_1 D_2}$$

Now, $R_0 < 1$ implies

$$\begin{aligned} D_1 D_2 - \beta_0 D_2 - \alpha\beta &> 0 \\ D_1 + D_2 - \beta_0 &> 0 \end{aligned}$$

The positivity of these two values and Eq. 6.7 implies that the other two eigenvalues of Jacobian matrix $M$ are negative.

Hence, the disease free equilibrium $E_0$ of the model system (6.1) is locally asymptotically stable if $R_0 \leq 1$ and unstable for $R_0 > 1$.

## *Global Stability of Disease Free Equilibrium*

In order to obtain the conditions for the global stability for $E_0$, we have used the approach set out in [14], which states that if the model system can be written in the following form

$$\begin{aligned} \frac{dX}{dt} &= F(X, Z) \\ \frac{dZ}{dt} &= G(X, Z),\ G(X, 0) = 0 \end{aligned} \tag{6.8}$$

here $X \in R^n$ are the uninfected individuals and $Z \in R^m$ describes the infected individuals. According to this notation, the disease free equilibrium is given by $Q_0 = (X_0, 0)$.

K1: For $\frac{dX}{dt} = F(X, 0)$, $X_0$ is globally asymptotically stable.
K2: $G(X, Z) = BZ - \hat{G}(X, Z)$ where $\hat{G}(X, Z) \geq 0$ for $X, Z \in \Omega$.

here $B = D_z G(X_0, 0)$ is a $M$-matrix (matrix whose non-diagonal elements belongs to $R^+ \cup 0$) and $\Omega$ is the feasible of the model. Now, the following theorem establishes the global stability of the disease free equilibrium $E_0$ for our proposed model system.

**Theorem 4** *The disease free equilibrium point is globally asymptotically stable provided* $R_0 \leq 1$.

***Proof*** First we will prove $K_1$ as

$$F(X,0) = \begin{bmatrix} \mu - \mu S \\ -\mu R \end{bmatrix}$$

The characteristic polynomial of system is given by

$$(-\mu - \Gamma)(-\mu - \Gamma) = 0$$

Hence, there are two negative roots $\Gamma = -\mu, -\mu$. Hence, $X = X_0$ is globally asymptotically stable.

Now, we have $G(X,Z) = BZ - \hat{G}(X,Z)$

$$= \begin{bmatrix} \beta_0 S_0 - (\alpha + \alpha_1 + \mu) & \beta S_0 \\ \alpha & -(\theta + \alpha_2 + \mu) \end{bmatrix} \begin{bmatrix} E \\ I \end{bmatrix} - \begin{bmatrix} \beta(S_0 - S)I + \beta_0(S_0 - S)E \\ 0 \end{bmatrix}$$

Here, $B = \begin{bmatrix} \beta_0 S_0 - (\alpha + \alpha_1 + \mu) & \beta S_0 \\ \alpha & -(\theta + \alpha_2 + \mu) \end{bmatrix}$ is an $M$ matrix.

Also, $\hat{G}(X,Z) = \begin{bmatrix} \beta(S_0 - S)I + \beta_0(S_0 - S)E \\ 0 \end{bmatrix} \geq 0 \;; \quad \forall (X,Z) \in \Omega.$

Hence $K_1$ and $K_2$ are satisfied, which proves our theorem.

## *Local Stability of Endemic Equilibrium*

**Theorem 5** *When $R_0 > 1$ the endemic equilibrium $E^*$ is locally asymptotically stable under the condition $A_j > 0 \;; \; \forall \; j = 1, 2, 3$ and 4 and*

$$A_1 A_2 A_3 > A_3^2 + A_1^2 A_4 \;, \quad A_1 A_2 > A_3$$

*expression of $A_1, A_2, A_3$ and $A_4$ are given in the proof.*

***Proof*** The Jacobian matrix about $E^*$ is given as

$$J_0 = \begin{bmatrix} a_{11} & a_{12} & a_{13} & 0 \\ a_{21} & a_{22} & a_{23} & 0 \\ 0 & a_{32} & a_{33} & 0 \\ 0 & a_{42} & a_{43} & a_{44} \end{bmatrix}$$

Here,

$$a_{11} = -\beta I^* - \beta_0 E^* - \mu \quad a_{12} = -\beta_0 S^* \quad a_{13} = -\beta S^* \quad a_{21} = \beta I^* + \beta_0 E^*$$
$$a_{22} = \beta_0 S^* - (\alpha + \alpha_1 + \mu) \quad a_{23} = \beta S^* \quad a_{32} = \alpha \quad a_{33} = -(\theta + \alpha_2 + \mu)$$
$$a_{42} = \alpha_1 \quad a_{43} = \alpha_2 \quad a_{44} = -\mu$$

The characteristic equation of $J_0$ is

$$\Gamma^4 + A_1\Gamma^3 + A_2\Gamma^2 + A_3\Gamma + A_4 = 0$$

Here,

$$\begin{aligned}
A_1 &= -(a_{11} + a_{22} + a_{33} + a_{44}) \\
A_2 &= (a_{11}a_{22} - a_{12}a_{21} + a_{11}a_{33} + a_{11}a_{44} + a_{22}a_{33} - a_{23}a_{32} + a_{22}a_{44} \\
&\quad + a_{33}a_{44}) \\
A_3 &= (-a_{11}a_{22}a_{33} + a_{11}a_{23}a_{32} + a_{12}a_{21}a_{33} - a_{13}a_{21}a_{32} - a_{11}a_{22}a_{44} \\
&\quad + a_{12}a_{21}a_{44} - a_{11}a_{33}a_{44} - a_{22}a_{33}a_{44} + a_{23}a_{32}a_{44}) \\
A_4 &= (a_{11}a_{22}a_{33}a_{44} - a_{11}a_{23}a_{32}a_{44} - a_{12}a_{21}a_{33}a_{44} + a_{13}a_{21}a_{32}a_{44})
\end{aligned}$$

Thus, under the conditions stated in the theorem, the local stability of the endemic equilibrium is guaranteed by the Routh-Hurwitz criterion.

## Numerical Simulation and Model Fitting

According to the data collected from [11], the total number of COVID-19 cases has crossed 227364 as of May 20. There has been more than 17669 deaths in the country due to this epidemic. A study is needed to observe that what will be the situation in the near future in terms of daily active cases. Therefore, the aim of the proposed model is to predict the future scenario of daily active cases of the COVID-19 epidemic in Italy by analyzing its present state in the country. In this section, we perform rigorous numerical simulations to get an insight into the pandemic in Italy (Fig. 6.3).

We calibrated the model (6.1) for daily active cases of COVID-19 in Italy and its four province namely Lombardia, Veneto, Emilia Romagna and Piemonte. For simulation, data for daily active cases has been taken from [15] for the period of 68 days (March 14 to May 20, 2020). We fit the model system (6.1) with daily active cases for the whole country as well as for the four provinces. We use in-built function *lsqcurvefit* in MATLAB (Mathworks, R2017a) to fit the model. The parameters $A$, $\beta$, $\beta_0$, $\alpha$, $\alpha_1$ (see Table 6.2) and initial conditions of the human population $(S(0), E(0), I(0), R(0))$ are estimated. Other parameters, involved in the model system (6.1) are taken from the literature (Table 6.3).

According to the situation report-11 available on the official website of WHO [17], the first two COVID-19 positive cases in Italy were reported on January 31. Both the infected individuals had a travel history to the city of Wuhan, China. This was the ini-

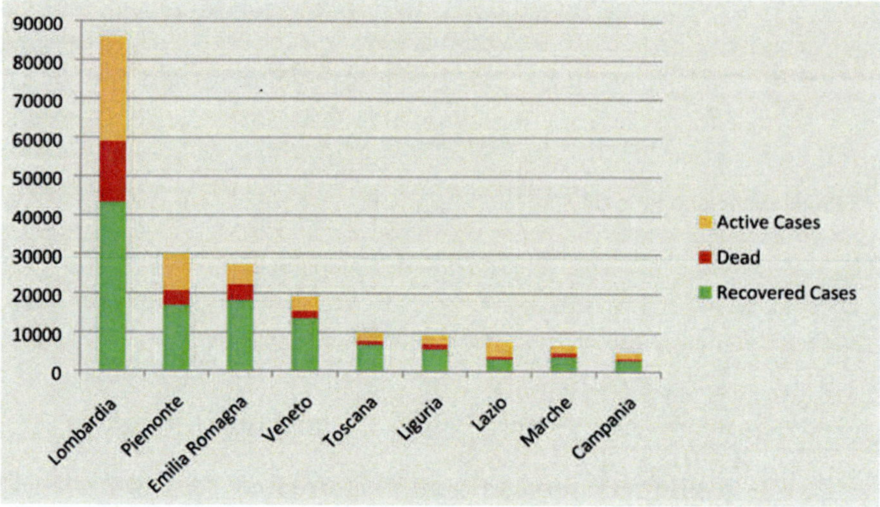


**Fig. 6.3** Current status of COVID-19 in various regions on Italy. (as on 20th May, 2020)

**Table 6.3** Estimated initial conditions of human population for Lombardia, Veneto, Emilia Romagna, Piemonte and Italy

| | Lombardia | Veneto | Emilia romagna | Piemonte | Italy |
|---|---|---|---|---|---|
| S(0) | 6,630,487.2250 | 4,463,338.6370 | 4,297,884.3050 | 3,899,297.0840 | 35,906,707.6100 |
| E(0) | 505,834.0860 | 148,574.0557 | 123,220.8240 | 151,910.3186 | 2,715,742.7120 |
| I(0) | 9905.8333 | 1771.0768 | 2833.7313 | 1410.4594 | 10,542.2400 |
| R(0) | 978.6327 | 199.5888 | 52.0534 | 199.1392 | 422.8864 |

tial phase of the epidemic in the country. Due to delayed imposition of lockdown, the epidemic started to grow slowly but significantly. By the end of February, the number of cases reached 1000 [11]. We fitted the proposed model system (6.1) for Italy and its four provinces with the official data available on [15] for the month of March, April and May. It is observed that the proposed model is able to predict the epidemic closely (see Figs. 6.6, 6.7, 6.8, 6.9 and 6.10). Also, the corresponding residual is also given in Fig. 6.5, which shows the efficiency of our model with the real data (Fig. 6.4).

Using the parameters given in Tables 6.1 and 6.2, we evaluated basic reproduction number ($R_0$) for Lombardia, Veneto, Emilia Romagna, Piemonte and Italy (see Table 6.4). As the Lombardia was worst affected province of Italy and thus recorded highest value of $R_0$ among others. The fitting of model to daily active cases data is represented in Figs. 6.7a and 6.10a for Lombardia, Veneto, Emilia Romagna, Piemonte and Italy, respectively. The corresponding residuals for these regions are given in Figs. 6.7b and 6.10b. Further, we tried to predict the situation of daily active cases of COVID-19 in near future. Graphs of prediction are shown in Figs. 6.7c and 6.10c for Lombardia, Veneto, Emilia Romagna, Piemonte and Italy, respectively.

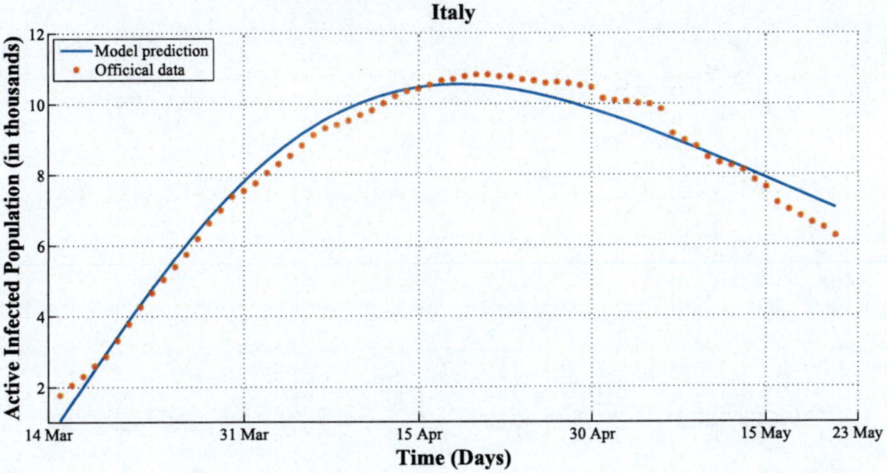


**Fig. 6.4** Fitted model to daily active cases data of Italy for the period of March 14 to May 20, 2020

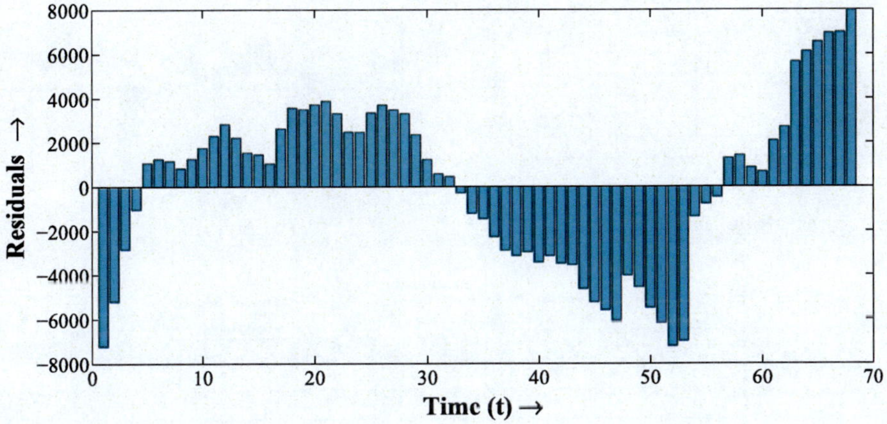


**Fig. 6.5** Residuals for the data fitting of Italy

**Table 6.4** Total population (N) and evaluated basic reproduction number ($R_0$) for Lombardia, Veneto, Emilia Romagna, Piemonte and Italy.

| | Lombardia | Veneto | Emilia Romagna | Piemonte | Italy |
|---|---|---|---|---|---|
| N | 10,000,000 | 5,000,000 | 4,500,000 | 4,400,000 | 60,000,000 |
| $R_0$ | 2.1382 | 1.7512 | 1.6331 | 1.9099 | 2.0683 |

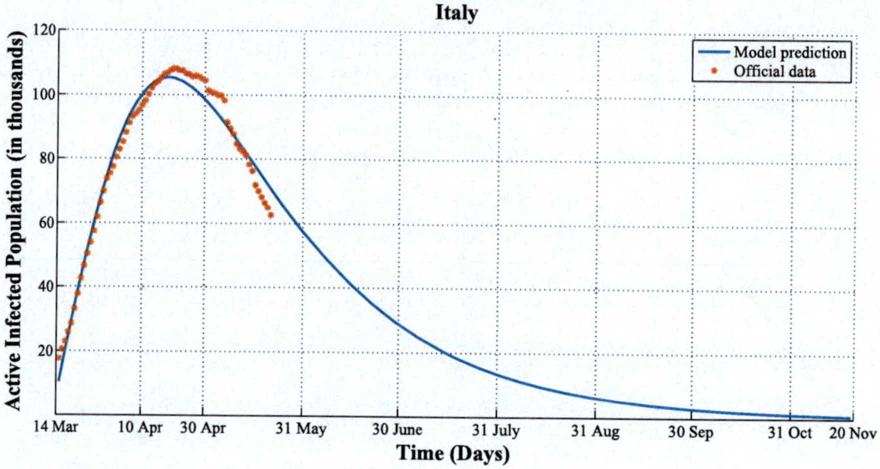


**Fig. 6.6** Prediction of the model for Italy

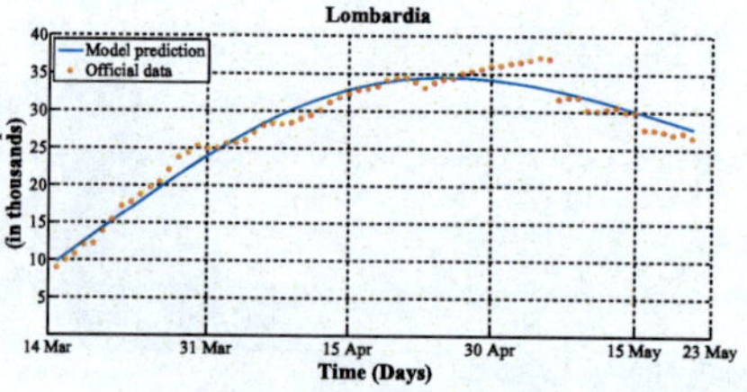


(a) Model fitting to data

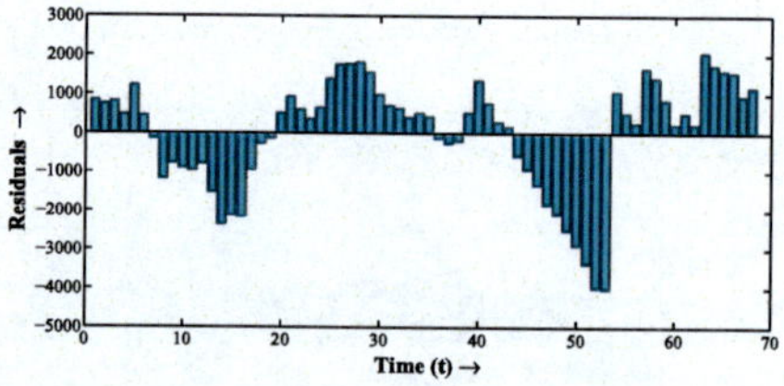


(b) Residual

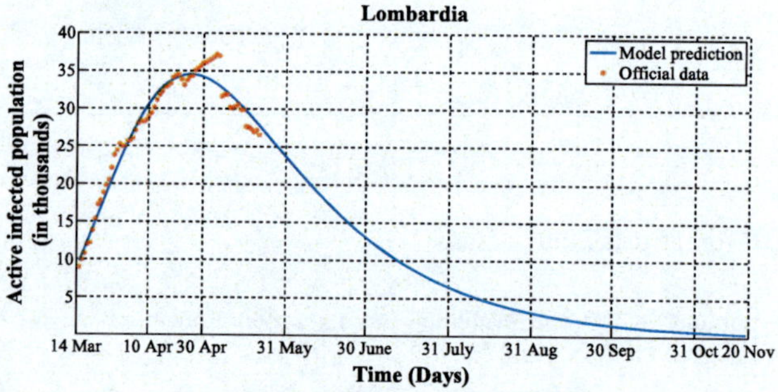


(c) Prediction of the model

**Fig. 6.7** Fitted model to daily active cases data of Lombardia for the period of March 14 to May 20, 2020

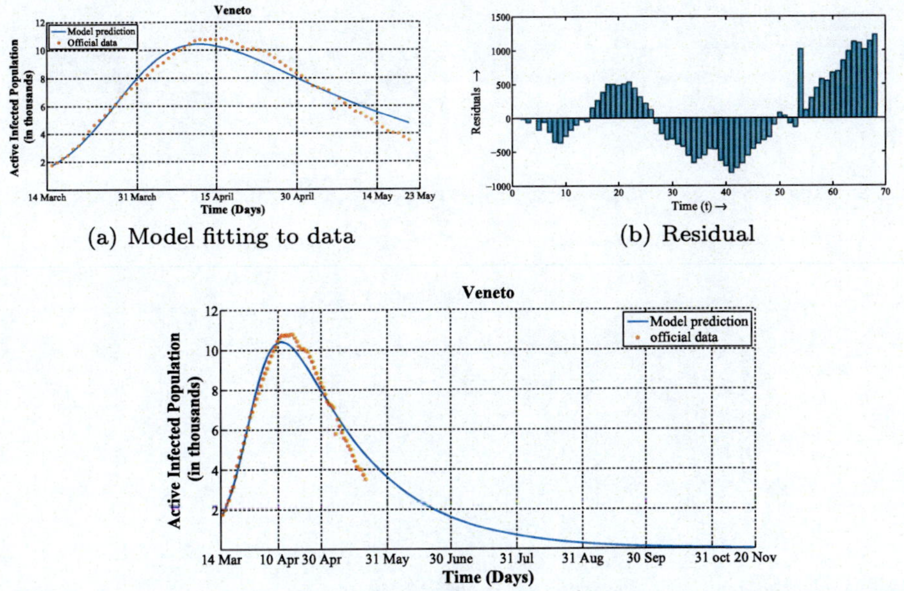


(a) Model fitting to data (b) Residual

(c) Prediction of the model

**Fig. 6.8** Fitted model to daily active cases data of Veneto for the period of March 14 to May 20, 2020. **b** Residual **c** Model prediction

It can be seen from Figures that the proposed model is fitted well with the actual active cases of COVID-19 for all provinces Lombardia, Veneto, Emilia Romagna, Piemonte, and Italy. In each case, first, we fit the model to the actual data from March 14 to May 20, and estimated parameter values. After that, we used these parameter values for prediction of the daily active cases for all regions considered in the present work. The prediction of the active cases for Lombardia is shown in Fig. 6.7c. From the Figure, it can be concluded that the epidemic will be over approximately by November 19, 2020 (250 days from 14 March). Next, the model fitting for Veneto and Emilia Romagna is performed and plotted in Figs. 6.8 and 6.9 and prediction for these regions is given in Figs. 6.8c and 6.9c, respectively. From the Figures, we conclude that the active cases will near to be eliminated in approximately 200 days (near about September 30, 2020). In continuation of this, Figs. 6.10 and 6.6 depicts model fitting and prediction of daily active cases of COVID-19 for Piemonte and whole country Italy, respectively. Again, it is observed that both the regions model show a good fit with actual daily active cases. One can see the model prediction in Figs. 6.10c and 6.6 for Piemonte and whole country Italy, respectively that the daily active cases will be annihilated in approximately 225 days (near about October 25, 2020).

Why Lombardia is epicenter of COVID-19 in Italy? Our model has shown that it has the highest infectives in Italy (Fig. 6.7). Besides, there may be other reasons as well; the initial case was found here and he had travel history to Wuhan and in

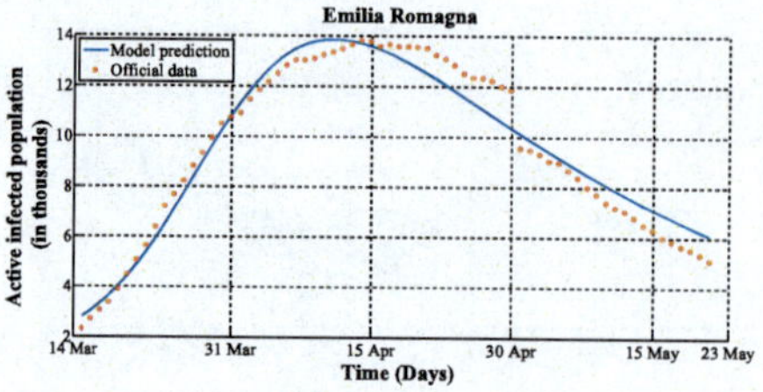


(a) Model fitting to data

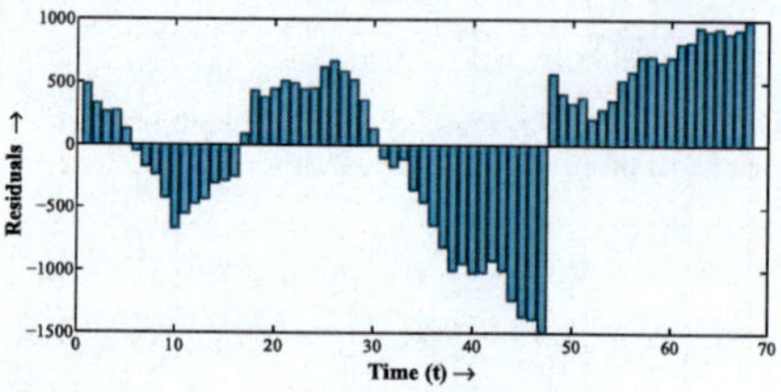


(b) Residual

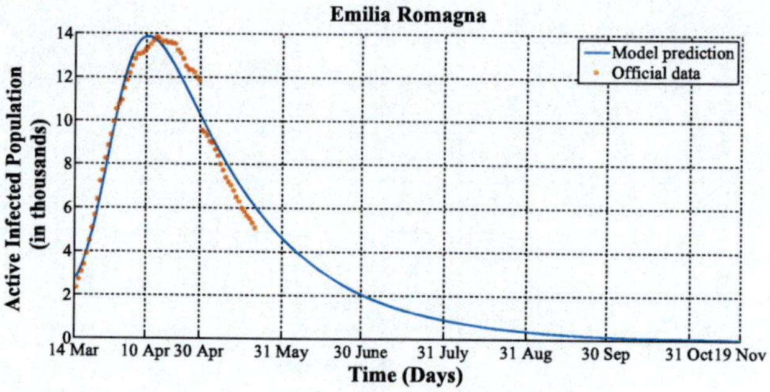


(c) Prediction of the model

**Fig. 6.9** Fitted model to daily active cases data of Emilia Romagna for the period of March 14 to May 20, 2020. **b** Residual **c** Model prediction

Wuhan L and S type of strains have been documented [18]. It has also been proposed by some researchers that L type is more deadly. It has one-sixth of Italy's 60 million population and is a region with one of the most high density of population. It also has more people over 65 years of age than any other region. In an interesting study, high association with pollution in this region has also been pointed out as one of the reasons in Lombardia [19].

## Game Changers

In this section, we will discuss the factors which can significantly control the spread of pandemic. The two major factors discussed here are (a) Early Lock down and (b) Rapid Isolation.

### *Impact of Early Lock Down*

On 9th March 2020, in response to the increasing COVID-19 pandemic in the region, the Government of Italy enforced a national quarantine restricting population move-

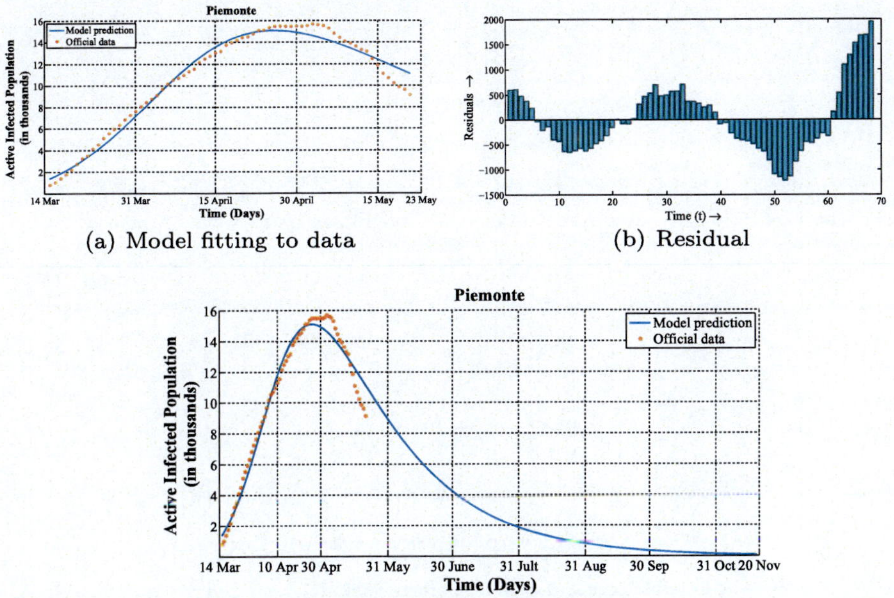


(a) Model fitting to data (b) Residual

(c) Prediction of the model

**Fig. 6.10** Fitted model to daily active cases data of Piemonte for the period of March 14 to May 20, 2020. **b** Residual **c** Model prediction

ment, with the exception of necessity, work and health circumstances. In this section, we will investigate the effect of early lock down on the spread of pandemic in Italy.

It is clear from Fig. 6.11 that the epidemic could have been controlled at a very early stage if the lockdown had been imposed early in Italy. Figure 6.11 shows three different scenarios of the epidemic in Italy

Figure 6.11a shows the current scenario of Italy. However, if lock down would have been imposed prior to 9th March 2020, the number of susceptible would have been significantly low. In Fig. 6.11b, we have considered a case of lock-down with 50% efficiency. It can be seen from the plot that, it has not only significantly reduce the number of infections, but also caused the overall death of pandemic by 31st August, 2020. Also, Fig. 6.11c indicate that the epidemic could have been eliminated even more earlier if the the efficiency of lock-down would have been 80%.

## *Impact of Rapid Isolation on Infected Individuals*

COVID-19 is a pandemic which is spreading all across the globe. Early research shows that the disease transmission rate from an infected individual to a susceptible is very high [20]. The transmission rate can be reduced by isolating the infected

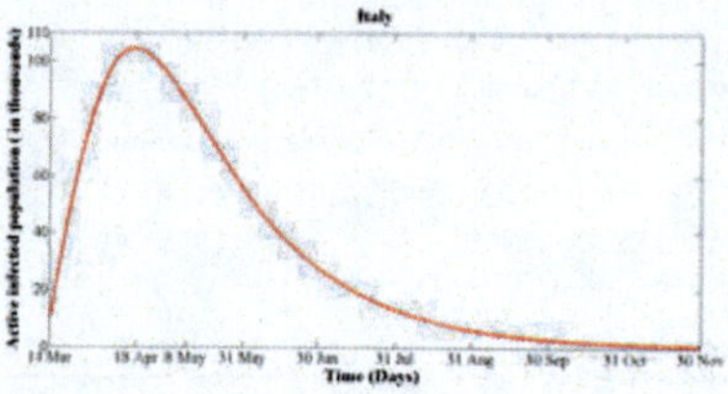


(a) Scenario of Italy with current precautionary measures

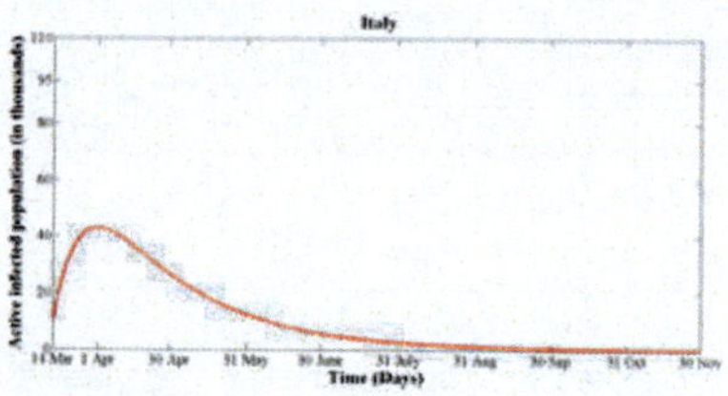


(b) Possible scenario with 50 percent lock down efficiency

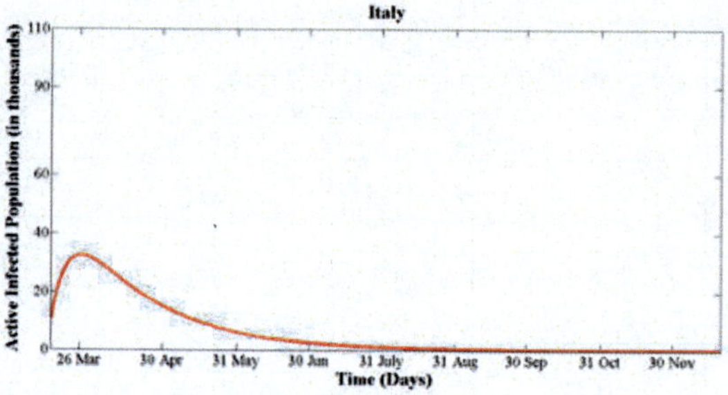


(c) Possible scenario with 80 percent lock down efficiency

**Fig. 6.11** Various possible cases of Italy corresponding to different lock down efficiency rate.

individuals as quickly as possible. In this subsection, with the help of numerical simulations we will show the variations in infected population for different values of $\beta$, disease transmission rate from infected individual to susceptible individuals.

Figure 6.12 shows various scenarios of the epidemic in Italy in case disease transmission rate would have been timely controlled. A rapid isolation of infected population will lead to reduce the disease transmission rate, $\beta$. From Fig. 6.12, we see that as disease transmission rate, $\beta$ is reduced by 75%, it not only decrease the active number of infections from 1.5 lacks to 35000, but also the overall lifespan of pandemic reduced from November 30th to July 15, 2020.

## Conclusion

A SEIR type compartmental model is proposed to study the current scenario of COVID-19 in Italy. Our proposed model accurately fits the officially available data of the pandemic in Italy. Also, we have discussed how the lock down that was imposed on 9th March, 2020 was a good but a delayed decision of the government of Italy. Through simulations, we have shown that a rapid isolation of the infective individuals and early lock down in the country are two of the most efficient procedures to terminate the spread of COVID-19. Our simulations shows that the pandemic in Italy will last till November, 2020. As of now, the vaccination of COVID-19 have not

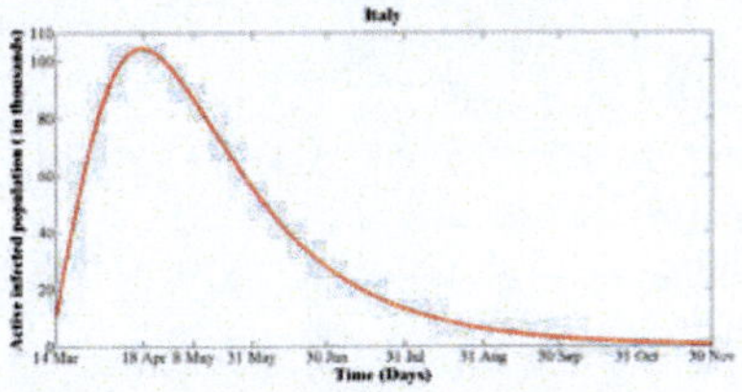


(a) Scenario of Italy with current precautionary measures

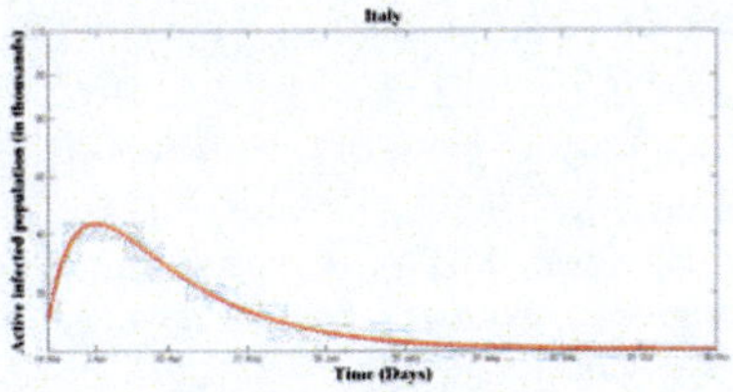


(b) Scenario with 50 percent rapid isolation

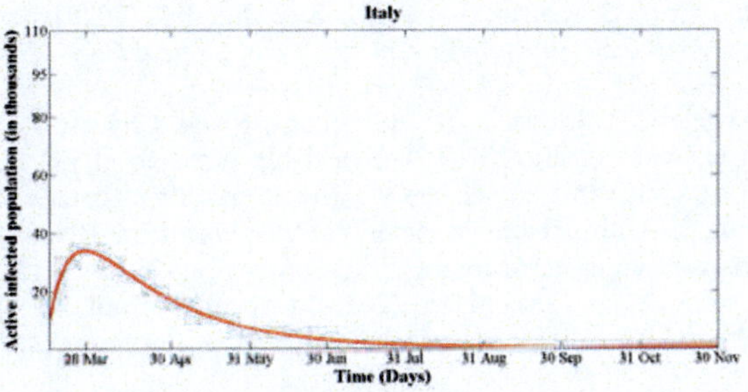


(c) Possible Scenario with 75 percent rapid isolation of infected population

**Fig. 6.12** Variation in infected population for different isolation of infected population.

been discovered. Hence, this research can also be beneficial for the countries which are in the initial stage of the pandemic, as our research describes two of the most effective procedures to counter the spread of the pandemic and its long term impact on the spread of disease.

We have also estimated the basic reproduction number $R_0$ for the disease. As our proposed model involves several parameters, we have shown the sensitivity of these parameters via numerical simulations. It is clear from the simulations (see Fig. 6.2) that the transmission rates, $\beta$ and $\beta_0$ are the most sensitive parameters. The reproduction number can be minimized if we can reduce these two parameters. Also, we have derived the value of the basic reproduction number for Italy and some of its highly effected regions. This research can be extended in various ways. One can refine the model by introducing new compartments in order to examine the epidemic more precisely. There are certain assumptions which we have made while constructing this model because of the limited data and short onset time. As more data will be available in the future, this model can be trained with more real data to increase its efficiency. Useful future directions have been proposed Ndairou et al. (2020) in their compartment model on Wuhan, China. Besides, future mathematical models may take virus strains prevailing in a region also in account.

**Acknowledgements** The research of the corresponding author (Nitu Kumari) is funded by Science and Engineering Research Board (SERB), under three separate grants with grant numbers MSC/2020/000369, MTR/2018/000727 and EMR/2017/ 005203.

## References


1. Dominguez, S. R., O'Shea, T. J., Oko, L. M., & Holmes, K. V. (2007). Detection of group 1 coronaviruses in bats in North America. *Emerging infectious diseases, 13*(9), 1295.
2. Li, Q., Guan, X., Wu, P., Wang, X., Zhou, L., Tong, Y., et al. (2020). Early transmission dynamics in Wuhan, China, of novel coronavirus-infected pneumonia. *New England Journal of Medicine*,.
3. Wang, D., Hu, B., Hu, C., Zhu, F., Liu, X., Zhang, J., et al. (2020). *Clinical characteristics of 138 hospitalized patients with 2019 novel coronavirus-infected pneumonia in wuhan*. China: JAMA.
4. Carlos, W. G., Dela Cruz, C. S., Cao, B., Pasnick, S., & Jamil, S. (2020). Novel Wuhan (2019-ncov) coronavirus. *American Journal of Respiratory and Critical Care Medicine, 201*(4), P7–P8.
5. Biscayart, C., Angeleri, P., Lloveras, S., do Socorro Souza Chaves, T., Schlagenhauf, P., Rodríguez-Morales, A. J., et al. (2020). The next big threat to global health? 2019 novel coronavirus (2019-ncov): What advice can we give to travellers?–interim recommendations January 2020, from the Latin-American society for travel medicine (slamvi).
6. https://www.who.int/emergencies/diseases/novel-coronavirus-2019/events-as-they-happen.
7. Grasselli, G., Pesenti, A., & Cecconi, M. (2020). Critical care utilization for the covid-19 outbreak in lombardy, italy: Early experience and forecast during an emergency response. *JAMA*.
8. Remuzzi, A., & Remuzzi, G. (2020). Covid-19 and Italy: What next? *The Lancet*.
9. Lazzerini, M., & Putoto, G. (2020). Covid-19 in Italy: Momentous decisions and many uncertainties. *The Lancet Global Health*.
10. Vattay, G. (2020). Predicting the ultimate outcome of the covid-19 outbreak in Italy. *arXiv preprint* arXiv: 2003.07912.
11. https://www.worldometers.info/coronavirus/country/italy/.
12. Diekmann, O., Heesterbeek, J. A. P., & Roberts, M. G. (2010). The construction of next-generation matrices for compartmental epidemic models. *Journal of the Royal Society Interface, 7*(47), 873–885.
13. Castillo-Chavez, C., & Song, B. (2004). Dynamical models of tuberculosis and their applications. *Mathematical Biosciences and Engineering, 1*(2), 361.
14. Castillo-Chavez, C., Blower, S., van den Driessche, P., Kirschner, D., & Yakubu, A.-A. (2002). *Mathematical approaches for emerging and reemerging infectious diseases: an introduction*, vol. 1. Springer Science & Business Media.
15. http://www.salute.gov.it/portale/nuovocoronavirus/archivioNotizieNuovoCoronavirus.jsp 2020.
16. https://knoema.com/legal/termsofuse. April 2020.
17. https://www.who.int/emergencies/diseases/novel-coronavirus-2019/situation-reports.
18. Tang, X., Wu, C., Li, X., Song, Y., Yao, X., & Wu, X., (2020). On the origin and continuing evolution of SARS-CoV-2. *National Science Review*, 03 2020. nwaa036.
19. Coccia, M. (2020). Factors determining the diffusion of covid-19 and suggested strategy to prevent future accelerated viral infectivity similar to covid. *Science of the Total Environment*, pp. 138474.
20. Liu, Y., Gayle, A. A., Wilder-Smith, A., & Rocklöv, J. (2020). The reproductive number of covid-19 is higher compared to sars coronavirus. *Journal of travel medicine*.

# Chapter 7
# Effective Lockdown and Plasma Therapy for COVID-19

**Nita H. Shah, Nisha Sheoran, and Ekta N. Jayswal**

**Abstract** COVID-19 is a major pandemic threat of 2019–2020 which originated in Wuhan. As of now, no specific anti-viral medication is available. Therefore, many countries in the world are fighting to control the spread by various means. In this chapter, we model COVID-19 scenario by considering compartmental model. The set of dynamical system of nonlinear differential equation is formulated. Basic reproduction number $R_0$ is computed for this dynamical system. Endemic equilibrium point is calculated and local stability for this point is established using Routh-Hurwitz criterion. As COVID-19 has affected more than 180 countries in several ways like medically, economy, etc. It necessitates the effect of control strategies applied by various government worldwide to be analysed. For this, we introduce different types of time dependent controls (which are government rules or social, medical interventions) in-order to control the exposure of COVID-19 and to increase recovery rate of the disease. By using Pontryagins maximum principle, we derive necessary optimal conditions which depicts the importance of these controls applied by the government during this epidemic.

**Keywords** COVID-19 · Basic reproduction number · Local stability · Optimal control

**Mathematics Subject Classification** 37Nxx

N. H. Shah · N. Sheoran (✉) · E. N. Jayswal
Department of Mathematics, Gujarat University, Ahmedabad, Gujarat, India
e-mail: sheorannisha@gmail.com

N. H. Shah
e-mail: nitahshah@gmail.com

E. N. Jayswal
e-mail: jayswal.ekta1993@gmail.com



N. H. Shah and M. Mittal (eds.), *Mathematical Analysis for Transmission of COVID-19*, Mathematical Engineering, https://doi.org/10.1007/978-981-33-6264-2_7

## Introduction

As of 3 May 2020, the countries affected by COVID-19 are suffering major loss in terms of economy (globally) and also many workers are losing their jobs. So far, the number of cases reported on 3 May 2020 are more than 3.24 million across 187 countries and territories, resulting in more than 243,000 deaths [1]. COVID-19 is type of virus that infects the respiratory system of humans. It originated in Wuhan (China) on 31 December 2019. It is highly contagious with the reproduction number 6.47 calculated by Tang et al. [2] (as on 22 January 2020). Being a major public health threat declared by WHO [3], it is necessary to control the pandemic by understanding early dynamics of transmission of disease in china which has been discussed by Kucharski et al. [4]. Since no pharmaceutical treatment is available, interventions such as complete ban on air travel, shutting down of educational institutions, enforcing lockdown in the entire country, social distancing as studied by Prem et al. [5], random testing at large scale studied by Mueller et al. [6] and by isolating cases of COVID-19 and there contacts (Hellewell et al. [7]) have helped some of the countries like China, Hong Kong to control the transmission of COVID-19.

Further to understand the spread of COVID-19 and to study the effect of various interventions measures adopted by individuals and government, compartmental modelling is significant. Some authors like Toda [8] developed basic SIR model to study the effectiveness of social distancing in reducing the spread, Peng et al. [9] developed SEIR compartmental model to study epidemics of COVID-19 in China. Also Tang et al. [10] modified SEIR model for new prediction of COVID-19. Piguillem et al. [11] extended standard SIR model to study the importance of rigours testing and concluded mandatory quarantine can bring world close to what is considered as optimal. Some of the early research work with modelling of COVID-19 to understand disease dynamics in various countries includes: study by Sun et al. [12], discussed the various characteristic to COVID-19 situation in china which helps in understanding the fatality rate and transmission rate of COVID-19 so as to help in controlling the epidemic spread, the importance of travel quarantine or travel restriction in Wuhan was studied by Chinazzi et al. [13]. Other related researchers include Zhao and Chen [14], Xu et al. [15], Yang et al. [16] etc.

Now, as the world is very well aware of COVID-19 and everywhere the respective government is carrying out necessary measures to control the spread or human to human transmission of COVID-19. The best way to visualize the importance of measures been taken is to analyse it by introducing optimal control theory using Pontryagins maximum principle [17] into the model. Some of the previous research includes: Sharomi and Malik [18] have discussed very nicely optimal control in epidemiology by considering various compartmental models, Lemos-Paião et al. [19] have also applied optimal control theory showing treatment of cholera with quarantine effects, Tilahun et al. [20] applied optimal control to pneumonia disease, etc. Similarly, in COVID-19 scenario also optimal control is applied by various authors like Djidjou-Demassea et al. [21] formulated a model to minimize the death and the cost by applying control until the vaccines arrives as it will take near about 18 months.

Mallela [22] also applied optimal control theory by taking social distancing as the control in his model. Also, Tsay et al. [23] have modelled COVID-19 outbreak in USA with optimal control theory, etc.

In this chapter, our target is to predict the importance of various control strategies such as lockdown, curfew, viral load testing, plasma therapy, etc., adopted by the government in COVID-19 environment, by introducing these, measures as time dependent controls into the model and using Pontryagins theory, we will be obtaining optimal control conditions. We will also simulate through trajectories the situation with and without control in an exposed environment.

This chapter is organized as follows: Sect. Formulation of Mathematical Model describes the formulation of mathematical model and calculation of its equilibrium points. In Sect. Basic Reproduction Number, basic reproduction number is computed. In Sect. Stability Analysis, local stability of the equilibrium point is established. In Sect. Optimal Control, we develop optimal control theory by taking various controls into the model and calculate optimality conditions. The results of optimal control and other numerical simulation are discussed in Sect. Numerical Simulation. Finally, the findings are summarized with conclusion in last Sect. Conclusion.

## Formulation of Mathematical Model

This study considers formulation of mathematical model of COVID-19 dividing human population into eight mutually exclusive compartmental model. The compartments taken into account are exposed class $E_{CO}$, identified population $I_F$, isolated population $I_{SO}$, test $T_E$—it is taken as the number of test done so far including both positive and negative test, population in COVID-19 care centre $C$, population with COVID-19 in hospital $H$, Home quarantined population $Q$ and recovered population $R$.

The parametric definitions and values used in formulation of this dynamical system are given by Table 7.1.

Here, we develop a mathematical model starting with the exposure stage of COVID-19, i.e. individuals those who are exposed to COVID-19 or are in surrounding of COVID-19 infectives are considered to be in this compartment also new recruitments to this class occur at the rate $B$. Out of this exposed class, COVID-19 infected individuals (both symptomatic and asymptomatic, where asymptomatic are those with less clinical symptoms such as fever, fatigue etc.) are identified joining the compartment $I_F$ at the rate $\beta_1$. After this, the identified population is isolated ($I_{SO}$) at the rate $\beta_2$. Isolation of asymptomatic infectives is a vital strategy in containing the spread of COVID-19. Next, isolated population is then tested through viral load test for COVID-19 by laboratories and this tested population is contained in $T_E$ class at the rate $\beta_3$. Here, if the population is tested positive for COVID-19 then, we again sub-divide this positive tested population into two classes as population in COVID-19 care centre $C$ and hospitalized population $H$. Here, we assume that if the positive tested population is not in need of emergency medical treatment and is not severe it

**Table 7.1** Parametric definitions and its values [*source* own]

| Notations | Description | Parametric values |
|---|---|---|
| $B$ | Birth rate | 0.01 |
| $\beta_1$ | Rate at which population exposed to COVID-19 is been identified | 0.0009 |
| $\beta_2$ | Rate at which identified population is isolated | 0.0086 |
| $\beta_3$ | Rate at which isolated population is tested | 0.0059 |
| $\beta_4$ | Rate at which individuals joins COVID-19 care centre | 0.0046 |
| $\beta_5$ | Rate at which individuals get admitted to hospital | 0.0024 |
| $\beta_6$ | Rate at which individuals are quarantined after tested | 0.0076 |
| $\beta_7$ | Rate at which individuals in COVID-19 care centre gets recovered | 0.00006 |
| $\beta_8$ | Rate at which individuals in hospital gets recovered | 0.007 |
| $\beta_9$ | Rate at which quarantined individual gets recovered | 0.0001 |
| $\mu$ | Natural morbidity rate | 0.00009 |
| $\mu_{CO}$ | Morbidity rate due to COVID-19 | 0.00029 |

goes to COVID-19 care centre with the rate $\beta_4$ and emergency situations get hospitalized at the rate $\beta_5$. The negative tested population is asked to home quarantine themselves ($Q$) at the rate $\beta_6$ which what the government is doing. Next, population from COVID-19 care centre, hospital and home quarantine are recovered at the rate $\beta_7$, $\beta_8$ and $\beta_9$, respectively. Also, $\mu$, $\mu_{CO}$ are taken as the morbidity rates.

The following set of nonlinear differential equations is established form the Fig. 7.1.

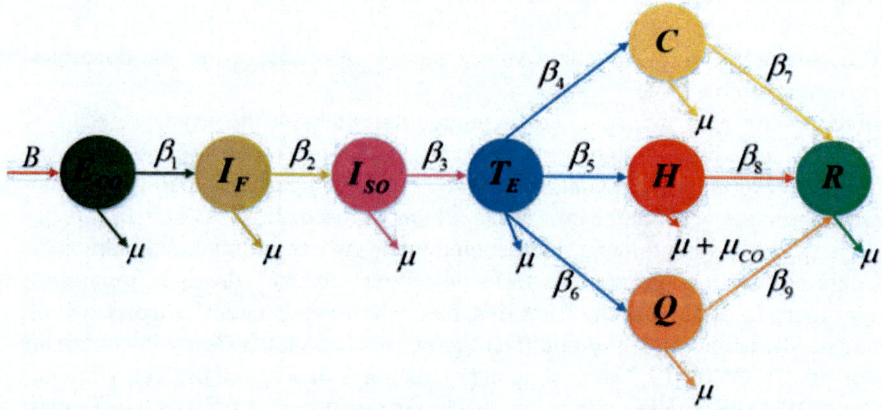


**Fig. 7.1** Compartmental diagram showing flow of human population through different compartments [*source* own]

$$\begin{aligned}
\frac{\mathrm{d}E_{CO}}{\mathrm{d}t} &= B - \beta_1 E_{CO} I_F - \mu E_{CO} \\
\frac{\mathrm{d}I_F}{\mathrm{d}t} &= \beta_1 E_{CO} I_F - (\beta_2 + \mu) I_F \\
\frac{\mathrm{d}I_{SO}}{\mathrm{d}t} &= \beta_2 I_F - (\beta_3 + \mu) I_{SO} \\
\frac{\mathrm{d}T_E}{\mathrm{d}t} &= \beta_3 I_{SO} - (\beta_4 + \beta_5 + \beta_6 + \mu) T_E \\
\frac{\mathrm{d}C}{\mathrm{d}t} &= \beta_4 T_E - (\beta_7 + \mu) C \\
\frac{\mathrm{d}H}{\mathrm{d}t} &= \beta_5 T_E - (\beta_8 + \mu + \mu_{CO}) H \\
\frac{\mathrm{d}Q}{\mathrm{d}t} &= \beta_6 T_E - (\beta_9 + \mu) Q \\
\frac{\mathrm{d}R}{\mathrm{d}t} &= \beta_7 C + \beta_8 H + \beta_9 Q - \mu R
\end{aligned} \tag{7.1}$$

where, $N = E_{CO} + I_F + I_{SO} + T_E + C + H + Q + R$.

The feasible region for the solutions of the system (7.1) is given by

$$\Lambda = \left\{ \begin{array}{l} (E_{CO}, I_F, I_{SO}, T_E, C, H, Q, R);\ E_{CO} + I_F + I_{SO} + T_E + C + H + Q + R \le \dfrac{B}{\mu}, \\ E_{CO} > 0, I_F > 0, I_{SO} > 0, C > 0, H > 0, Q > 0, R > 0 \end{array} \right\}$$

## *Equilibrium Solutions*

Solving above system of equation, we get following equilibrium point

1. Disease-free equilibrium point $E^0\left(\frac{B}{\mu}, 0, 0, 0, 0, 0, 0, 0\right)$
2. Endemic equilibrium point $E^*(E^*_{CO}, I^*_F, I^*_{SO}, T^*_E, C^*, H^*, Q^*, R^*)$

where

$$E_{CO}^* = \frac{\beta_2 + \mu}{\beta_1}$$

$$I_F^* = \frac{B\beta_1 - \mu(\beta_2 + \mu)}{\beta_1(\beta_2 + \mu)}$$

$$I_{SO}^* = \frac{\beta_2(B\beta_1 - \mu(\beta_2 + \mu))}{\beta_1(\beta_2 + \mu)(\beta_3 + \mu)}$$

$$T_E^* = \frac{\beta_2\beta_3(B\beta_1 - \mu(\beta_2 + \mu))}{\beta_1(\beta_4 + \beta_5 + \beta_6 + \mu)(\beta_2 + \mu)(\beta_3 + \mu)}$$

$$C^* = \frac{\beta_2\beta_3\beta_4(B\beta_1 - \mu(\beta_2 + \mu))}{\beta_1(\beta_4 + \beta_5 + \beta_6 + \mu)(\beta_2 + \mu)(\beta_3 + \mu)(\beta_7 + \mu)}$$

$$H^* = \frac{\beta_2\beta_3\beta_5(B\beta_1 - \mu(\beta_2 + \mu))}{\beta_1(\beta_4 + \beta_5 + \beta_6 + \mu)(\beta_2 + \mu)(\beta_3 + \mu)(\beta_8 + \mu + \mu_{CO})}$$

$$Q^* = \frac{\beta_2\beta_3\beta_6(B\beta_1 - \mu(\beta_2 + \mu))}{\beta_1(\beta_4 + \beta_5 + \beta_6 + \mu)(\beta_2 + \mu)(\beta_3 + \mu)(\beta_9 + \mu)}$$

$$R^* = \frac{\begin{array}{c}\beta_2\beta_3(B\beta_1 - \mu(\beta_2 + \mu))((\beta_9 + \mu)(\beta_4\beta_7(\mu + \mu_{CO}) + \beta_5\beta_8(\beta_7 + \mu)) \\ + \beta_6\beta_9\mu_{CO}(\beta_7 + \mu) + (\beta_8 + \mu)(\beta_7\beta_9(\beta_4 + \beta_6) + \beta_6\beta_9\mu))\end{array}}{\beta_1\mu(\beta_4 + \beta_5 + \beta_6 + \mu)(\beta_2 + \mu)(\beta_3 + \mu)(\beta_7 + \mu)(\beta_8 + \mu + \mu_{CO})(\beta_9 + \mu)}$$

## Basic Reproduction Number

Basic reproduction number $R_0$ is defined as number of secondary infections produced due to a single infection in a completely susceptible population. Basic reproduction number plays a significant role in determining the disease spread and in developing control strategies.

Basic reproduction is derived using next-generation matrix method by Diekmann et al. [24]. Here, $F$ is the Jacobian matrix of the new recruitments in the population and $V$ is the Jacobian matrix of the new transfer of exposed individuals from one compartment to another.

$$F = \begin{bmatrix} \beta_1 E_{CO} & 0\,0\,0\,0\,0\,0 & \beta_1 I_F \\ 0 & 0\,0\,0\,0\,0\,0 & 0 \\ 0 & 0\,0\,0\,0\,0\,0 & 0 \\ 0 & 0\,0\,0\,0\,0\,0 & 0 \\ 0 & 0\,0\,0\,0\,0\,0 & 0 \\ 0 & 0\,0\,0\,0\,0\,0 & 0 \\ 0 & 0\,0\,0\,0\,0\,0 & 0 \\ 0 & 0\,0\,0\,0\,0\,0 & 0 \end{bmatrix}$$

$$V = \begin{bmatrix} -\beta_2-\mu & 0 & 0 & 0 & 0 & 0 & 0 & 0 \\ -\beta_2 & \beta_3+\mu & 0 & 0 & 0 & 0 & 0 & 0 \\ 0 & -\beta_3 & \beta_4+\beta_5+\beta_6+\mu & 0 & 0 & 0 & 0 & 0 \\ 0 & 0 & -\beta_4 & \beta_7+\mu & 0 & 0 & 0 & 0 \\ 0 & 0 & -\beta_5 & 0 & \beta_8+\mu+\mu_{CO} & 0 & 0 & 0 \\ 0 & 0 & -\beta_6 & 0 & 0 & \beta_9+\mu & 0 & 0 \\ 0 & 0 & 0 & -\beta_7 & -\beta_8 & -\beta_9 & \mu & 0 \\ \beta_1 E_{CO} & 0 & 0 & 0 & 0 & 0 & 0 & \beta_1 I_F+\mu \end{bmatrix}$$

The reproduction number $R_0$ is the spectral radius of $FV^{-1}$ evaluated at $E^0\left(\frac{B}{\mu}, 0, 0, 0, 0, 0, 0, 0\right)$ and is given by the expression $R_0 = \frac{B\beta_1}{(\beta_2+\mu)\mu}$.

Here, $R_0$ gives the number of newly exposed individual to COVID-19 due to single exposure in a population which has been calculated as 11.5 using data from Table 7.1. This shows that an individual who is exposed to COVID-19 through any mode of transmission of disease via infected individual exposes 12 more individuals. Here, if $R_0 < 1$, it means that the exposure to COVID-19 is deteriorating which indicates the die out situation of COVID-19. This is stage which the world has not yet achieved. And $R_0 > 1$ shows the existence of endemic equilibrium point. Which is the scenario as off 3rd May 2020. In the next section, we will discuss the local stability of endemic point only.

## Stability Analysis

Here, we study local stability of endemic equilibrium point using Routh-Hurwitz criterion.

**Theorem 1** *The equilibrium point $E^*$ is locally asymptotically stable if* $(\beta_2+\mu) > \beta_1 E^*_{CO}$.

***Proof*** The Jacobian matrix for the dynamical system (7.1) is given by

$$J^* = \begin{bmatrix} -a_{11} & -\beta_1 E^*_{CO} & 0 & 0 & 0 & 0 & 0 & 0 \\ \beta_1 I^*_F & -a_{22} & 0 & 0 & 0 & 0 & 0 & 0 \\ 0 & \beta_2 & -a_{33} & 0 & 0 & 0 & 0 & 0 \\ 0 & 0 & \beta_3 & -a_{44} & 0 & 0 & 0 & 0 \\ 0 & 0 & 0 & \beta_4 & -a_{55} & 0 & 0 & 0 \\ 0 & 0 & 0 & \beta_5 & 0 & -a_{66} & 0 & 0 \\ 0 & 0 & 0 & \beta_6 & 0 & 0 & -a_{77} & 0 \\ 0 & 0 & 0 & 0 & \beta_7 & \beta_8 & \beta_9 & -a_{88} \end{bmatrix}$$

where

$a_{11} = \beta_1 I^*_F + \mu$, $a_{22} = -\beta_1 E^*_{CO} + (\beta_2 + \mu)$,

$a_{33} = \beta_3 + \mu$, $a_{44} = \beta_4 + \beta_5 + \beta_6 + \mu$, $a_{55} = \beta_7 + \mu$,

$a_{66} = \beta_8 + \mu + \mu_{CO}$, $a_{77} = \beta_9 + \mu$, $a_{88} = \mu$.

Here, trace$(J^*) = -(a_{11} + a_{22} + a_{33} + a_{44} + a_{55} + a_{66} + a_{77} + a_{88}) < 0$ and $\det(J^*) > 0$ if $(\beta_2 + \mu) > \beta_1 E^*_{CO}$. Hence, by Routh-Hurwitz criterion [25], the endemic equilibrium point is locally asymptotically stable if $(\beta_2 + \mu) > \beta_1 E^*_{CO}$.

## Optimal Control

In this section, we consider different measures adopted by government as control in order to study the effectiveness of this model. Here, we apply Pontryagins maximum principle [17] in order to determine necessary conditions for optimality by introducing time dependent controls in the system (7.1). The introduced controls as follows can be observed in Fig. 7.2: $u_1$ and $u_2$ are taken as lock down and curfew control in order to restrict the exposure to COVID-19, $u_3$ as viral load test which detects and measures virus level consistently in COVID-19 infected patient, $u_4$ is a control which allows more and more individuals to opt for COVID-19 care centre, $u_5$ a control which allows only emergency medical patients get into the hospital and $u_6$ as plasma therapy in-order to increase recovery rate of hospitalized patients.

And the modified system (7.1) with controls is rewritten as

$$\frac{\mathrm{d}E_{CO}}{\mathrm{d}t} = B - \beta_1 E_{CO} I_F - \mu E + u_1 I_F$$

$$\frac{\mathrm{d}I_F}{\mathrm{d}t} = \beta_1 E_{CO} I_F - (\beta_2 + \mu) I_F - u_1 I_F + u_2 I_{SO}$$

$$\frac{\mathrm{d}I_{SO}}{\mathrm{d}t} = \beta_2 I_F - (\beta_3 + \mu) I_{SO} - u_2 I_{SO} - u_3 I_{SO}$$

$$\frac{\mathrm{d}T_E}{\mathrm{d}t} = \beta_3 I_{SO} - (\beta_4 + \beta_5 + \beta_6 + \mu) T_E - u_4 T_E + u_5 H + u_3 I_{SO}$$

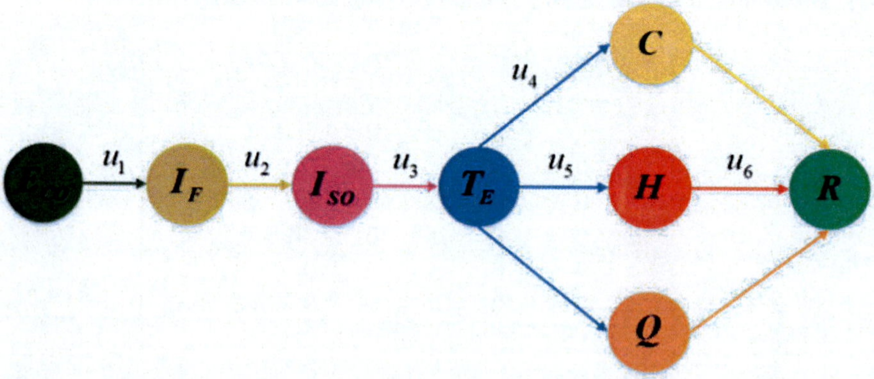


**Fig. 7.2** Optimal controls applied to the Fig. 7.1 [*source* own]

$$\frac{dC}{dt} = \beta_4 T_E - (\beta_7 + \mu)C + u_4 T_E$$
$$\frac{dH}{dt} = \beta_5 T_E - (\beta_8 + \mu + \mu_{CO} + u_5 + u_6)H$$
$$\frac{dQ}{dt} = \beta_6 T_E - (\beta_9 + \mu)Q$$
$$\frac{dR}{dt} = \beta_7 C + \beta_8 H + \beta_9 Q - \mu R + u_6 H$$

For this, we consider following objective function

$$J(c_i, \Lambda) = \int_0^T \big(W_1 E_{CO}^2 + W_2 I_F^2 + W_3 I_{SO}^2 + W_4 T_E^2 + W_5 C^2 + W_6 H^2 + W_7 Q^2 + W_8 R^2 + v_1 u_1^2 + v_2 u_2^2 + v_3 u_3^2 + v_4 u_4^2 + v_5 u_5^2 + v_6 u_6^2\big)\, dt$$

The control functions $u_1, u_2, u_3, u_4, u_5$ and $u_6$ are bounded, Lebesgue integrable functions. Here, $\Lambda$ denotes the set of all compartmental variables. The coefficients $W_1, W_2, W_3, W_4, W_5, W_6, v_1, v_2, v_3, v_4, v_5, v_6$ are the balancing cost functions.

Now, we seek to find out $u_1^*, u_2^*, u_3^*, u_4^*, u_5^*, u_6^*$ for the time $t = 0$ to $t = T$ such that

$$J(u_i(t)) = \min\{J(u_i^*, \Lambda)/(u_i) \in \phi\}, \quad i = 1, 2, 3, 4, 5, 6$$

where $\phi$ is a smooth function on the interval [0, 1].

Next, we introduce the Lagrangian function as follows

$$L(u, \lambda) = W_2 I_F^2 + W_3 I_{SO}^2 + W_4 T_E^2 + W_5 C^2 + W_6 H^2 + W_7 Q^2 + W_8 R^2 + v_1 u_1^2 + v_2 u_2^2 + v_3 u_3^2 + v_4 u_4^2 + v_5 u_5^2 + v_6 u_6^2$$

To obtain the value of Lagrangian function, we define Hamiltonian $H$ for the optimal control as

$$\begin{aligned} H = {} & W_2 I_F^2 + W_3 I_{SO}^2 + W_4 T_E^2 + W_5 C^2 + W_6 H^2 + W_7 Q^2 \\ & + W_8 R^2 + v_1 u_1^2 + v_2 u_2^2 + v_3 u_3^2 + v_4 u_4^2 + v_5 u_5^2 \\ & + v_6 u_6^2 + \lambda_1 (B - \beta_1 E_{CO} I_F - \mu E + u_1 I_F) \\ & + \lambda_2 (\beta_1 E_{CO} I_F - (\beta_2 + \mu) I_F - u_1 I_F + u_2 I_{SO}) + \lambda_3 (\beta_2 I_F \\ & - (\beta_3 + \mu) I_{SO} - u_2 I_{SO} - u_3 I_{SO}) \\ & + \lambda_4 (\beta_3 I_{SO} - (\beta_4 + \beta_5 + \beta_6 + \mu) T_E - u_4 T_E + u_5 H + u_3 I_{SO}) \\ & + \lambda_5 (\beta_4 T_E - (\beta_7 + \mu) C + u_4 T_E) \\ & + \lambda_6 (\beta_5 T_E - (\beta_8 + \mu + \mu_{CO} + u_5 + u_6) H) \end{aligned}$$

$$+\lambda_7(\beta_6 T_E - (\beta_9 + \mu)Q)$$
$$+\lambda_8(\beta_7 C + \beta_8 H + \beta_9 Q - \mu R + u_6 H)$$

Now using Pontryagins maximum principle and existence condition discussed by Fleming and Rishel [26], we obtain adjoint equations for the adjoint variable $\lambda_i = (\lambda_1, \lambda_2, \lambda_3, \lambda_4, \lambda_5, \lambda_6, \lambda_7, \lambda_8)$ associated with the state variables $(E_{CO}, I_F, I_{SO}, T_E, C, H, Q, R)$

$$\dot{\lambda}_1 = -2W_1 E_{CO} + (\lambda_1 - \lambda_2)\beta_1 I_F + \lambda_1 \mu$$

$$\dot{\lambda}_2 = -2W_2 I_F + (\lambda_1 - \lambda_2)\beta_1 E_{CO} + (\lambda_2 - \lambda_1)u_1 + (\lambda_2 - \lambda_3)\beta_2 + \lambda_2 \mu$$

$$\dot{\lambda}_3 = -2W_3 I_{SO} + (\lambda_3 - \lambda_2)u_2 + (\lambda_3 - \lambda_4)(\beta_3 + u_3) + \lambda_3 \mu$$

$$\dot{\lambda}_4 = -2W_4 T_E + \beta_5(\lambda_4 - \lambda_6) + \beta_6(\lambda_4 - \lambda_7) + (\lambda_4 - \lambda_5)(\beta_4 + u_4) + \lambda_4 \mu$$

$$\dot{\lambda}_5 = -2W_5 C + (\lambda_5 - \lambda_8)\beta_7 + \lambda_5 \mu$$

$$\dot{\lambda}_6 = -2W_6 H + (\lambda_6 - \lambda_4)u_5 + (\lambda_6 - \lambda_8)(u_6 + \beta_8) + (\mu + \mu_C)\lambda_6$$

$$\dot{\lambda}_7 = -2W_7 Q + (\lambda_8 - \lambda_7)\beta_9 + \lambda_7 \mu$$

$$\dot{\lambda}_8 = -2W_8 R + \lambda_8 \mu$$

The optimality conditions for control are given by
$u_1^* = \max\left(a_1, \min\left(b_1, \frac{I_F(\lambda_2-\lambda_1)}{2v_1}\right)\right)$, $u_2^* = \max\left(a_2, \min\left(b_2, \frac{I_{SO}(\lambda_3-\lambda_2)}{2v_2}\right)\right)$,
$u_3^* = \max\left(a_3, \min\left(b_3, \frac{I_{SO}(\lambda_3-\lambda_4)}{2v_3}\right)\right)$, $u_4^* = \max\left(a_4, \min\left(b_4, \frac{T_E(\lambda_4-\lambda_4)}{2v_4}\right)\right)$,
$u_5^* = \max\left(a_5, \min\left(b_5, \frac{H(\lambda_6-\lambda_4)}{2v_5}\right)\right)$ and $u_6^* = \max\left(a_6, \min\left(b_6, \frac{H(\lambda_6-\lambda_8)}{2v_6}\right)\right)$.
In next section, we will discuss these controls through plot.

## Numerical Simulation

In this section, simulation is carried out to understand the compartmental model of COVID-19. Here, we also discuss the various controlling interventions applied to the system (7.1). The data taken in Table 7.1 has been calculated and assumed

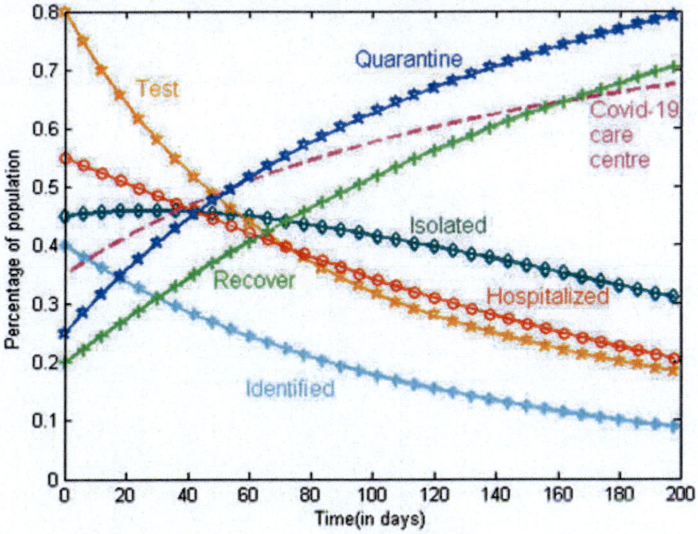


**Fig. 7.3** Transmission dynamics of COVID-19 [*source* own]

accordingly to the current scenario and can be found on https://ourworldindata.org/covid-testing [27].

In Fig. 7.3, we plot trajectory of all the compartments taken in our model. The path of various compartment can be observed showing behaviour of the model. Numerically, we observe that approximately 40% of population exposed to COVID-19 when tested shows positive report and is hospitalized within 72 days. About 35% of identified population adopts quarantine law. It is also seen that individual hospitalized gets recover at a faster rate than the individuals in COVID-19 care centre. Again, approximately 49% of exposed population when tested, shows negative report and is asked to home quarantine themselves. From Fig. 7.3, we also observe with time hospitalization decreases and individuals in quarantine and care centres increases.

The system (7.1) is said to exhibit oscillatory behaviour observed in Fig. 7.4. We see that Fig. 7.4a–c, i.e. exposed to COVID-19, identified and isolation, respectively, these compartments gets stabilized in due course of time. Exposed population initially decreases as seen in Fig. 7.4a which leads increase in identified (Fig. 7.4b) and isolation (Fig. 7.4b) population. Figure 7.4d–f oscillates as number of tests, individuals in COVID-19 care centre and individuals in hospital fluctuates with time. Quarantine population with the fluctuation decreases with time observed in Fig. 7.4g and recovery rate increases with time as seen in Fig. 7.4h.

Next, we discuss the simulation of the model after applying various controls.

In this model, we have applied 6 controls namely $u_1, u_2, u_3, u_4, u_5, u_6$ as lock down control, curfew, viral load test, control to increase population in care centres and to decrease population in hospital and plasma therapy as a medical intervention, respectively. Optimal control conditions are developed using iterative method. We start with solving state equations with a guess for controls within a simulated time and

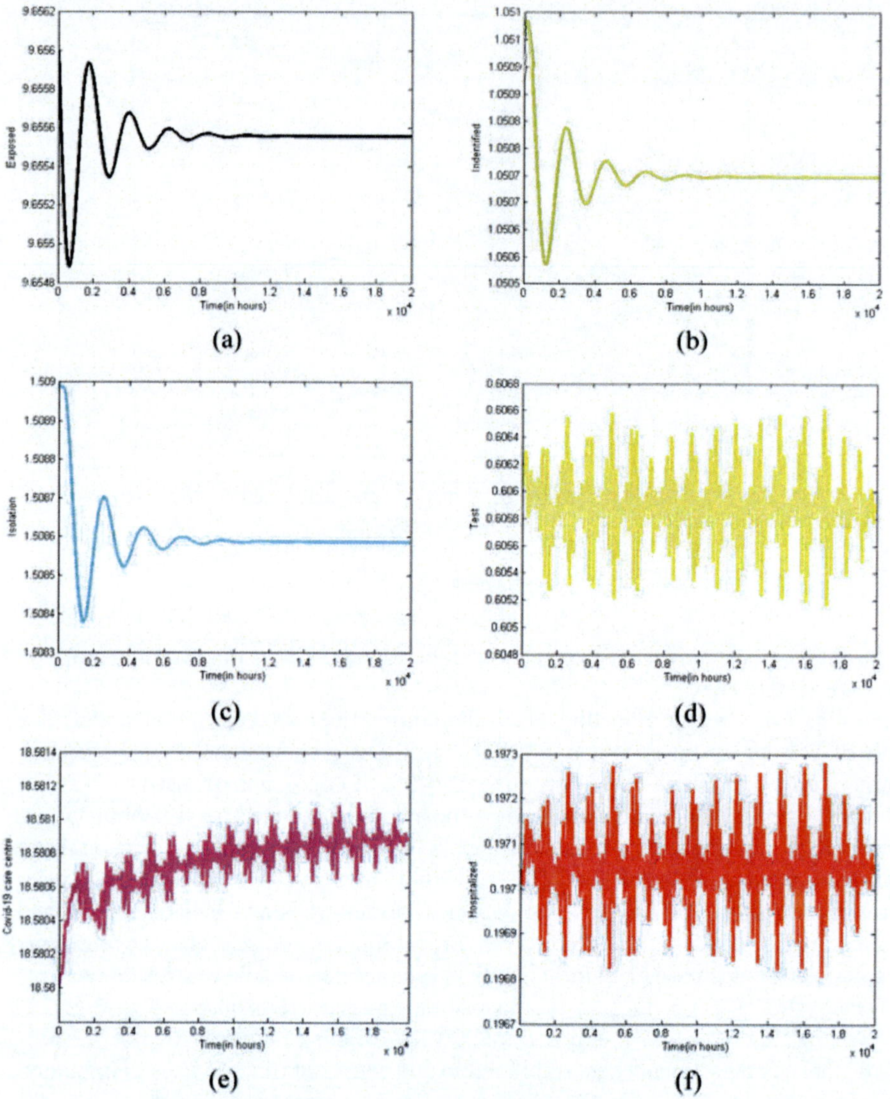


**Fig. 7.4** Oscillation observed in different compartments [*source* own]

apply fourth order Runge–Kutta method. Here, we plot each compartment against all the controls applied to the system (7.1) which is observed in Fig. 7.5. From Fig. 7.5a, we observe decrease in exposure to COVID-19. Showing importance of lockdown and curfew control in the model. Similarly, identification of COVID-19 cases increases initially for 2 weeks but then with time and with all the controls it decreases as observed in Fig. 7.5b. Similarly, decrease in number of isolated cases is observed in Fig. 7.5c. After encouraging individuals for the test as shown in Fig. 7.5d,

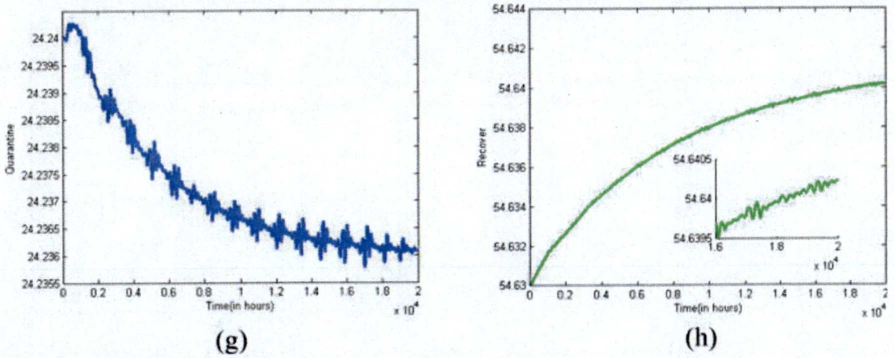


**Fig. 7.4** (continued)

first 4 weeks helps to reduce catching infection. Again, number of individuals in COVID-19 care centre increases seen in Fig. 7.5e and number of COVID-19 cases decreases in hospital (Fig. 7.5f) under the influence of all the controls. Quarantine population decreases (Fig. 7.5g) when control is applied as compared to when no control is applied. This happens due to decrease in exposure cases which indirectly decreases identification and isolation population. Recovery (Fig. 7.5h) increases with the decrease in hospitalized human population which shows the positive effect of plasma therapy used as one of the controls.

Figure 7.6 shows the plot of objective function under the influence of all the controls taken in this chapter. Here, we observe that implementing all these controls strictly can end this pandemic within 2 months.

## Conclusion

Observing current situation of the world, a huge proportion is infected with COVID-19. There are several social, medical interventions taken up by the government throughout the world to fight against the transmission of COVID-19 in absence of vaccination. In this chapter, we constructed a model considering exposure stage of COVID-19. Here we have computed basic reproduction number and showed it to be greater than 1 indicating current scenario of various countries. We have also calculated endemic equilibria and has shown it to be locally stable using Routh-Hurwitz criterion. Now until vaccine arrives, COVID-19 pandemic will have caused huge loss. Therefore, it is for our safety to follow various measures adopted by different government. In this chapter, we applied optimal control theory using social, medical measures as a control in a COVID-19 exposure scenario. It is observed that using all the controls strictly life can come back to normal within 2 months. This model suggest testing of COVID-19 at large scale also plays important role in combating

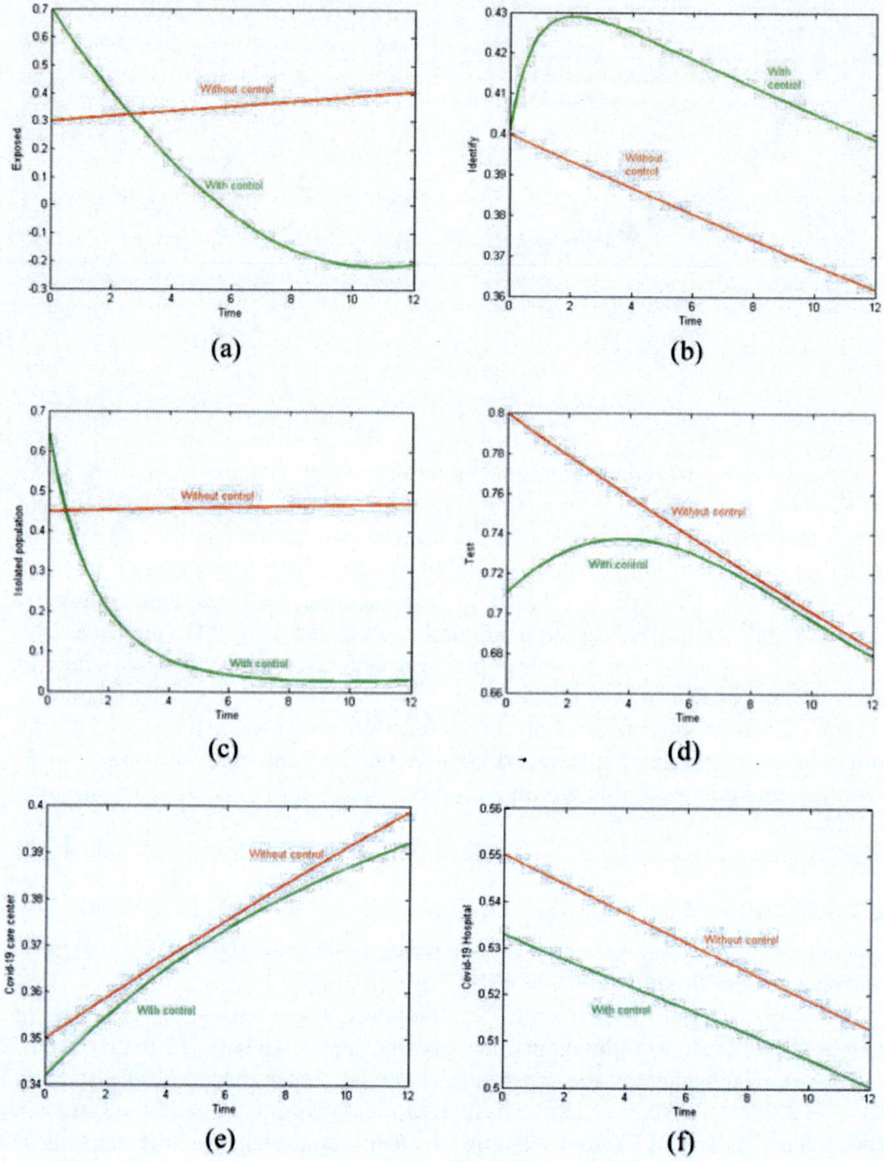


**Fig. 7.5** Effect of various controls applied to the model [*source* own]

COVID-19. Also, plasma therapy used as one of the controls plays vital role in increasing recovery rate of infective's.

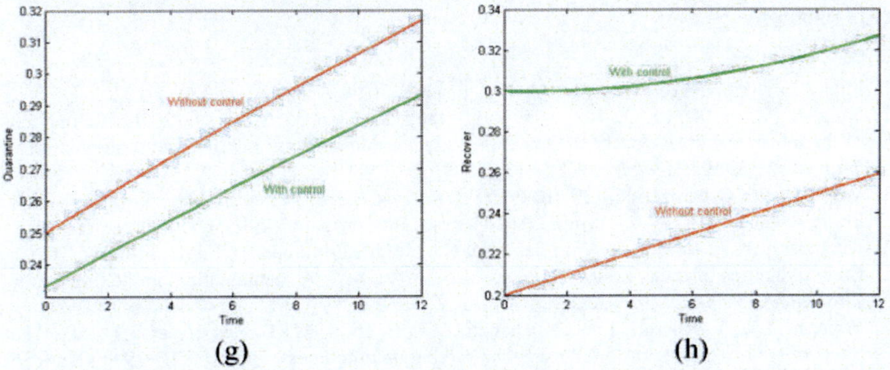


**Fig. 7.5** (continued)

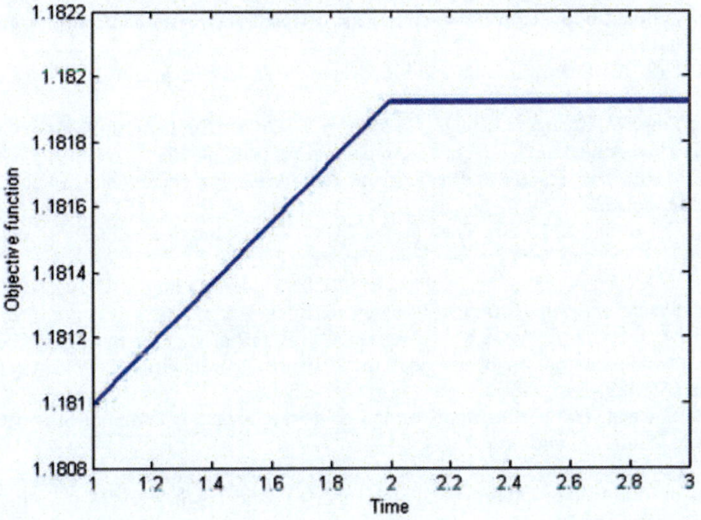


**Fig. 7.6** Plot of objective function [*source* own]

One of the limitations of this model is we have not taken into account cost-effectiveness strategies while applying controls.

**Acknowledgements** The authors thank reviewers for their constructive comments. The authors thank DST-FIST file # MSI-097 for technical support to the department. Second author (NS) would like to extend sincere thanks to the Education Department, Gujarat State for providing scholarship under ScHeme Of Developing High quality research (SHODH). Third author (ENJ) is funded by UGC granted National Fellowship for Other Backward Classes (NFO-2018-19-OBC-GUJ-71790).

**Data Availability** The data used to support the findings of this study are included within the article.

**Conflict of Interest** The authors do not have conflict of interest.

## References


1. https://en.wikipedia.org/wiki/Coronavirus_disease_2019. Accessed on 3 May 2020.
2. Tang, B., Wang, X., Li, Q., Bragazzi, N. L., Tang, S., Xiao, Y., & Wu, J. (2020). Estimation of the transmission risk of the 2019-nCoV and its implication for public health interventions. *Journal of Clinical Medicine, 9*(2), 462.
3. World Health Organization. (2020). *WHO Director-General's opening remarks at the media briefing on COVID-19—11 March 2020*. Geneva, Switzerland: World Health Organization.
4. Kucharski, A. J., Russell, T. W., Diamond, C., Liu, Y., Edmunds, J., Funk, S., et al. (2020). Early dynamics of transmission and control of COVID-19: A mathematical modelling study. *The Lancet Infectious Diseases*.
5. Prem, K., Liu, Y., Russell, T. W., Kucharski, A. J., Eggo, R. M., Davies, N., et al. (2020). The effect of control strategies to reduce social mixing on outcomes of the COVID-19 epidemic in Wuhan, China: A modelling study. *The Lancet Public Health*.
6. Mueller, M., Derlet, P. M., Mudry, C., & Aeppli, G. (2020). Using random testing to manage a safe exit from the COVID-19 lockdown. *arXiv preprint* arXiv:2004.04614.
7. Hellewell, J., Abbott, S., Gimma, A., Bosse, N. I., Jarvis, C. I., Russell, T. W., et al. (2020). Feasibility of controlling COVID-19 outbreaks by isolation of cases and contacts. *The Lancet Global Health*.
8. Toda, A. A. (2020). Susceptible-infected-recovered (sir) dynamics of COVID-19 and economic impact. *arXiv preprint* arXiv:2003.11221.
9. Peng, L., Yang, W., Zhang, D., Zhuge, C., & Hong, L. (2020). Epidemic analysis of COVID-19 in China by dynamical modeling. *arXiv preprint* arXiv:2002.06563.
10. Tang, Z., Li, X., & Li, H. (2020). Prediction of new coronavirus infection based on a modified SEIR model. *medRxiv*.
11. Piguillem, F., & Shi, L. (2020). The optimal COVID-19 quarantine and testing policies (No. 2004). Einaudi Institute for Economics and Finance (EIEF).
12. Sun, P., Lu, X., Xu, C., Sun, W., & Pan, B. (2020). Understanding of COVID-19 based on current evidence. *Journal of Medical Virology, 92*(6), 548–551.
13. Chinazzi, M., Davis, J. T., Ajelli, M., Gioannini, C., Litvinova, M., Merler, S., et al. (2020). The effect of travel restrictions on the spread of the 2019 novel coronavirus (COVID-19) outbreak. *Science, 368*(6489), 395–400.
14. Zhao, S., & Chen, H. (2020). Modeling the epidemic dynamics and control of COVID-19 outbreak in China. *Quantitative Biology*, 1–9.
15. Xu, T., Chen, C., Zhu, Z., Cui, M., Chen, C., Dai, H., & Xue, Y. (2020). Clinical features and dynamics of viral load in imported and non-imported patients with COVID-19. *International Journal of Infectious Diseases*.
16. Yang, C., & Wang, J. (2020). A mathematical model for the novel coronavirus epidemic in Wuhan, China. *Mathematical Biosciences and Engineering, 17*(3), 2708–2724.
17. Pontryagin, L. S. (2018). *Mathematical theory of optimal processes*. Routledge.
18. Sharomi, O., & Malik, T. (2017). Optimal control in epidemiology. *Annals of Operations Research, 251*(1–2), 55–71.
19. Lemos-Paião, A. P., Silva, C. J., & Torres, D. F. (2017). An epidemic model for cholera with optimal control treatment. *Journal of Computational and Applied Mathematics, 318*, 168–180.
20. Tilahun, G. T., Makinde, O. D., & Malonza, D. (2017). Modelling and optimal control of pneumonia disease with cost-effective strategies. *Journal of Biological Dynamics, 11*(sup2), 400–426.
21. Djidjou-Demasse, R., Michalakis, Y., Choisy, M., Sofonea, M. T., & Alizon, S. (2020). Optimal COVID-19 epidemic control until vaccine deployment. *medRxiv*.
22. Mallela, A. (2020). Optimal Control applied to a SEIR model of 2019-nCoV with social distancing. *medRxiv*.
23. Tsay, C., Lejarza, F., Stadtherr, M. A., & Baldea, M. (2020). Modeling, state estimation, and optimal control for the US COVID-19 outbreak. *arXiv preprint* arXiv:2004.06291.

24. Diekmann, O., Heesterbeek, J. A. P., & Roberts, M. G. (2010). The construction of next-generation matrices for compartmental epidemic models. *Journal of the Royal Society Interface, 7*(47), 873–885.
25. Routh, E. J. (1877). *A treatise on the stability of a given state of motion: Particularly steady motion.* Macmillan and Company.
26. Fleming, W. H., & Rishel, R. W. (2012). *Deterministic and stochastic optimal control* (Vol. 1). Springer Science & Business Media.
27. https://ourworldindata.org/covid-testing. Accessed on 25 April 2020.

# Chapter 8
# Controlling the Transmission of COVID-19 Infection in Indian Districts: A Compartmental Modelling Approach

**Ankit Sikarwar, Ritu Rani, Nita H. Shah, and Ankush H. Suthar**

**Abstract** The widespread of the novel coronavirus (2019-nCoV) has adversely affected the world and is treated as a Public Health Emergency of International Concern by the World Health Organization. Assessment of the basic reproduction number with the help of mathematical modeling can evaluate the dynamics of virus spread and facilitate critical information for effective medical interventions. In India, the disease control strategies and interventions have been applied at the district level by categorizing the districts as per the infected cases. In this study, an attempt has been made to estimate the basic reproduction number $R_0$ based on publically available data at the district level in India. The susceptible-exposed-infected-critically infected-hospitalization-recovered (SEICHR) compartmental model is constructed to understand the COVID-19 transmission among different districts. The model relies on the twelve kinematic parameters fitted on the data for the outbreak in India up to May 15, 2020. The expression of basic reproduction number $R_0$ using the next-generating matrix is derived and estimated. The study also employs three time-dependent control strategies to control and minimize the infection transmission from one district to another. The results suggest an unstable situation of the pandemic that can be minimized with the suggested control strategies.

**Keywords** COVID-19 · India · Districts · Compartmental model · Basic reproduction number · Optimal control theory

A. Sikarwar (✉) · R. Rani
Department of Development Studies, International Institute for Population Sciences, Mumbai 400088, Maharashtra, India
e-mail: anks.sik@gmail.com

R. Rani
e-mail: rituroy4@yahoo.com

N. H. Shah · A. H. Suthar
Department of Mathematics, Gujarat University, Ahmedabad 380009, Gujarat, India
e-mail: nitahshah@gmail.com

A. H. Suthar
e-mail: ankush.suthar1070@gmail.com



N. H. Shah and M. Mittal (eds.), *Mathematical Analysis for Transmission of COVID-19*, Mathematical Engineering, https://doi.org/10.1007/978-981-33-6264-2_8

**Mathematics Subject Classification** 37Nxx

## Introduction

A novel coronavirus has caused an outbreak of atypical pneumonia, now formally called COVID-19 by the World Health Organization, first in Wuhan in December 2019 and then rapidly spread out in the whole of China and the other parts of the world. The emergence of novel coronavirus has threatened the world, leading to a public health crisis and unprecedented challenges. According to WHO estimates on July 25, 2020, 15,538,736 confirmed cases and 634,325 deaths so far have been reported globally [23]. The escalated spreading, the virulence of the pandemic, and the alarming case fatality rates even in the western countries with efficient healthcare and medical facilities have raised concerns for the immediate control of the infection [24]. The remarkable variability of the infection numbers and mortality in different nations and settings has also captured the focus of scientists and researchers worldwide [3, 11]. Due to the complex nature of the virus, non-availability of any standard treatment, and delay in the vaccines, the pandemic has challenged the humans in a very serious and critical manner [18]. To control the spread of the virus decision-makers have come up with non-pharmacological measures and controlling interventions such as social distancing, quarantining, use of face mask, isolation, contact tracing, and lockdown. Many countries have efficiently controlled the spread of the virus by implementing these measures.

As per July 25, 2020, India has the third-highest number of COVID-19 cases in the world with a total of 1,336,861 confirmed cases and 31,358 deaths [23]. Even though the proportion of the total population infected is low compared with other countries; India is at a higher risk of community transmission because of crowded living conditions, congested cities, high slum-dwelling population, poor health care facilities, and high levels of poverty [12]. India was quick to close its international borders and impose an immediate lockdown, which WHO praised as "tough and timely". The lockdown provided the government time to prepare for a possible surge in cases when the pandemic was forecasted to peak in the coming weeks. Still, India's population of 1.3 billion across diverse states, health inequalities, widening economic and social disparities, and distinct cultural values present unique challenges [9]. As the spread of the virus has begun from large cities to smaller towns and rural areas, the detrimental health effects are likely to be more severe for poorer people from socioeconomically disadvantaged settings.

The role of mathematical models analyzing the transmission of various infectious diseases has been recognized by many researchers [7, 8, 19]. Due to the inclusion of realistic factors, the use of such models has been remained constructive for healthcare professionals and policymakers [10, 20]. There has been a sudden peak in the studies with mathematical modeling since the emergence of coronavirus [5, 6, 14, 16, 17, 21]. Compartment models are constructed for epidemiological analysis by classifying the population into relevant compartments. The most commonly used of these models is the Susceptible-Infected-Recovered (SIR) model. These models are adopted by the scientific community to understand the transmission dynamics of COVID-19 [16, 22]. The optimal control theory plays an important role to develop control strategies and to analyze their impact on the transmission of infectious diseases [1, 13]. The compartmental models with objective specific algorithms were used to predict the regional peaks of the pandemics and to evaluate the effectiveness of various control measures implemented during pandemics.

Considering the applicability of compartment models to understand the district level transmission of COVID-19 infections, in this study, an attempt has been made to estimate the basic reproduction number R0 based on the publically available data at the district level in India. The susceptible-exposed-infected-critically infected-hospitalized-recovered (SEICHR) compartmental model for COVID-19 transmission among different districts is constructed. To calculate the threshold value for the infection the basic reproduction number is formulated. Moreover, the stability of the model is optimized by applying optimal control theory to the SEICHR-model. The control theory aims to find the optimal treatment and control strategies such that it improves the recovery rate of infected individuals, as well as reduce the transmission of COVID-19 by reducing the contact rate.

## Model Development

A compartmental model for COVID-19 transmission dynamics among different districts is constructed. District-wise data of COVID-19 infections till May 15, 2020, is acquired from the Ministry of Health and Family Welfare (MoHFW) portal of COVID-19 (https://www.mohfw.gov.in/). According to the proposed model, we divided 693 districts of India into six compartments, class of susceptible district ($S$), class of exposed district ($E$), class of district infected by COVID-19 ($I$), class of critically infected district ($C$), class of districts with hospitalization facilities ($H$), and class of recovered districts ($R$).

The model considers the susceptible district as the one which does not have any case of infection until the considered date. Moreover, the district is not surrounded by the districts having infection cases. The exposed districts are surrounded by the districts having infections. A connection between susceptible and exposed districts is considered. Infected and critically infected districts are categorized on the bases of severity in the infected cases in the districts. The model assumes that the susceptible districts will become exposed over time and then will become infected or critically infected based on the number of cases. These districts with the treatment in hospitalization facilities may be declared as recovered over time. These recovered districts again have the probability of becoming exposed ones. A detailed diagram of the flow of the model with compartments is shown in Fig. 8.1.

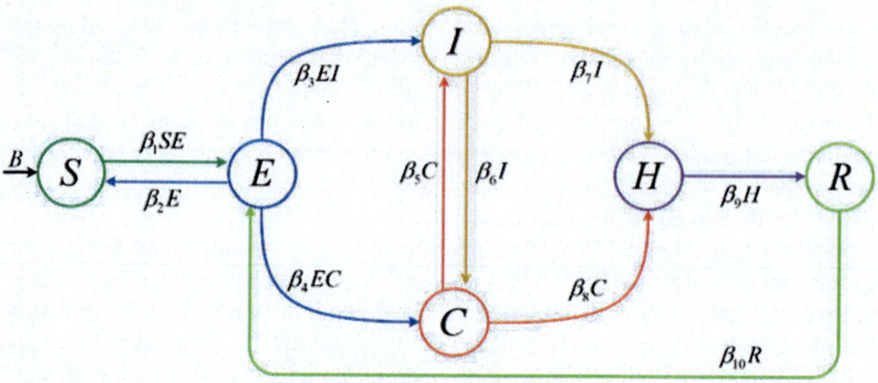


**Fig. 8.1** System diagram of the COVID-19 transmission model

Using this flow diagram of the COVID-19 transmission model, a formulated system of nonlinear differential equations is showed by Eq. 8.1.

$$\begin{aligned}
\frac{dS(t)}{dt} &= B - \beta_1 SE + \beta_2 E \\
\frac{dE(t)}{dt} &= \beta_1 SE - \beta_2 E - \beta_3 EI - \beta_4 EC + \beta_{10} R \\
\frac{dI(t)}{dt} &= \beta_3 EI + \beta_5 C - (\beta_6 + \beta_7 + \mu_d) I \\
\frac{dC(t)}{dt} &= \beta_4 EC + \beta_6 I - (\beta_5 + \beta_8 + \mu_d) C \\
\frac{dH(t)}{dt} &= \beta_7 I + \beta_8 C - (\beta_9 + \mu_d) H \\
\frac{dR(t)}{dt} &= \beta_9 H - \beta_{10} R
\end{aligned} \tag{8.1}$$

The estimation of parameters and parametric values is a crucial part of compartment models. Optimal estimation of these values ensures the validity of the model to describe the transmission of the pandemic. The data for the infected, critically infected, and the recovered area is taken by observing the current situation of deviation in different districts in India as per the MoHFW statistics. The mortality rate due to COVID-19 is calculated by observing data from the Worldometer (https://www.worldometers.info/coronavirus/). Approximate parametric values are calculated using the data available and some are assumed. The parameters used in the model are listed in Table 8.1.

Note that, in this model, we have considered only the human population. Hence, only non-negative initial conditions are used.

**Table 8.1** Parameters and their value used in the model

| Parameters | Description | Parametric values |
|---|---|---|
| $B$ | Rate by which any new area becomes susceptible | 0.6 |
| $\beta_1$ | Rate at which susceptible area becomes exposed | 0.3533 |
| $\beta_2$ | Rate at which exposed area becomes susceptible | 0.0321 |
| $\beta_3$ | Rate by which exposed area gets infected | 0.3814 |
| $\beta_4$ | Rate by which exposed area becomes critically infected | 0.1914 |
| $\beta_5$ | Rate at which critically infected area recover from the critical situation and becomes infected | 0.0821 |
| $\beta_6$ | Rate at which critically infected area becomes critically infected | 0.4577 |
| $\beta_7$ | Hospitalization rate in an infected area | 0.8232 |
| $\beta_8$ | Hospitalization rate in a critically infected area | 0.9532 |
| $\beta_9$ | Recovery rate of the hospitalized area | 0.2487 |
| $\beta_{10}$ | Rate at which recovered area again becomes exposed | 0.1211 |
| $\mu_d$ | Mortality rate due to COVID-19 | 0.25 |

*Source* Estimated by authors

## *Positivity and Boundedness of the Solution*

Note that, the solutions of the system (8.1) are non-negative and bounded if initial conditions are non-negative.

**Theorem 1** *Let* $(S(t), E(t), I(t), C(t), H(t), R(t))$ *be the solution of the system (8.1), with initial condition* $(S(0), E(0), I(0), C(0), H(0), R(0))$ *and the compact set* $\Lambda$ *is a bounded and positively invariant set where all solutions in* $R_+^6$ *converge.*

Where

$$\Lambda = \{(S, E, I, C, H, R) \in R_+^6 : N' \leq B\} \tag{8.2}$$

***Proof*** Define $N(t) = S(t) + E(t) + I(t) + C(t) + H(t) + R(t)$, by differentiating it with respect to $t$, we get $N' = B - \mu_d(I + C + H)$.

Also, all the parameters used in the model are considered non-negative. Based on the system (8.1), as $t \to \infty$, we have $N' \leq B$, since $\mu_d \geq 0$. Hence, $\Lambda$ is a bounded and positively invariant set.

$N' \leq B$ gives $N'(t) = B + N(0)e^{-t}$, where $N(0)$ are the initial conditions.

## Basic Reproduction Number and Equilibrium Point

The basic reproduction number $R_0$ is defined as the average number of secondary infected cases rising from an average primary case in an entirely exposed/susceptible population.

The optimum solution of the system (8.1), the endemic equilibrium point is as follow:

$$E_p^* = \left(S^*, E^*, I^*, C^*, H^*, R^*\right) \tag{8.3}$$

where $S^* = \frac{\beta_2 r+B}{r\beta_1}$, $E^* = r$, $I^* = \frac{B\beta_5(\beta_9+\mu_d)}{\mu_d k}$, $C^* = \frac{B}{\mu_d k}(\beta_9 + \mu_d)(\beta_6 + \beta_7 - \beta_3 r + \mu_d)$, $H^* = \frac{B}{\mu_d k}(\beta_8(\beta_6 + \mu_d) + \beta_8(\beta_7 - \beta_3 r) + \beta_7\beta_5)$, $R^* = \frac{\beta_9 B}{\mu_d k\beta_{10}}(\beta_8(\beta_6 + \mu_d)+\beta_8(\beta_7 - \beta_3 r) + \beta_7\beta_5)$.

Here, $k = (\beta_9 + \mu_d)(\beta_5 + \beta_6 + \beta_7 + \mu_d - \beta_3 r) + \beta_7(\beta_5 + \beta_8) + \beta_8(\mu_d - \beta_3 r)$ and $r$ is the highest root of the polynomial $p(z) = s_2 z^2 - s_1 z + s_0 = 0$, coefficients of the polynomial are:$s_2 = \beta_3\beta_4$, $s_1 = (\mu_d(\beta_3 + \beta_4) + \beta_3(\beta_5 + \beta_8) + \beta_4(\beta_6 + \beta_7))$ and $s_0 = (\beta_8 + \mu_d)(\beta_6 + \beta_7 + \mu_d) + \beta_5(\beta_7 + \mu_d)$.

This $R_0$ based on endemic equilibrium point can be calculated using the next-generation matrix method [2, 4]. The above dynamical system (8.1) can be re-formed and separate in the following two matrices.

$$f = \begin{bmatrix} \beta_1 SE \\ \beta_3 EI \\ \beta_4 EC \\ 0 \\ 0 \\ 0 \end{bmatrix} \text{ and } v = \begin{bmatrix} \beta_2 E + \beta_3 EI + \beta_4 EC - \beta_{10} R \\ -\beta_5 C + (\beta_6 + \beta_7 + \mu_d)I \\ -\beta_6 I + (\beta_5 + \beta_8 + \mu_d)C \\ -B + \beta_1 SE - \beta_2 E \\ -\beta_7 I - \beta_8 C + (\beta_9 + \mu_d)H \\ -\beta_9 H + \beta_{10} R \end{bmatrix}$$

The Jacobian matrix of $f$ and $v$ around endemic equilibrium point (3) is $F$ and $V$. Note that, the matrix $F$ shows the new infectious rates and the matrix $V$ shows other rates transferred in between the compartments, are given, respectively, by:

$$F = \begin{bmatrix} \beta_1 S^* & 0 & 0 & \beta_1 E^* & 0 & 0 \\ \beta_3 I^* & \beta_3 E^* & 0 & 0 & 0 & 0 \\ \beta_4 C^* & 0 & \beta_4 E^* & 0 & 0 & 0 \\ 0 & 0 & 0 & 0 & 0 & 0 \\ 0 & 0 & 0 & 0 & 0 & 0 \\ 0 & 0 & 0 & 0 & 0 & 0 \end{bmatrix}$$

$$V = \begin{bmatrix} \beta_2 + \beta_3 I^* + \beta_4 C^* & \beta_3 E^* & \beta_4 E^* & 0 & 0 & -\beta_{10} \\ 0 & \beta_6 + \beta_7 + \mu_d & -\beta_5 & 0 & 0 & 0 \\ 0 & -\beta_6 & \beta_5 + \beta_8 + \mu_d & 0 & 0 & 0 \\ \beta_1 S^* - \beta_2 & 0 & 0 & \beta_1 E^* & 0 & 0 \\ 0 & -\beta_7 & -\beta_8 & 0 & \beta_9 + \mu_d & 0 \\ 0 & 0 & 0 & 0 & -\beta_9 & \beta_{10} \end{bmatrix}$$

The basic reproduction number ($R_0$) is given by the spectral radius of the matrix ($FV^{-1}$).

$$FV^{-1} = \begin{bmatrix} x_1 & x_2 & x_3 & 1 & x_4 & x_1 \\ x_5 & x_6 & x_7 & 0 & x_8 & x_5 \\ x_9 & x_{10} & x_{11} & 0 & x_{12} & x_9 \\ 0 & 0 & 0 & 0 & 0 & 0 \\ 0 & 0 & 0 & 0 & 0 & 0 \\ 0 & 0 & 0 & 0 & 0 & 0 \end{bmatrix}$$

where $x_1 = \frac{\mu_d k \beta_2 r}{r a_1}$, $x_2 = \frac{-\mu_d k a_2 \beta_2 r}{r a_1 a_5 (\beta_9 + \mu_d)}$, $x_3 = \frac{-\mu_d k a_3 \beta_2 r}{r a_1 a_5 (\beta_9 + \mu_d)}$, $x_4 = \frac{\mu_d k \beta_9 \beta_2 r}{r a_1 (\beta_9 + \mu_d)}$, $x_5 = \frac{B\beta_3\beta_5(\beta_9 + \mu_d)}{a_1}$, $x_6 = \frac{\beta_3 r (\beta_5 + \beta_8 + \mu_d)}{a_5} - \frac{B\beta_3\beta_5 a_2}{a_1 a_5}$, $x_7 = \frac{\beta_3\beta_5 r}{a_5} - \frac{B\beta_3\beta_5 a_3}{a_1 a_5}$, $x_8 = \frac{B\beta_3\beta_5\beta_9}{a_1}$, $x_9 = \frac{B\beta_4 a_4}{a_1}$, $x_{10} = \frac{\beta_4\beta_6 r}{a_5} - \frac{B\beta_4 a_2 a_4}{a_1 a_5 (\beta_9 + \mu_d)}$, $x_{11} = \frac{\beta_4 r (\beta_6 + \beta_7 + \mu_d)}{a_5} - \frac{B\beta_4 a_3 a_4}{a_1 a_5 (\beta_9 + \mu_d)}$, $x_{12} = \frac{B\beta_4 a_4 a_9}{a_1 (\beta_9 + \mu_d)}$,

$$a_1 = (\mu_d + \beta_9)(B\beta_3(\beta_5 - \beta_4 r) + \beta_4\beta_7 B) + (\beta_6 + \mu_d)(\mu_d B\beta_4 + \beta_4\beta_9 B) + \mu_d\beta_2 k$$

$$a_2 = (\mu_d + \beta_9) r (\beta_4\beta_6 + \beta_3\mu_d) - (\beta_8 + \mu_d)\beta_7\beta_9 + \mu_d\beta_3 r(\beta_5 + \beta_8) \\ + (\beta_3 r - \beta_7)\beta_5\beta_9 + (\beta_3 r - \beta_6)\beta_8\beta_9$$

$$a_3 = (\mu_d + \beta_6) r \mu_d \beta_4 + (\beta_3\beta_5 + \beta_4\beta_7)\mu_d r + (\beta_4 r - \beta_8)\beta_9(\mu_d + \beta_7) \\ + (\beta_4 r - \beta_9)\beta_9\beta_6 + (\beta_3 r - \beta_7)\beta_9\beta_5$$

$a_4 = (\mu_d + \beta_9)(\beta_6 + \beta_7 + \mu_d - \beta_3 r)$ and $a_5 = (\mu_d + \beta_8)(\beta_6 + \beta_7 + \mu_d) + (\mu_d + \beta_7)\beta_5$.

Reproduction number is:

$$R_0 = \frac{1}{6} m^{1/3} - \frac{6n}{m^{1/3}} + \frac{1}{3}(x_{11} + x_6 + x_1) \tag{8.4}$$

where $n = \frac{1}{9}x_1(x_{11} + x_6 - x_1) + \frac{1}{9}x_6(x_{11} - x_6) - \frac{1}{3}(x_{10}x_7 + x_5x_2 + x_9x_3) - \frac{1}{9}x_{11}^2$ and

$$m = (x_3x_9 - x_6^2)(6x_1x_6(x_{10}x_7 - 4x_3x_9) - 6x_{11}x_6(x_{11}^2 + 4x_3x_9) \\ +6x_3x_9(5x_1x_{11} - 2x_3x_9 + 2x_6x_9))$$

$$
\begin{aligned}
&+ (x_3x_9 + x_6^2)(3x_{11}x_6(2x_{10}x_7 - x_{11}x_6) \\
&-6x_2x_5(x_{11}^2 + 6x_3x_9) - 3x_{11}^2x_3x_9) \\
&- 3x_1^4(4x_{10}x_7 + x_{11}^2) \\
&+ (x_3x_9 + x_2x_5)\big(30x_1^2(x_{11}x_6 - 2x_{10}x_7) - 3x_1x_5(x_1x_2 - 18x_{10}x_3)\big) \\
&+ 12x_1^3x_{10}(2x_{11}x_7 + x_3x_5 + 2x_6x_7) \\
&+ 3x_2x_9(3x_2x_7 - 4x_3x_6)(2x_5x_6 + 9x_7x_9) \\
&- 3x_1^4x_6(x_6 - 2x_{11}) + 6x_1^3x_{11}^3(1 - x_6) - 12x_3x_6^4x_9 \\
&+ 6x_1^3x_2(x_5x_6 + 2x_7x_9) - 6x_1^3x_6(x_3x_9 - x_6) \\
&+ 6x_1^2x_{10}x_7(4x_{10}x_7 - x_{11}) - 18x_1^2x_{10}x_3x_5(x_6 + x_{11}) \\
&- 3x_1^2x_{11}^3(x_{11} + 2x_6) - 6x_1^2x_{11}^2(x_2x_5 + 4x_3x_9) \\
&- 6x_1^2x_{11}x_6^2(3x_{11} - x_6) - 3x_1^2x_2x_9(x_3x_5 + 6x_{11}x_7) \\
&- 6x_{11}^2x_2x_9(x_3x_5 + 3x_6x_7) - 12x_{11}x_2^2x_5(2x_5x_6 + 9x_7x_9) \\
&+ 6x_2x_3x_9x_{11}(19x_5x_6 + 9x_7x_9) \\
&- 18x_1^2x_2x_6(x_5x_6 + x_7x_9) - 3x_1^2x_1^6(x_6^2 + x_2x_5) \\
&- 3x_1^2x_3x_9(x_3x_9 + 2x_6^2) + 54x_{10}x_{11}x_2x_7^2x_9 \\
&- 24x_1x_{10}^2x_7(x_6x_7 + 4x_3x_5) - 6x_1x_{10}x_{11}^2(x_{11}x_7 + 3x_3x_5) \\
&+ 6x_1x_{10}x_{11}x_7(5x_{11}x_6 + 19x_2x_5) + 6x_1x_{10}x_{11}x_3(x_7x_9 + 7x_5x_6) \\
&+ 30x_1x_{10}x_{11}x_6(x_3x_5 + x_6x_7) + 6x_1x_{10}x_5x_6(x_2x_7 - 3x_3x_6) \\
&+ 6x_{10}x_{11}x_2x_5(x_6x_7 - 18x_3x_5) + 6x_{10}x_{11}x_3x_5(9x_3x_9 - 3x_6^2) \\
&+ 6x_{10}x_2x_5^2(9x_3x_6 - 6x_2x_7) + 6x_2x_5x_7x_{10}(5x_3x_9 - x_6^2) \\
&+ 6x_2x_7x_9x_{10}(10x_3x_5 + 9x_6x_7) + 6x_1x_{11}x_2x_6(5x_5x_6 + 12x_7x_9) \\
&+ 108x_1x_{10}x_7x_9(x_3x_6 - x_2x_7) \\
&+ 6x_1x_{11}^2x_2(3x_7x_9 - 10x_5x_6) \\
&+ 6x_1x_{11}^2x_6(5x_3x_9 - x_6^2) + 6x_1x_{11}^3(x_3x_9 + 4x_2x_5) \\
&+ 6x_1x_{11}^3x_6(x_{11} - x_6) + 6x_1x_{11}x_2x_5(x_3x_9 - 4x_2x_5) \\
&- 6x_1x_{11}x_6^2(5x_3x_9 - x_6^2) + 6x_1x_2^2x_5(5x_5x_6 + 9x_7x_9) \\
&+ 6x_1x_2x_3x_9(x_5x_6 + 4x_7x_9) + 9x_{10}^2x_3x_5(9x_3x_5 + 6x_6x_7) \\
&- 3x_{10}^2x_7^2(x_{11}^2 + 4x_{10}x_7) + 6x_{10}^2x_{11}x_7(5x_6x_7 + 3x_3x_5) \\
&- 3x_{10}^2x_7^2(12x_3x_9 + x_6^2) + 6x_{10}x_{11}^2(x_6x_7 + 2x_3x_5) \\
&- 6x_{10}x_{11}^2x_5(10x_2x_7 + 3x_3x_6) + 12x_{10}x_3x_6^2(x_5x_6 - 5x_7x_9) \\
&- 3x_{11}^4(x_6^2 + 4x_2x_5) + 12x_{11}^3x_2(x_7x_9 + 2x_5x_6) \\
&- 6x_{11}x_2x_6^2(x_5x_6 + 3x_7x_9) - 3x_2^2x_5x_6(x_5x_6 - 12x_7x_9) \\
&- 6x_1^2x_{10}x_7(10x_3x_9 + x_6^2) - 6x_1^3x_{11}(x_2x_5 + x_3x_9 + x_6^2)
\end{aligned}
$$

$$- 12x_1x_{10}^2x_7(2x_{11}x_7 + x_3x_5) - 36x_{10}x_3^2x_9(x_7 + 3x_5x_6)$$
$$+ 6x_1x_2x_6^2(x_5x_6 - 3x_7x_9) + 36x_{10}^2x_5x_7(x_{11}x_3 - x_2x_7)$$
$$- 6x_{10}x_{11}^2x_7(x_3x_9 + 4x_6^2) - 12x_2^2x_5^2(x_2x_5 + 3x_3x_9)$$
$$+ 24x_{11}^2x_2x_5(x_2x_5 - 2x_3x_9)$$

After substitution of parametric values given in Table 8.1 in Eq. 8.4, the resultant value of the reproduction number is 1.7976. It represents the average number of secondarily infected districts generated by an infected district in a completely susceptible region.

## Optimal Control Theory

Control measures play a vital role to minimize the spread of an infectious virus. In this section, we extend the system (8.1) to include three time-dependent control strategies,$u_1(t)$, $u_2(t)$ and $u_3(t)$ which could be the best control measures to restrict the spread of COVID-19 infections from one district to another (Fig. 8.2). The first control strategy $u_1(t)$ is lockdown to minimize the chances of the exposed district to be converted into infected districts. Implementation of effective lockdown restricts the intra-state and inter-state movement of individuals. By this measure, there are very fewer chances of an exposed district becoming infected even if it is neighboring a district with considerable cases. The second control strategy $u_2(t)$ is applied between infected to the critically infected district. These measures include social isolation (quarantine) of infected individuals, precautionary measures taken by the family members of an infected individual, and contact tracing. These strategies will reduce the risk of the infected district to become critically infected. The third control strategy $u_3(t)$ is applied to increase the recovery rate in the districts having responsive medical

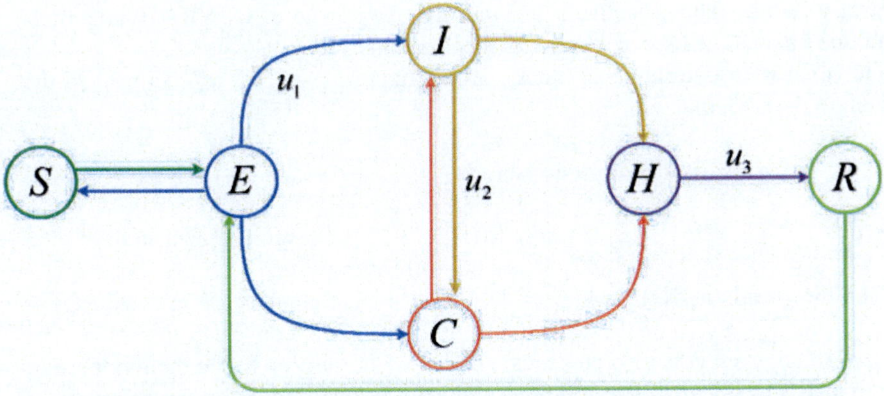


Fig. 8.2 COVID-19 model with control variables

or hospital facilities for COVID-19 cases. The study has considered an increase in hospital beds and ventilators as the third controlling strategy to increase the recovery rates in a district. The effect of these control strategies on the transmission of the virus is explained in the numerical simulation section.

According to the above assumptions, the COVID-19 model (8.1) is modified by including control variables as follow:

$$
\begin{aligned}
\frac{dS(t)}{dt} &= B - \beta_1 SE + \beta_2 E \\
\frac{dE(t)}{dt} &= \beta_1 SE - \beta_2 E - \beta_3 EI - \beta_4 EC + \beta_{10} R + u_1 E \\
\frac{dI(t)}{dt} &= \beta_3 EI + \beta_5 C - (\beta_6 + \beta_7 + \mu_d) I - u_1 E + u_2 I \\
\frac{dC(t)}{dt} &= \beta_4 EC + \beta_6 I - (\beta_5 + \beta_8 + \mu_d) C - u_2 I \\
\frac{dH(t)}{dt} &= \beta_7 I + \beta_8 C - (\beta_9 + \mu_d) H - u_3 H \\
\frac{dR(t)}{dt} &= \beta_9 H - \beta_{10} R + u_3 H
\end{aligned}
\tag{8.5}
$$

According to this extended model, the optimal control situation with the objective function is formulated by

$$
\begin{aligned}
\text{Minimise } J(u_1, u_2, u_3) = \int_0^T & (A_1 S(t) + A_2 E(t) + A_3 I(t) + A_4 C(t) + A_5 H(t) \\
& + A_6 R(t) + w_1 u_1^2 + w_2 u_2^2 + w_3 u_3^2) d
\end{aligned}
\tag{8.6}
$$

In the above objective function (8.6), $A_i, \ i = 1, 2, ...6$, are weight constants of the respective compartments and $w_j, \ j = 1, 2, 3$ are weight constant of respective control variables. The objective is to minimize the spread of COVID-19 infections from one infected district to another susceptible district.

Our goal is to determine optimal control functions $(u_1^*, u_2^*, u_3^*)$, subject to the system (8.5), such that

$$
J\left(u_1^*, u_2^*, u_3^*\right) = \text{minimise}(J(u_1, u_2, u_3)/(u_1, u_2, u_3) \in \phi) \tag{8.7}
$$

where $\phi = \{(u_1, u_2, u_3)/u_i(t)$ is Lebesgue measurable on $[0, T], \ 0 \le u_i(t) \le 1, \ i = 1, 2, 3\}$ is a control strategy set.

The integrand, $A_1 S(t) + A_2 E(t) + A_3 I(t) + A_4 C(t) + A_5 H(t) + A_6 R(t) + w_1 u_1^2 + w_2 u_2^2 + w_3 u_3^2$ of the objective function (8.6) is convex in the set $\phi$. The control strategy set $\phi$ is also close and convex by definition. Since the model (8.5) is bounded and linear in the control variables, the conditions for the existence of optimal control are satisfied. Hence, here exist $(u_1^*, u_2^*, u_3^*) \in \phi$ such that,

$$J\left(u_1^*, u_2^*, u_3^*\right) = \text{minimise}(J(u_1, u_2, u_3)/(u_1, u_2, u_3) \in \phi).$$

Let convert this optimality problem into a problem of maximizing a Langrangian function $L$, with respect to all control variables. We use Pontryagins maximum principle for the necessary condition of an optimal control problem [15].

$$\begin{aligned}
L = {} & A_1 S^2(t) + A_2 E^2(t) + A_3 I^2(t) + A_4 C^2(t) + A_5 H^2(t) \\
& + A_6 R^2(t) + w_1 u_1^2 + w_2 u_2^2 + w_3 u_3^2 \\
& + \lambda_1(B - \beta_1 SE + \beta_2 E) + \lambda_2(\beta_1 SE - \beta_2 E - \beta_3 EI - \beta_4 EC + \beta_{10} R + u_1 E) \\
& + \lambda_3(\beta_3 EI + \beta_5 C - (\beta_6 + \beta_7 + \mu_d)I - u_1 E + u_2 I) \\
& + \lambda_4(\beta_4 EC + \beta_6 I - (\beta_5 + \beta_8 + \mu_d)C - u_2 I) \\
& + \lambda_5(\beta_7 I + \beta_8 C - (\beta_9 + \mu_d)H - u_3 H) + \lambda_6(\beta_9 H - \beta_{10} R + u_3 H)
\end{aligned}$$

For the given optimal control $u^* = (u_1^*, u_2^*, u_3^*)$ and corresponding state solutions of the corresponding system (8.5), there exist adjoint functions, $\lambda_i$, $i = 1, 2, \ldots 6$, as follow:

$$\lambda_1' = \frac{-\partial L}{\partial S} = -2A_1 S + (\lambda_1 - \lambda_2)\beta_1 E$$

$$\begin{aligned}
\lambda_2' = \frac{-\partial L}{\partial E} = {} & -2A_2 E + (\lambda_1 - \lambda_2)(\beta_1 S - \beta_2) + (\lambda_2 - \lambda_3)(\beta_3 I - u_1) \\
& + (\lambda_2 - \lambda_4)\beta_4 C
\end{aligned}$$

$$\lambda_3' = \frac{-\partial L}{\partial I} = -2A_3 I + (\lambda_2 - \lambda_3)\beta_3 E + (\lambda_3 - \lambda_4)(\beta_6 - u_2) + (\lambda_3 - \lambda_5)\beta_7 + \lambda_3 \mu_d$$

$$\lambda_4' = \frac{-\partial L}{\partial C} = -2A_4 C + (\lambda_2 - \lambda_4)\beta_4 E - (\lambda_3 - \lambda_4)\beta_5 + (\lambda_4 - \lambda_5)\beta_8 + \lambda_4 \mu_d$$

$$\lambda_5' = \frac{-\partial L}{\partial H} = -2A_5 H + (\lambda_5 - \lambda_6)(\beta_9 + u_3) + \lambda_5 \mu_d$$

$$\lambda_6' = \frac{-\partial L}{\partial R} = -2A_6 R - (\lambda_2 - \lambda_6)\beta_{10}$$

Using terminal condition $\lambda_i(T) = 0$, for $i = 1, 2, \ldots 6$ and optimality condition, $-\frac{\partial L}{\partial u_i} = 0$, for $i = 1, 2, 3$, the optimal control variables $u_1^*$, $u_2^*$ and $u_3^*$ are solved.

$$u_1 = \frac{(\lambda_3 - \lambda_2)E}{2w_1}, \; u_2 = \frac{(\lambda_4 - \lambda_3)I}{2w_2} \text{ and } u_3 = \frac{(\lambda_5 - \lambda_6)H}{2w_3} \tag{8.8}$$

Moreover, optimal control strategies $u_1^*$, $u_2^*$ and $u_3^*$ are given by:

$$u_1^* = \max\left\{0, \min\left\{1, \frac{(\lambda_3 - \lambda_2)E}{2w_1}\right\}\right\} \tag{8.9}$$

$$u_2^* = \max\left\{0, \min\left\{1, \frac{(\lambda_4 - \lambda_3)I}{2w_2}\right\}\right\} \tag{8.10}$$

$$u_3^* = \max\left\{0, \min\left\{1, \frac{(\lambda_5 - \lambda_6)H}{2w_3}\right\}\right\} \tag{8.11}$$

## Numerical Simulation

This section attempts to describe the graphical representation of variations in the compartmental model. Also change in compartments under influence of optimal control strategies is observed and analyzed. Parametric values used for simulation are as given in Table 8.1.

Figure 8.3 shows the variation in each compartment of the model with time. The initialization of susceptible, exposed, infected, critically infected, hospitalized, and recovered districts are given by $S(0) = 8$, $E(0) = 6$, $I(0) = 3$, $C(0) = 1.3$, $H(0) = 4$ and $R(0) = 1.03$, respectively. It can be observed in the graph that, hospitalization or medication is required at a very high level in the initial phase of the outbreak, after that as infected and critically infected area decreases after the second week, hospitalization is also decreased subsequently. But then again augmentation in exposed class is observed after the third week, which shows a virus survive in nature for a long time and the possibility of periodic nature of the infection.

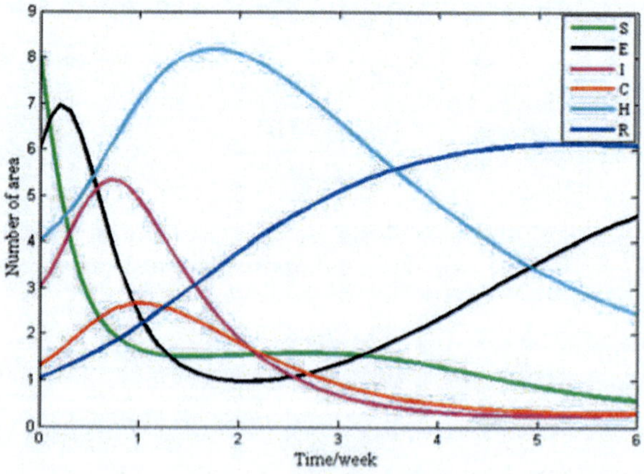


**Fig. 8.3** Oscillations in the compartments during the COVID-19 outbreak

Figure 8.4a shows that the susceptible districts are transformed into exposed districts at a high intensity. Figure 8.4b shows that the exposed district moves toward infected districts which means the exposed district becomes infected quickly as the virus spreads quickly from one district to another district. Figure 8.4c shows the periodic nature of COVID-19, as a critically infected district again becomes exposed after medication and exposed district moves to the critically infected district repeatedly.

Figure 8.5 portraits the three-dimensional relationship between various compartments of the model. Figure 8.5a, b shows the intensity at which infected and critically infected district gets recover after hospitalization and medication respectively. The light and dense graph show hospitalization in the infected area varies repeatedly from high to low but the requirement of hospitalization in the critically infected

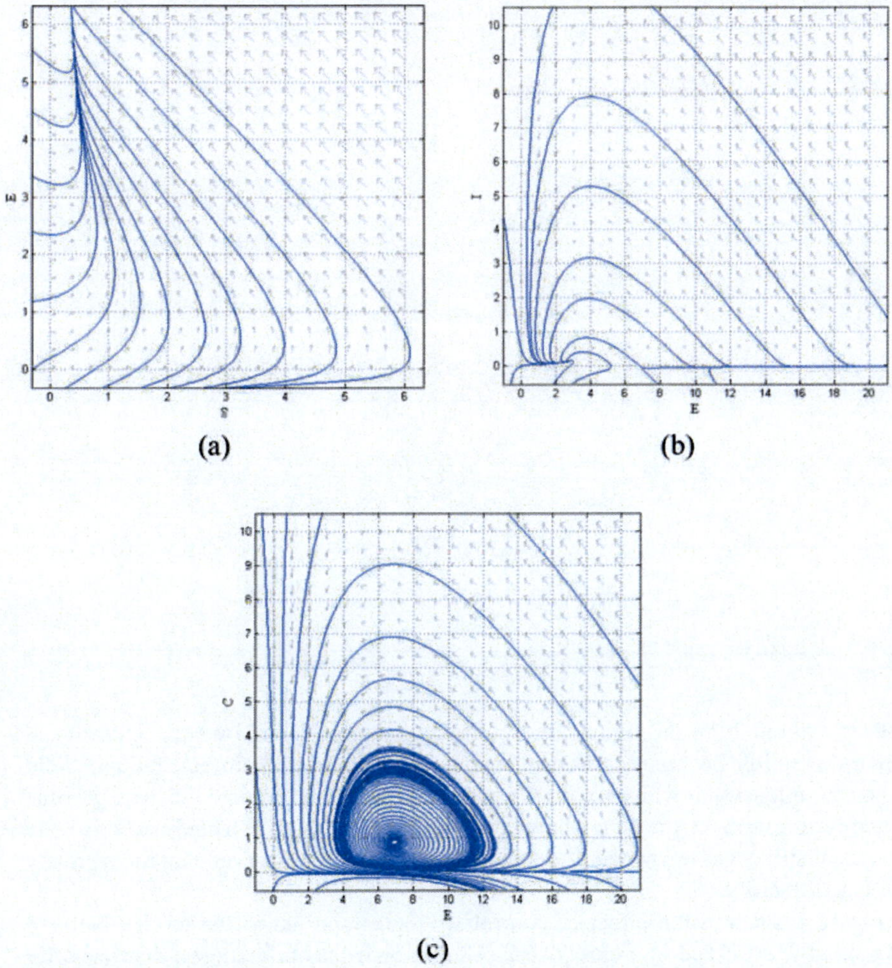


**Fig. 8.4** Planner graph shows the intensity of transmission of the infection

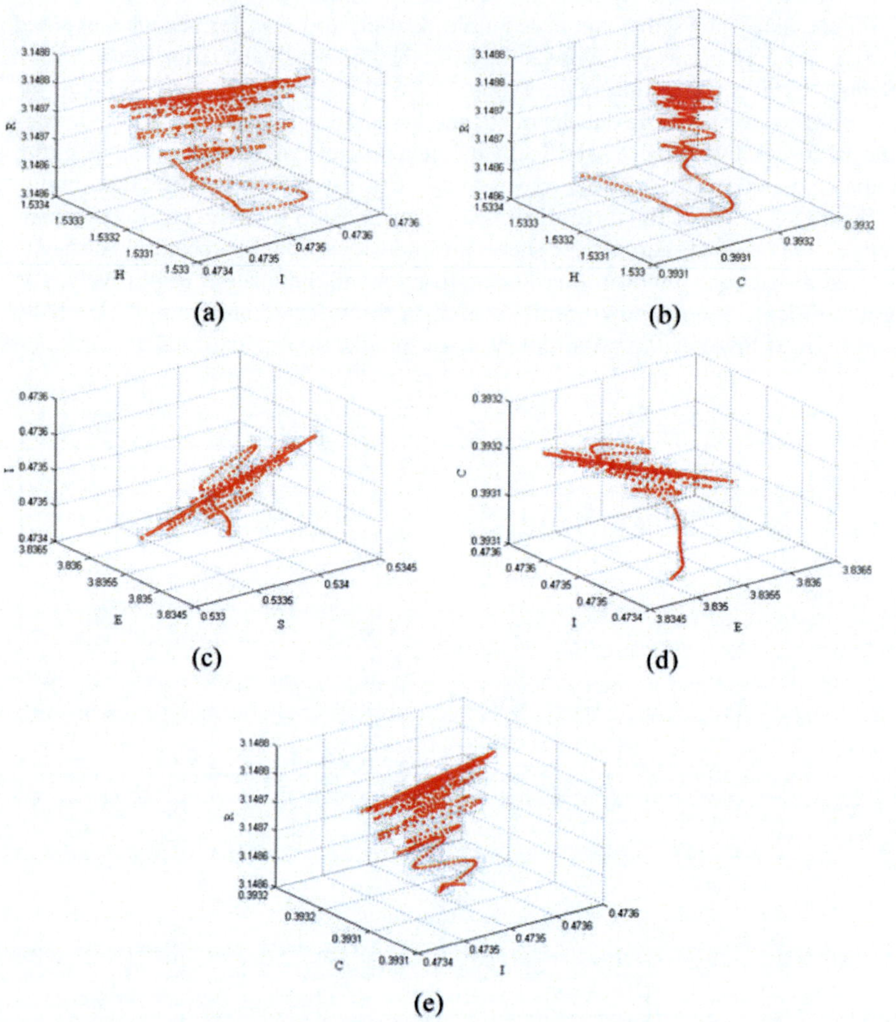


**Fig. 8.5** Scatter diagram of COVID-19 model

district remains high throughout the outbreak. Figure 8.5c shows the intensity at which susceptible become exposed and exposed district become infected. Figure 8.5d shows the intensity at which exposed become infected and infected district become critically infected. Figure 8.5e shows the intensity at which infected and critically infected districts get recovered after medication. It can be observed that the recovery rate is satisfactory.

Figure 8.6 shows the effect of control strategies applied to the model. Notable decrement is observed in a class of susceptible districts and exposed districts after applying control strategies in Fig. 8.6a, b, respectively. Due to the long incubation

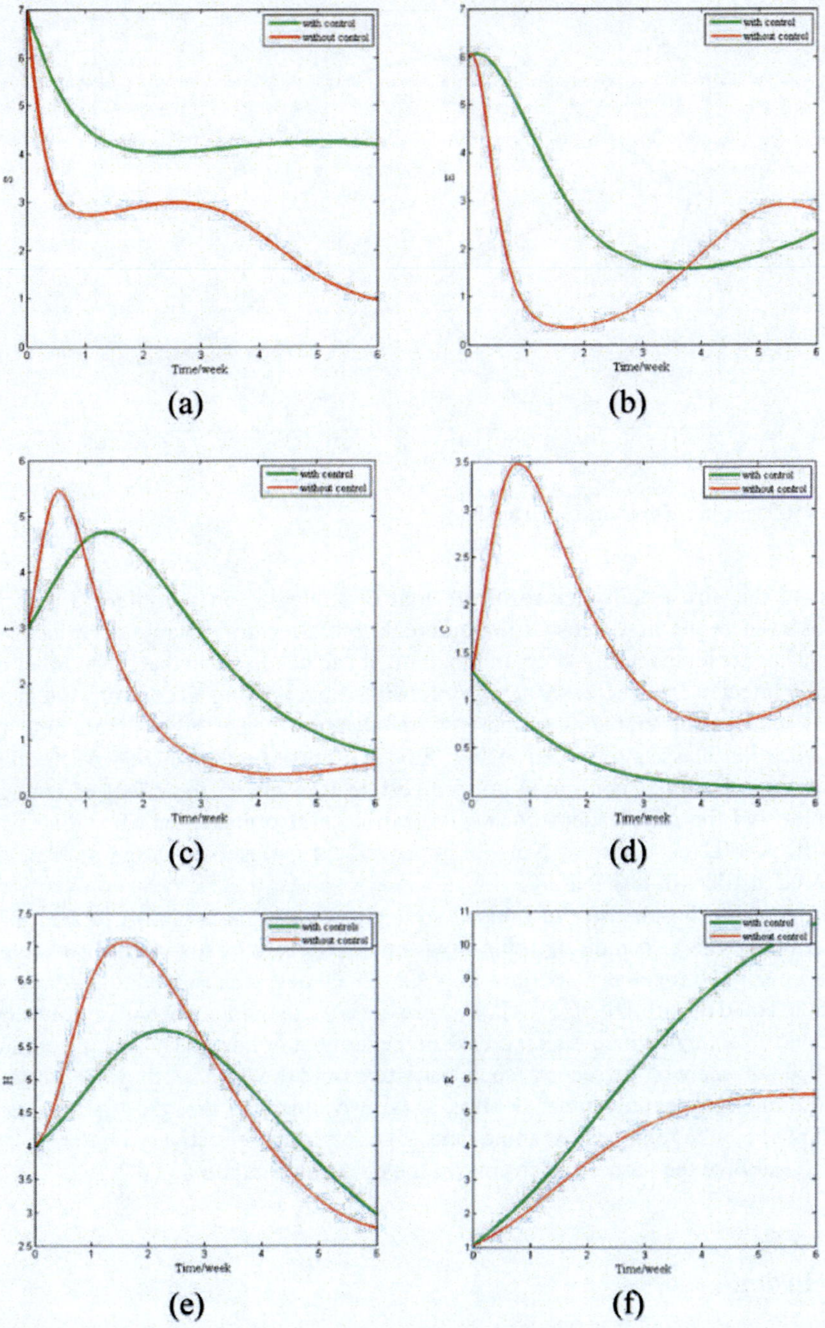


**Fig. 8.6** Variation in each compartment with and without control strategies

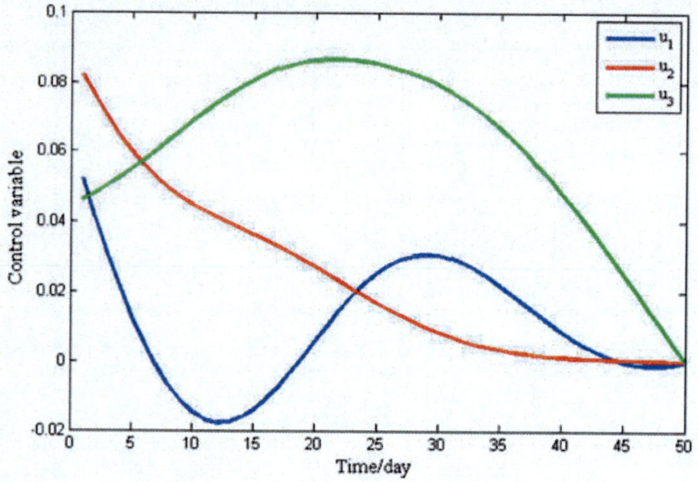


**Fig. 8.7** Change in control variables with time

period of the virus, the quick result of control strategies on the infected area is not observed in the first week of the outbreak, but after that notable decrement is observed as presented in Fig. 8.6c. In Fig. 8.6d, it can be observed that the number of critically infected area decreases at high intensity after applying all control strategies. One of the control strategies supports hospitalization hence highly increment in hospitalization class is observed in Fig. 8.6e during the initial three weeks, after that situation becomes controlled up to a certain level under the effect of control strategies and the requirement of hospitalization is also decreased after the third week of COVID-19 outbreak. Notable improvement is observed in the number of recovered districts in Fig. 8.6f.

Deviation in the intensity of control strategies over time is shown in Fig. 8.7. Furthermore, it is also indicated that how much intensity of the control variables should be applied together to control the COVID-19 outbreak in around 50 days. It should be noted that initially for 25 days, there is a need to implement and improve the third control strategy which is an increase in the number of hospital beds and ventilators. A requirement of the second (social isolation) and the third strategy is observed until the outbreak comes under control. It is also observed that the first strategy (lockdown) can be removed for some days (5–6 days) and effectively implemented again to stabilize the model and to control the epidemic situation.

## Conclusion

In this chapter, a mathematical compartment model is constructed to study and analyze the transmission dynamics of COVID-19 across the districts of India. The study attempts to formulate a basic reproduction number at the district level to depict the effect of one infected districts on the other susceptible districts. The results show that every infected district is infecting almost two other susceptible districts. The basic reproduction number is greater than one, hence it shows the epidemic nature of the disease. Considering the unstable situation of the spread of the virus from one district to another, this study has applied optimal control theory to the proposed model. Among many control strategies, this study specifies the importance of measures like lockdown, quarantine, contact tracing, and health care facilities at various stages of the COVID-19 transmission. The effectiveness of these control measures is analyzed separately and in combination with all six compartments and explained graphically by simulation of the SEICHR model. Numerical simulations of the model suggest that the control strategies play a significant role to minimize the transmission of COVID-19 cases from one district to the other.

**Acknowledgements** All the authors are thankful to DST-FIST file # MSI-097 for technical support to the Department of Mathematics, Gujarat University. The fourth author (AHS) is funded by a Junior Research Fellowship from the Council of Scientific & Industrial Research (file no.-09/070(0061)/2019-EMR-I).

## References

1. Biswas, M. H. A., Haque, M. M., & Mallick, U. K. (2019). Optimal control strategy for the immunotherapeutic treatment of HIV infection with state constraint. *Optimal Control Applications and Methods, 40*(4), 807–818.
2. Diekmann, O., Heesterbeek, J. A. P., & Metz, J. A. (1990). On the definition and the computation of the basic reproduction ratio $R_0$ in models for infectious diseases in heterogeneous populations. *Journal of Mathematical Biology, 28*(4), 365–382.
3. Duhoe, A. A. A., & Toffa, B. A. (2020). COVID-19: A blessing or curse on affected countries and its citizens. *Research Journal in Advanced Social Sciences, 1,* 32–39.
4. Garba, S. M., Gumel, A. B., & Bakar, M. A. (2008). Backward bifurcations in dengue transmission dynamics. *Mathematical Biosciences, 215*(1), 11–25.
5. Giordano, G., Blanchini, F., Bruno, R., Colaneri, P., Di Filippo, A., Di Matteo, A., & Colaneri, M. (2020). Modelling the COVID-19 epidemic and implementation of population-wide interventions in Italy. *Nature Medicine, 26,* 1–6. https://doi.org/10.1038/s41591-020-0883-7.
6. Ivorra, B., Ferrández, M. R., Vela-Pérez, M., & Ramos, A. M. (2020). Mathematical modeling of the spread of the coronavirus disease 2019 (COVID-19) taking into account the undetected infections. The case of China. *Communications in Nonlinear Science and Numerical Simulation, 88,* 105303. https://doi.org/10.1016/j.cnsns.2020.105303
7. Kao, R. R. (2002). The role of mathematical modelling in the control of the 2001 FMD epidemic in the UK. *Trends in Microbiology, 10*(6), 279–286.
8. Keeling, M. J., & Danon, L. (2009). Mathematical modelling of infectious diseases. *British Medical Bulletin, 92*(1), 33–42.
9. Lancet, T. (2020). India under COVID-19 lockdown. *Lancet, 395*(10233), 1315. https://doi.org/10.1016/S0140-6736(20)30938-7
10. Lloyd, A. L. (2001). Destabilization of epidemic models with the inclusion of realistic distributions of infectious periods. *Proceedings of the Royal Society of London. Series B: Biological Sciences, 268*(1470), 985–993.

11. Miller, L. E., Bhattacharyya, R., & Miller, A. L. (2020). Spatial analysis of global variability in Covid-19 burden. *Risk Management and Healthcare Policy., 13,* 519–522.
12. Mohanty, S. K. (2020). Contextualising geographical vulnerability to COVID-19 in India. *The Lancet Global Health, 8*(9), 1104–1105.
13. Mondal, M. K., Hanif, M., & Biswas, M. H. A. (2017). A mathematical analysis for controlling the spread of Nipah virus infection. *International Journal of Modelling and Simulation., 37*(3), 185–197.
14. Peirlinck, M., Linka, K., Costabal, F. S., & Kuhl, E. (2020). Outbreak dynamics of COVID-19 in China and the United States. *Biomechanics and Modeling in Mechanobiology.* https://doi.org/10.1007/s10237-020-01332-5.
15. Pontryagin, L. S. (2018). *Mathematical theory of optimal processes.* Routledge.
16. Postnikov, E. B. (2020). Estimation of COVID-19 dynamics "on a back-of-envelope": Does the simplest SIR model provide quantitative parameters and predictions? *Chaos, Solitons & Fractals, 135,* 109841.
17. Prem, K., Liu, Y., Russell, T. W., Kucharski, A. J., Eggo, R. M., Davies, N., et al. (2020). The effect of control strategies to reduce social mixing on outcomes of the COVID-19 epidemic in Wuhan, China: A modelling study. *The Lancet Public Health, 5*(5), 261–270.
18. Ranney, M. L., Griffeth, V., & Jha, A. K. (2020). Critical supply shortages—The need for ventilators and personal protective equipment during the Covid-19 pandemic. *New England Journal of Medicine, 382*(18), e41. https://doi.org/10.1056/NEJMp2006141.
19. Reiner, R. C., Jr., Perkins, T. A., Barker, C. M., Niu, T., Chaves, L. F., Ellis, A. M., et al. (2013). A systematic review of mathematical models of mosquito-borne pathogen transmission: 1970–2010. *Journal of the Royal Society Interface., 10*(81), 20120921.
20. Sattenspiel, L., Lloyd, A. (2009). *The geographic spread of infectious diseases: models and applications* (Vol. 5). Princeton University Press.
21. Tuite, A. R., Fisman, D. N., & Greer, A. L. (2020). Mathematical modelling of COVID-19 transmission and mitigation strategies in the population of Ontario, Canada. *CMAJ, 192*(19), E497–E505.
22. Wang, M., & Flessa, S. (2020). Modelling Covid-19 under uncertainty: What can we expect? *The European Journal of Health Economics, 21,* 665–668. https://doi.org/10.1007/s10198-020-01202-y.
23. World Health Organization. (2020). WHO coronavirus disease (COVID-19) dashboard. https://covid19.who.int/. Accessed 25 July 2020.
24. Zhang, X., Ma, R., & Wang, L. (2020). Predicting turning point, duration and attack rate of COVID-19 outbreaks in major Western countries. *Chaos, Solitons & Fractals, 135,* 109829 (2020). https://doi.org/10.1016/j.chaos.2020.109829.

# Chapter 9
# Fractional SEIR Model for Modelling the Spread of COVID-19 in Namibia

**Samuel M. Nuugulu, Albert Shikongo, David Elago, Andreas T. Salom, and Kolade M. Owolabi**

**Abstract** In this chapter, a fractional SEIR model and its robust first-order unconditionally convergent numerical method is proposed. Initial conditions based on Namibian data as of 4 July 2020 were used to simulate two scenarios. In the first scenario, it is assumed that the proper control mechanisms for kerbing the spread of COVID-19 are in place. In the second scenario, a worst-case scenario is presented. The worst case is characterised by ineffective COVID-19 control mechanisms. Results indicate that if proper control mechanisms are followed, Namibia can contain the spread of COVID-19 in the country to a lowest level of 1, 800 positive cases by October 2020. However, if no proper control mechanisms are followed, Namibia can hit a potentially unmanageable level of over 14, 000 positive cases by October 2020. From a mathematical point of view, results indicate that the fractional SEIR model and the proposed method are appropriate for modelling the chaotic nature observed in the spread of COVID-19. Results herein are fundamentally important to policy and decision-makers in devising appropriate control and management strategies for curbing further spread of COVID-19 in Namibia.

**Keywords** Fractional calculus · Fractional SEIR model · COVID-19 · Well-posedness · Numerical methods · Stability analysis

S. M. Nuugulu (✉) · A. Shikongo · D. Elago
Department of Mathematics, University of Namibia, Private Bag 13301, Windhoek, Namibia
e-mail: snuugulu@unam.na

A. Shikongo
e-mail: ashikongo@unam.na

D. Elago
e-mail: delago@unam.na

A. T. Salom
CERENA-Polo FEUP, Faculty of Engineering, University of Porto, Porto, Portugal
e-mail: andreas.s@fe.up.pt

K. M. Owolabi
Department of Mathematical Sciences, Federal University of Technology, Akure, Ondo State, Nigeria
e-mail: kmowolabi@futa.edu.ng



N. H. Shah and M. Mittal (eds.), *Mathematical Analysis for Transmission of COVID-19*, Mathematical Engineering, https://doi.org/10.1007/978-981-33-6264-2_9

**Mathematics Subject Classification** 92B05 · 65D15 · 92D25 · 45F05 · 91G50 · 74S40

## Introduction

It is now common knowledge that towards the end of the year 2019 a severe acute respiratory syndrome coronavirus 2 (SARS-CoV-2) emerged in a Wuhan City of China. In [1, 2], for example, it is mentioned that COVID-19 infections are transmitted from one person to the other primarily through saliva droplets or charges from the noses of infected persons when coughing or sneezing, or physical contacts with an infected person and touching of infected surfaces. The World Health Organisation (WHO) declared the new coronavirus disease a public health emergency of international concern on 31 January 2020. By March 2020, the disease had already spread to the rest of the world in a very shorter period of time. Thus, the WHO on the 11 March 2020 declared COVID-19 a global pandemic. Since then, COVID-19 has been claiming many lives on a daily basis and currently (as of 11 July 2020 (14:00 GMT +8)) the hard-hit countries are San Marino with more than 12,380 deaths, Belgium with over 843 deaths, Andorra with over 672 deaths, followed by the UK with over 656 deaths.

COVID-19 is a highly contagious viral disease that is perpetually imposing severe burdens on public health and economies and has thus created chaos across the globe see for example [2–4] and references therein. Since there exists no clinically tested vaccine or proper medication for treating COVID-19, most governments across the world have drawn their attentions to stiffening of precautionary measures such as lockdowns, social distancing protocols, self-isolations, quarantines as well as enforcing basic public health practices of regular washing and sanitisation of hands to help control further spread of the virus, see for example [2, 5]. In Namibia, President Hage Geingob declared a state of emergency on the 17 March 2020 and stipulated some unprecedented measures to help curb further spreading of COVID-19 in the country.

In view of the above-mentioned developments, models for studying dynamical behaviour of COVID-19 have been developed. For instance, Kassa et al. in [6] formulated and analysed a COVID-19 mathematical model with model parameters estimated from available COVID-19 data. The authors' investigation involved a backward bifurcation analysis which is believed to arise when recovered individuals do not develop permanent immunity for the disease, i.e., ($R_o = 1$) and disappear in the absence of re-infection ($R_o < 1$). Fanelli and Piazza in [7] analysed and forecasted the spread of COVID-19 in China, Italy and France using a simple day-lag map points of a simple susceptible-infected-recovery-death (SIRD) integer-orders model.

Mishra et al. in [2] developed a three special compartmental quarantine models, susceptible-immigrant-home isolation-infectious-hospital quarantine-recovered (SIR) model. The authors performed numerical simulations and concluded that hospitals quarantine and home isolations are indispensable forces to control spread of the virus in the absence of treatment and vaccine. Postnikov in [4] studied the dynamics of COVID-19 using a simple susceptible-infectious-recovered (SIR) integer-order model proposed by Kermack and McKendrick in [4]. The author applied the sequential reduction of the SIR to a logistic regression-based equation and applied the resultant model to validating the COVID-19 recent data reported by the European Centre for Disease Prevention and Control.

Ullah and Khan [8] also developed a new mathematical transmission model to explore the transmission dynamics and impact of non-pharmaceutical control of the COVID-19 pandemic in Pakistan. In the first instance, they developed their model without optimal control variables and estimated the model parameters from the reported cases using a nonlinear least square curve fitting technique. In the second case, after reformulating the model, they added two time-dependent control variables, i.e., quarantine, hospitalisation and self-isolation interventions for the infected individual. They reported that their model outputs are in good agreement with the COVID-19 confirmed cases in Pakistan. The infection horizons of COVID-19 estimation using a data-driven approach and an SIR model with a time varying disease transmission rate are studied in [9] and [10], respectively. Moreover, Manotosh et al. in [5] formulated a model of integer order to study the dynamical behaviour of the spread of COVID-19 by quantifying the basic reproductive number to help predict and control further the spread of COVID-19 by introducing quarantine and governmental measures components in the model.

The theory of fractional calculus is more than three centuries old just like classical integer-order calculus, but it was not popular in science [11] and engineering field [12–14] and recent discoveries of fractal geometries in different scientific fields of applications such as finance, love dynamics as well as disease modelling, see for example [15–20] and references therein have justified that fractional calculus-based models are appropriate for handling real-world phenomenons better than their integer-order counterparts.

Fractional calculus epidemiological models provide a general version of the integer-order epidemiological models by replacing integer-order derivatives with corresponding non-integers-order derivatives. Since fractional differential operators are non-local and are often characterised by power processes [18], fading memories [15] as well cross-overs [14] such features make them appropriate in dealing with dynamical real-life phenomena, see for example [15, 17, 18]. Though these features until recently have only been commonly observed in stock price dynamics, underground water modelling, etc., (see for example [17, 18, 21–23] and reference therein), the extension of fractional calculus in this chapter and also in other recent work in [15, 16, 24, 25] to modelling the chaotic nature of the spread of COVID-19 is well appropriate.

Zhang et al. in [25] applied fractional differential equations in modelling the dynamics and mitigation scenarios of COVID-19 for the first time in China. The authors proposed and applied a time-dependent susceptible-exposed-infectious-recovered (SEIR) model to fit and predict the time series of COVID-19 for three months (22 December 2019 to 22 March 2020) data from China. Their validated

SEIR model was then applied further to predict the dynamic behaviour in Japan, Italy, South Korea and the USA.

Atangana in [16] also applied new fractional operators to modelling the spread of COVID-19 as well as to investigating the effect of governments' lockdown protocols. The author considered the possibility of infection of medical personnels by dead bodies during the postmortem procedures or direct contact during the funeral arrangements, removed the transmission rate from dead bodies and incorporated lockdown and social distancing effects using the next generation matrix. Their results indicate that zero basic reproductive number can be achieved if lockdown recommendations are observed properly.

In [15], Alkahtani and Alzaid investigated the stability conditions for a numerical method based on Lagrange polynomial for fractional-orders epidemiological model with eight compartments. Another fractional derivative-based model was also formulated by Khan and Atangana [24] to describe the dynamics of COVID-2019. They applied an Atangana-Baleanu derivative operator to their model as they believe that many properties such as the kernel which is nonlocal and non-singular, and the crossover behaviours within the model are best to explain using Atangana-Baleanu. Their results indicate that the virus is locally asymptotically stable when the reproduction less than a unit. With the given data, they estimated that the basic reproduction number for the given data is $R_o \approx 2.4829$. The Atangana-Baleanu fractional derivative operator-based model was also used by Khan et al. in [26] in studying the dynamics of COVID-19 by incorporating the quarantine and isolations principles in the model formulations.

The exponential growth of the pandemic and chaotic situations it caused globally in the health sector, international trades, travel and tourism, as well as energy sector is undoubtebly significant, see Nuugulu et al. in [23]. The unprecedented effects of the virus have forced governments and private institutions globally to take drastic measures to contain further spread of the virus. Some of those measures implemented globally are travel restrictions, lockdowns, physical distancing and self-isolations among others. While these measures are in place, researchers around the world are now working on different theory and models to understand the dynamics of the spread and possible impact of COVID-19. The primary motivation therein is to assist the policy and decision-makers as well as health officials to plan for healthcare needs and craft out appropriate mitigation strategies as epidemic unfold.

Namibia being a semi-arid country situated in Southern Africa has also not being spared, by the time of finalising this manuscript, Namibia has recorded a total of 25 positive cases, a majority of which are all travel related. Compared to other countries in the region, Namibia has been applauded of its effective efforts in cushioning the unanticipated impacts of the outbreak. Declaration of a state of emergency on the outbreak earlier and locking down the entire country was some of the best strategies the country used in curbing further spread of the virus.

This chapter therefore firstly serves to propose a fractional calculus-based model and its robust numerical method for modelling the spread of COVID-19 in Namibia. Secondly, the study serves to illustrate why there is a need for Namibia to continue on its best trajectory in controlling and managing further spread of COVID-19 as well as set out some policy recommendations for managing the spread of COVID-19 in the country. The rest of the chapter is organised as follows. In Sect. Conceptual Model, we present the conceptual model under consideration, whereas in Sect. Mathematical Analysis of the Conceptual Model, we carry out the analysis of the conceptual model. The numerical method is derived in Sect. Construction of the Numerical Method and the results and discussions are presented in Sect. Results and Discussion. Section Concluding Remarks and Policy Recommendations presents the conclusion and policy recommendations.

## Conceptual Model

The model under consideration is a fractional susceptible-exposed-infected-recovered (SEIR) model. Our classes are defined as follow: susceptible ($S$)—those individuals who are at the risk of infection, exposed ($E$)—those individuals who are already exposed to the virus by being in contact with a person who is infected, infected ($I$)—comprising of individuals who are diagnosed and tested positive for COVID-19 and lastly the recovered individuals in class ($R$). Figure 9.1 shows a graphical representation of the dynamics of the considered SEIR model.

A fractional SEIR model is a generalisation of the well-known classical SEIR model. In this model, we assume that the change in the population follows a super-fractal diffusion process with fractional steps characterised by the Hurst parameter $\alpha \in [1/2, 1)$. The Hurst parameter $\alpha$ is very fundamental in effectively modelling the chaotic nature of the spread of the COVID-19. Assuming that the incubation period is a random variable with exponential distribution with parameter $\vartheta$ (i.e., the average incubation period is $\vartheta^{-1}$) and also assuming the presence of vital dynamics with birth rate $\Lambda$ equal to death rate $\mu$, we get the following model

$$\begin{cases} {}^cD_t^\alpha S_t = \Lambda - \mu S_t - \beta \frac{I_t}{N} S_t, \\ {}^cD_t^\alpha E_t = \beta \frac{I_t}{N} S_t - (\mu + \vartheta) E_t, \\ {}^cD_t^\alpha I_t = \vartheta E_t - (\pi + \mu) I_t, \\ {}^cD_t^\alpha R_t = \pi I_t - \mu R_t, \\ \text{with} \\ S_t(0) = S_0 \geq 0,\ E_t(0) = E_0 \geq 0,\ I_t(0) = I_0 \geq 0,\ R_t(0) = R_0 \geq 0, \ \text{on } [0, T], \end{cases} \tag{9.1}$$

where $t \leq T \in >, \alpha \in [1/2, 1)$, ${}^cD_t^\alpha, \Lambda, \mu, \beta, \vartheta, \pi$ denote fractional derivative in defined in the Caputo sense, influx rate, death rate, infection rate, incubation rate,

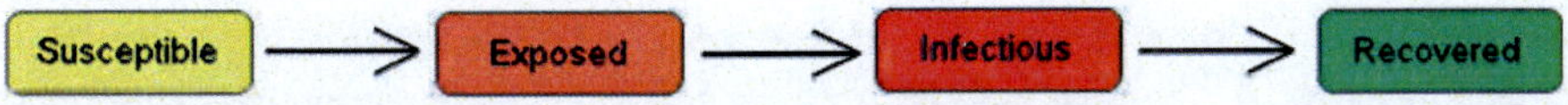


**Fig. 9.1** Graphical representation of the proposed spread of COVID-19. *Source* own

recovery rate, respectively. We further assume that $N = S_t + E_t + I_t + R_t$ denotes the total population, where subscript $t$ denotes time.

## Mathematical Analysis of the Conceptual Model

In this section, we assess the existence and uniqueness of solution to (9.1), as well as its continuous dependency on data, equilibrium points, reproductive number and stability conditions.

### *Existence of Uniqueness of Solution and Continuously Dependency on the Data*

Since the conceptual model in Eq. (9.1) is an initial value problem (IVP), it can be expressed as

$$\mathbf{u}(t) = \mathbf{u}(0) + \frac{1}{\Gamma(\alpha)} \int_0^T (t-s)^{\alpha-1} \mathbf{F}(s, \mathbf{u}(z)) ds, \tag{9.2}$$

where

$$\mathbf{u}(t) = (S_t, E_t, I_t, R_t, )', \mathbf{F} = (f_1, f_2, f_3, f_4)'$$

in which,

$$\begin{cases} f_1 = \Lambda - \mu S_t - \beta \frac{I_t}{m} S_t, \\ f_2 = \beta \frac{I_t}{m} S_t - (\mu + \vartheta) E_t, \\ f_3 = \vartheta E_t - (\pi + \mu) I_t, \\ f_4 = \pi I_t - \mu R_t. \end{cases}$$

Since the IVP in Eq. (9.1) is a system of continuous functions, then it is in the space of continuous function $C[0, T]$, thus, it suffices to define the operator

$$A\mathbf{u} := \mathbf{u}(0) + \frac{1}{\Gamma(\alpha)} \int_0^T (t-s)^{\alpha-1} \mathbf{F}(s, \mathbf{u}(z)) ds, \tag{9.3}$$

such that the following results hold.

**Theorem 1** *Let* $\mathbf{V}$ *denote a nonempty closed subset of a Banach space* $\mathbb{B}$, *such that for every* $M_j \geq 0$, $\sum_{j=0}^{\infty}$ *converges. Moreover, let* $A : \mathbf{V} \to \mathbf{V}$ *denotes a mapping satisfying*

$$\|A^{(j)}\mathbf{u} - A^{(j)}\mathbf{v}\| \leq M_j \|\mathbf{u} - \mathbf{v}\|,$$

*for every* $j \in \gg$ *and* $\mathbf{u}, \mathbf{v} \in \mathbf{V}$.

***Proof*** The proof to these results has already been established in [27], therefore, we conclude that there exist a unique solution to the IVP (9.1). □

Next, we establish results on the continuous dependency of the unique solution to the IVP in (9.1) on the data. To do so, we re-define the IVP in (9.1) as

**Definition 1**

$$D^\alpha(\mathbf{u} - T_0[\mathbf{u}])(t) = \mathbf{F}(t, \mathbf{u}(t)), \text{ with } \mathbf{u}(0) = \mathbf{u}_0, \tag{9.4}$$

where $T_0[\mathbf{u}]$ denotes Taylor polynomial of order 0 for $\mathbf{u}$ centred at origin.

**Theorem 2** *Let* $\mathcal{D} := [0, T] \times [\mathbf{u}_0 - \delta, \mathbf{u}_0 + \delta]$, *for some* $\delta > 0$. *Furthermore, let* $\mathbf{u}, \mathbf{v}$ *denote the unique solutions of*

$$D^\alpha(\mathbf{u} - T_0[\mathbf{u}])(t) = \mathbf{F}(t, \mathbf{u}(t)), \text{ with } \mathbf{u}(0) = \mathbf{u}_0,$$

*and*

$$D^\alpha(\mathbf{v} - T_0[\mathbf{v}])(t) = \mathbf{F}(t, \mathbf{v}(t)), \text{ with } \mathbf{v}(0) = \mathbf{v}_0,$$

*respectively. Then,*

$$\|\mathbf{u} - \mathbf{v}\|_\infty = \mathcal{O}\left(\| \max |\mathbf{u}_0 - \mathbf{v}_0| \|\right),$$

*over any compact interval in which* $\mathbf{u}$ *and* $\mathbf{v}$ *exist.*

**Lemma 1** *Let* $\mathcal{D} := [0, T] \times [\mathbf{u}_0 - \delta, \mathbf{u}_0 + \delta]$, *for some* $\delta > 0$ *such that* $\mathbf{F}, \tilde{\mathbf{F}}$ *are continuous on* $\mathcal{D}$ *and* $f$ *satisfy Lipschitz conditions with respect* $\mathbf{u}$. *Furthermore, let* $\mathbf{u}, \mathbf{v}$ *denote the unique solutions of*

$$D^\alpha(\mathbf{u} - T_0[\mathbf{u}])(t) = \mathbf{F}(t, \mathbf{u}(t)), \text{ with } \mathbf{u}(0) = \mathbf{u}_0,$$

*and*

$$D^\alpha(\mathbf{v} - T_0[\mathbf{v}])(t) = \mathbf{F}(t, \mathbf{v}(t)), \text{ with } \mathbf{v}(0) = \mathbf{v}_0,$$

*respectively. Then,*

$$\|\mathbf{u} - \mathbf{v}\|_\infty = \mathcal{O}\left(\|\mathbf{F} - \tilde{\mathbf{F}}\|\right),$$

*over any compact interval in which* $\mathbf{u}$ *and* $\mathbf{v}$ *exist.*

***Proof*** The proof to Theorem 2 and Lemma 1 has also been established in [27]. □

The uniqueness and data dependency of the model on the data has been demonstrated in the two considered scenarios in Sect. Results and Discussion.

## *Positivity of Solution*

In this section, we simply present that the solution to the IVP in equation (9.1) is biological feasible.

**Theorem 3** *The IVP in Eq.* (9.1) *possesses positive unique solution.*

***Proof*** In view of the total population, we see that

$$ {}^{c}D_t^{\alpha} N(t) = {}^{c}D_t^{\alpha} S_t + {}^{c}D_t^{\alpha} E_t + {}^{c}D_t^{\alpha} I_t + {}^{c}D_t^{\alpha} R_t, $$

which yield

$$ {}^{c}D_t^{\alpha} N(t) = \Lambda - \mu N \geq 0. $$

Therefore, the biological feasible region for the IVP in Eq. (9.1) is

$$ \Gamma := \{(S_t, E_t, I_t, R_t) \in \mathbb{R}_+^4 : S_t, E_t, I_t, R_t \geq 0; \quad S_t + E_t + I_t + R_t = N\}. $$

□

## *Equilibrium Points*

When

$$ {}^{c}D_t^{\alpha} S_t = {}^{c}D_t^{\alpha} E_t = {}^{c}D_t^{\alpha} I_t = {}^{c}D_t^{\alpha} R_t = 0, $$

the IVP in Eq. (9.1) yields the following results.

**Theorem 4** *The fractional SEIR model* (9.1) *possesses at most two equilibrium points, namely the disease free equilibrium point* $(\frac{\Lambda}{\mu}, 0, 0, 0)$ *and an endemic equilibrium point* $(S_t^\star, E_t^\star, I_t^\star, R_t^\star)$*, where*

$$ S_t^\star = \frac{(\vartheta + \mu)(\pi + \mu)}{\beta\vartheta}, E_t^\star = \left(\frac{\beta S_t^\star - (\pi + \mu)}{\mu}\right) I_t^\star, I_t^\star = \left(\frac{\vartheta}{\pi + \mu}\right) E_t^\star, R_t^\star = \left(\frac{\pi\vartheta}{\pi + \mu^2}\right) E_t^\star. $$

***Proof*** In view of the third and fourth equations in Eq. (9.1), we have

$$I_t^{\star} = \left(\frac{\vartheta}{\pi + \mu}\right) E_t^{\star} \text{ and } R_t^{\star} = \left(\frac{\pi}{\mu}\right) I_t^{\star}, \tag{9.5}$$

respectively. From equations in (9.19), we obtain

$$R_t^{\star} = \left(\frac{\pi\vartheta}{\pi + \mu^2}\right) E_t^{\star}, \tag{9.6}$$

in which we get the following cases.

- *Case 1:* When $E_t^{\star} = I_t^{\star} = R_t^{\star} = 0$.
- *Case 2:* When $E_t^{\star} \neq 0$, $I_t^{\star} \neq 0$, and $R_t^{\star} \neq 0$,.

In view of *Case 1:*, we see from the first equation one in (9.1) that

$$\Lambda - \mu S_t - \beta \frac{I_t}{N} S_t = 0, \tag{9.7}$$

whenever, ${}^{c}D_t^{\alpha} S_t = 0$, and $N_t = 1$. Solving for $S_t$ in Eq. (9.7) we find

$$S_t^{\star} = \frac{\Lambda}{\mu}. \tag{9.8}$$

Equation (9.8) imply that the disease free equilibrium point is $\left(\frac{\Lambda}{\mu}, 0, 0, 0\right) \in \Gamma$.

Similarly, for *Case 2:*, we are adding the second equation to the third equation in (9.1) to get

$$(\beta S_t^{\star} - (\pi + \mu))I_t^{\star} - \mu E_t^{\star} = 0. \tag{9.9}$$

Then, substituting equation in (9.19) into equation in (9.24) and simplify further, yields

$$\beta S_t^{\star} \vartheta = (\vartheta + \mu)(\pi + \mu),$$

which is equivalent to

$$S_t^{\star} = \frac{(\vartheta + \mu)(\pi + \mu)}{\beta\vartheta},$$

which conclude the prove. □

### *Reproductive Number*

We understand that the reproductive number, henceforth denoted by $R_o$ is defined as the number of secondary cases that one case would produce in completely susceptible individuals. Therefore, to determine it from the disease free situation

$$S_t^\star \geq \frac{\Lambda}{\mu}. \tag{9.10}$$

Since the endemic equilibrium $S_t^\star$ is

$$S_t^\star = \frac{(\vartheta + \mu)(\pi + \mu)}{\beta\vartheta}, \tag{9.11}$$

then substituting (9.11) into (9.10) we find

$$\frac{(\vartheta + \mu)(\pi + \mu)}{\beta\vartheta} \geq \frac{\Lambda}{\mu}, \tag{9.12}$$

which is equivalent to

$$1 \geq \frac{\Lambda\beta\vartheta}{\mu(\vartheta + \mu)(\pi + \mu)}.$$

Hence, the reproductive number is

$$R_o = \frac{\Lambda\beta\vartheta}{\mu(\vartheta + \mu)(\pi + \mu)}. \tag{9.13}$$

Equation in (9.13) imply that if $R_o < 1$, the disease free equilibrium point is

$$(S_t^\star, E_t^\star, I_t^\star, R_t^\star) = \left(\frac{\Lambda}{\mu}, 0, 0, 0\right) \geq 0,$$

whereas, if $R_o > 1$ then the endemic equilibrium point

$$(S_t^\star, E_t^\star, I_t^\star, R_t^\star) = \left(\frac{(\vartheta + \mu)(\pi + \mu)}{\beta\vartheta}, \left(\frac{\beta S_t^\star - (\pi + \mu)}{\mu}\right) I_t^\star, \left(\frac{\vartheta}{\pi + \mu}\right) E_t^\star, \left(\frac{\pi\vartheta}{\pi + \mu^2}\right) E_t^\star\right) > 0.$$

In the next section, we investigate the endemic equilibrium points with respect to $R_o$.

### *Stability Analysis*

Linearising the system in Eq. (9.1) at the equilibrium point $(S_t^\star, E_t^\star, I_t^\star, R_t^\star)$, we obtain the non-zero entries of the Jacobian matrix as

$$\begin{cases} J(S_t^\star, E_t^\star, I_t^\star, R_t^\star)_{(1,1)} = -\mu - \beta I_t^\star, J(S_t^\star, E_t^\star, I_t^\star, R_t^\star)_{(1,3)} = -\beta I^\star, \\ J(S_t^\star, E_t^\star, I_t^\star, R_t^\star)_{(2,2)} = -(\mu + \vartheta), J(S_t^\star, E_t^\star, I_t^\star, R_t^\star)_{(2,3)} = \beta S^\star, J(S_t^\star, E_t^\star, I_t^\star, R_t^\star)_{(3,2)} = \vartheta, \\ J(S_t^\star, E_t^\star, I_t^\star, R_t^\star)_{(3,3)} = -(\pi + \mu), J(S_t^\star, E_t^\star, I_t^\star, R_t^\star)_{(4,3)} = \pi, J(S_t^\star, E_t^\star, I_t^\star, R_t^\star)_{(4,4)} = -\mu. \end{cases} \tag{9.14}$$

The Jacobian matrix enables us to determine the nature of the disease free equilibrium and endemic free equilibrium. This we establish in the next section.

#### Stability of the Disease Free Equilibrium

At the disease free equilibrium,

$$(S_t^\star, E_t^\star, I_t^\star, R_t^\star) = \left(\frac{\Lambda}{\mu}, 0, 0, 0\right),$$

we see that characteristic equation associated with the Jacobian matrix in Eq. (9.27) is

$$P(\lambda) = \lambda^4 + A\lambda^3 + B\lambda^2 + C\lambda + \mu^2 D, \tag{9.15}$$

where

$$\begin{cases} A = (4\mu + \pi + \vartheta), B = (3\pi\mu + \pi\vartheta - \vartheta\beta + 6\mu^2 + 3\mu\vartheta), \\ C = (3\pi\mu^2 + 2\pi\mu\vartheta - 2\vartheta\beta\mu + 4\mu^3 + 3\mu^3\vartheta), \\ D = (\pi\mu + \pi\vartheta - \vartheta\beta + \mu^2 + \mu\vartheta). \end{cases}$$

Thus, the roots of characteristic polynomials in Eq. (9.15) are

$$\begin{cases} \lambda_1 = -\mu, \ \lambda_2 = -\mu, \\ \lambda_3 = -\frac{\pi}{2} - \mu - \frac{\vartheta}{2} + \frac{\sqrt{\pi^2 - 2\pi\vartheta + 4\vartheta\beta + \vartheta^2}}{2} \\ \lambda_4 = -\frac{\pi}{2} - \mu - \frac{\vartheta}{2} - \frac{\sqrt{\pi^2 - 2\pi\vartheta + 4\vartheta\beta + \vartheta^2}}{2}. \end{cases} \tag{9.16}$$

Since all roots have negative real parts, it implies that a disease free equilibrium point is locally stable.

#### Stability of the Endemic Free Equilibrium

For the endemic free equilibrium, we obtain the following roots to the associated characteristic equation

$$\begin{cases} \lambda_1 = -\mu, \ \lambda_2 = -\mu - \beta\left(\frac{\Lambda\beta\vartheta - \mu^3 - \mu^2\vartheta - \mu^2\pi - \mu\vartheta\pi}{\beta(\mu^2 + \mu\vartheta + \mu\pi + \vartheta\pi)}\right) \\ \lambda_3 = \frac{-\pi}{2} - \mu - \frac{\vartheta}{2} + \frac{\sqrt{\pi^2 - 2\pi\vartheta + 4\vartheta\beta(\frac{(\vartheta+\mu)(\pi+\mu)}{\beta\vartheta}) + \vartheta^2}}{2}, \\ \lambda_4 = \frac{-\pi}{2} - \mu - \frac{\vartheta}{2} - \frac{\sqrt{\pi^2 - 2\pi\vartheta + 4\vartheta\beta(\frac{(\vartheta+\mu)(\pi+\mu)}{\beta\vartheta}) + \vartheta^2}}{2}. \end{cases} \tag{9.17}$$

Since the roots of characteristic equation are real and negative with $R_o > 1$, then the endemic free equilibrium point is locally asymptotically stable. Therefore, the following results follow.

**Theorem 5** *The basic reproduction number $R_o < 1$ is globally stable in the feasible region, whereas, if $R_o > 1$ the unique endemic equilibrium is globally asymptotically stable in the interior of the feasible region.*

***Proof*** The proof to this theorem has already been established in [28]. □

## Construction of the Numerical Method

In this section, we design a robust numerical scheme for solving (9.1). Let $M$ be positive integers and define $k = T/M$ time step-size. Further denote $t_m = mk$; $m = 0, 1, 2, \ldots, M$, such that $t_m \in [0, T]$, then the fractional derivatives in (9.1) can be approximated using the following numerical quadrature as formulated in [17].

We will illustrate below using initial value problem (IVP)

$$\begin{cases} {}^cD_t^\alpha S_t = f(t, S(t)), \\ \\ S_t(0) = S_0, \end{cases} \tag{9.18}$$

where $f(t, S(t)) = \Lambda - \mu S_t - \beta \frac{I_t}{N} S_t$, the equations for the rest of the compartments will be given analogously.

$$
\begin{aligned}
{}^{c}D_t^{\alpha} S_t &= \frac{1}{\Gamma(1-\alpha)} \int_0^{t_m} \frac{\mathrm{d}S_t}{\mathrm{d}t} (t_m - \tau)^{-\alpha} \mathrm{d}\tau, \\
&= \frac{1}{\Gamma(1-\alpha)} \sum_{j=0}^{m} \int_{jk}^{jk} \left( \frac{S_{j+1} - S_j}{k} + \mathcal{O}(k) \right) (mk - \tau)^{-\alpha} \mathrm{d}\tau, \\
&= \frac{1}{\Gamma(1-\alpha)} \sum_{j=0}^{m} \left( \frac{S_{j+1} - S_j}{k} + \mathcal{O}(k) \right) \left( \frac{(mk - (j+1)k)^{1-\alpha} - (mk - jk)^{1-\alpha}}{1-\alpha} \right), \\
&= \frac{1}{\Gamma(1-\alpha)} \frac{1}{1-\alpha} \sum_{j=0}^{m} \left( \frac{S_{j+1} - S_j}{k} + \mathcal{O}(k) \right) \left[ (m-j+1)^{1-\alpha} - (m-j)^{1-\alpha} \right] k^{1-\alpha}, \\
&= \frac{1}{(\alpha-1)!} \frac{1}{k^{\alpha}} \sum_{j=0}^{m} [S_{j+1} - S_j] \left[ (m-j+1)^{1-\alpha} - (m-j)^{1-\alpha} \right] \\
&\quad + \frac{1}{(\alpha-1)!} \sum_{j=0}^{m} \left[ (m-j+1)^{1-\alpha} - (m-j)^{1-\alpha} \right] \mathcal{O}(k) k^{1-\alpha}, \\
&= \frac{1}{(\alpha-1)!} \frac{1}{k^{\alpha}} \sum_{j=0}^{m} (S_{j+1} - S_j) \left[ (m-j+1)^{1-\alpha} - (m-j)^{1-\alpha} \right] \\
&\quad + \frac{1}{(\alpha-1)!} \sum_{j=0}^{m} \left[ (m-j+1)^{1-\alpha} - (m-j)^{1-\alpha} \right] \mathcal{O}(k^{2-\alpha}).
\end{aligned} \tag{9.19}
$$

Shifting the indices in (9.19), we obtain

$$
\begin{aligned}
{}^{c}D_t^{\alpha} S_t = \frac{1}{(\alpha-1)!} \frac{1}{k^{\alpha}} \sum_{j=0}^{m} (S_{j+1} - S_j) \left[ (j+1)^{1-\alpha} - j^{1-\alpha} \right] \\
+ \frac{1}{(\alpha-1)!} \sum_{j=0}^{m} \left[ (j+1)^{1-\alpha} - j^{1-\alpha} \right] \mathcal{O}(k^{2-\alpha}).
\end{aligned} \tag{9.20}
$$

Lct,

$$
\Psi := \frac{1}{(\alpha-1)!} \frac{1}{k^{\alpha}}, \tag{9.21}
$$

and

$$
\gamma_j := (j+1)^{1-\alpha} - j^{1-\alpha}; \quad j = 0, 1, \ldots, m, \tag{9.22}
$$

such that $1 = \gamma_0 > \gamma_1 > \cdots > \gamma_m \to 0$. Substituting $\Psi$ and $\gamma_j$ into (9.20) yield

$$
\begin{aligned}
{}^{c}D_t^{\alpha}S_t &= \Psi\sum_{j=0}^{m}\gamma_j\left(S_{j+1}-S_j\right)+\frac{1}{(\alpha-1)!}\sum_{j=0}^{m}\gamma_j\mathcal{O}(k^{2-\alpha}),\\
&= \Psi\sum_{j=0}^{m}\gamma_j\left(S_{j+1}-S_j\right)+\frac{1}{(\alpha-1)!}n^{1-\alpha}\mathcal{O}(k^{2-\alpha}),\\
&= \Psi\sum_{j=0}^{m}\gamma_j\left(S_{j+1}-S_j\right)+\frac{1}{(\alpha-1)!}\left(\frac{t_m}{k}\right)^{1-\alpha}\mathcal{O}(k^{2-\alpha}),\\
&= \Psi\sum_{j=0}^{m}\gamma_j\left(S_{j+1}-S_j\right)+\frac{t_m^{1-\alpha}}{(\alpha-1)!}k.
\end{aligned}
\tag{9.23}
$$

Therefore, ${}^{c}D_t^{\alpha}S_t$ in (9.1)

$$
{}^{c}D_t^{\alpha}S_t = \Psi\sum_{j=0}^{m}\gamma_j(S_{j+1}-S_j)+\mathcal{O}(k). \tag{9.24}
$$

Similarly, the rest of the compartments are

$$
\begin{cases}
{}^{c}D_t^{\alpha}E_t = \Psi\sum_{j=0}^{m}\gamma_j(E_{j+1}-E_j)+\mathcal{O}(k),\\
{}^{c}D_t^{\alpha}I_t = \Psi\sum_{j=0}^{m}\gamma_j(I_{j+1}-I_j)+\mathcal{O}(k),\\
{}^{c}D_t^{\alpha}R_t = \Psi\sum_{j=0}^{m}\gamma_j(R_{j+1}-R_j)+\mathcal{O}(k).
\end{cases}
\tag{9.25}
$$

Substituting expressions in Eq. (9.24) into (9.1) and neglecting the error terms of $\mathcal{O}(k)$, equation in (9.1) simplifies to

$$
\begin{cases}
\sum_{j=0}^{m}\gamma_j(S_{j+1}-S_j) = \hat{\Lambda}-\hat{\mu}S_m-\hat{\beta}\frac{I_m}{N}S_m,\\
\\
\sum_{j=0}^{m}\gamma_j(E_{j+1}-E_j) = \hat{\beta}\frac{I_m}{N}S_m-(\hat{\mu}+\hat{\vartheta})E_m,\\
\\
\sum_{j=0}^{m}\gamma_j(I_{j+1}-I_j) = \hat{\vartheta}E_m-(\hat{\pi}+\hat{\mu})I_m,\\
\\
\sum_{j=0}^{m}\gamma_j(R_{j+1}-R_j) = \hat{\pi}I_m-\hat{\mu}R_m,
\end{cases}
\tag{9.26}
$$

whereby $\hat{\Lambda}=\frac{\Lambda}{\Psi}$, $\hat{\mu}=\frac{\mu}{\Psi}$, $\hat{\beta}=\frac{\beta}{\Psi}$, $\hat{\vartheta}=\frac{\vartheta}{\Psi}$, $\hat{\pi}=\frac{\pi}{\Psi}$.

By expanding the left hand-side of (9.26), and shifting indices yield to

$$\begin{cases} \sum_{j=0}^{m} \gamma_j (S_{j+1} - S_j) = S_{m+1} - \gamma_m S_0 + \sum_{j=0}^{m-1} \varphi_j S_{m-j}, \\ \sum_{j=0}^{m} \gamma_j (E_{j+1} - E_j) = E_{m+1} - \gamma_m E_0 + \sum_{j=0}^{m-1} \varphi_j E_{m-j}, \\ \sum_{j=0}^{m} \gamma_j (I_{j+1} - I_j) = I_{m+1} - \gamma_m I_0 + \sum_{j=0}^{m-1} \varphi_j I_{m-j}, \\ \sum_{j=0}^{m} \gamma_j (R_{j+1} - R_j) = R_{m+1} - \gamma_m R_0 + \sum_{j=0}^{m-1} \varphi_j R_{m-j}, \end{cases} \tag{9.27}$$

where $\varphi_j = \gamma_j - \gamma_{j+1}, \quad j = 0, 1, \cdots, m.$

***Remark 1*** The following observations are trivial to show.

$$\begin{cases} 1 = \gamma_0 > \gamma_1 > \cdots \gamma_m \to 0, \\ \varphi_0 = 1 - \gamma_1, \\ \sum_{j=0}^{m-1} \varphi_j = 1 + \gamma_m, \\ \sum_{j=0}^{\infty} \varphi_j = 1 > 1 - (2^{1-\alpha} - 1^{1-\alpha}) = 2 - 2^{1-\alpha} = \varphi_0 > \varphi_1 > \cdots \to 0. \end{cases} \tag{9.28}$$

The $\sum_{j=0}^{m-1}(.)$ components in (9.27) represent the long memory effects in the COVID-19 data, which is justified by the dependence between new and the existing cases. Therefore, substituting (9.27) into (9.26) yield

$$\begin{cases} S_{m+1} = \gamma_m S_0 + \hat{\Lambda} - \hat{\mu} S_m - \hat{\beta} \frac{I_m}{N} S_m - \sum_{j=0}^{m-1} \varphi_j S_{m-j}, \\ E_{m+1} = \gamma_m E_0 + \hat{\beta} \frac{I_m}{N} S_m - (\hat{\mu} + \hat{\vartheta}) E_m - \sum_{j=0}^{m-1} \varphi_j E_{m-j}, \\ I_{m+1} = \gamma_m I_0 + \hat{\vartheta} E_m - (\hat{\pi} + \hat{\mu}) I_m - \sum_{j=0}^{m-1} \varphi_j I_{m-j}, \\ R_{m+1} = \gamma_m R_0 + \hat{\pi} I_m - \hat{\mu} R_m - \sum_{j=0}^{m-1} \varphi_j R_{m-j}. \end{cases} \tag{9.29}$$

Equation (9.29) can be expanded into

$$\begin{cases} S_{m+1} = \gamma_m S_0 + \hat{\Lambda} - \varphi_0 \hat{\mu} S_m - \hat{\beta} \frac{I_m}{N} S_m - (\varphi_1 S_{m-1} + \cdots + \varphi_{m-2} S_2 + \varphi_{m-1} S_1), \\ E_{m+1} = \gamma_m E_0 + \hat{\beta} \frac{I_m}{N} S_m - (\hat{\mu} + \hat{\vartheta} + \varphi_0) E_m - (\varphi_1 E_{m-1} + \varphi_2 E_{m-2} \cdots + \varphi_{m-2} E_2 + \varphi_{m-1} E_1), \\ I_{m+1} = \gamma_m I_0 + \hat{\vartheta} E_m - (\hat{\pi} + \hat{\mu} + \varphi_0) I_m - (\varphi_1 I_{m-1} + \varphi_2 I_{m-2} \cdots + \varphi_{m-2} I_2 + \varphi_{m-1} I_1), \\ R_{m+1} = \gamma_m R_0 + \hat{\pi} I_m - (\hat{\mu} - \varphi_0) R_m - (\varphi_1 R_{m-1} + \varphi_2 R_{m-2} \cdots + \varphi_{m-2} R_2 + \varphi_{m-1} R_1). \end{cases} \tag{9.30}$$

The above scheme (9.30) was implemented as an iterative process using MATLAB.

## *Analysis of the Numerical Method*

The exact solution of this problem is not available and in order to calculate the maximum pointwise error and rate of convergence, we use the double mesh principle. We define the double mesh principle as follow;

$$\epsilon_C^M = \max_{x_i \in \bar{D}_C^M} \left| U^M(x_j) - U^{2M}(x_j) \right|, \tag{9.31}$$

and

$$\epsilon^M = \max_C \epsilon_C^M, \tag{9.32}$$

where $\bar{D}_C^M$ is the domain with $U^M(x_j)$ and $U^{2M}(x_j)$ denoting numerical solutions obtained using $M$ and $2M$ mesh intervals, respectively (Table 9.1).

Furthermore, the robust orders of convergence are computed using

$$r = \log_2 \left( \frac{\epsilon_C^M}{\epsilon_C^{2M}} \right). \tag{9.33}$$

We will use (9.32) and (9.33), respectively, to compute the maximum absolute errors and orders of convergence of the numerical scheme in (9.30). The stability and convergence results to the two scenarios presented in Sect. Results and Discussion are appearing in Tables 9.2, 9.3, 9.5, 9.6 and 9.4.

**Table 9.1** Parameter values for Example 1

| Birth rate ($\lambda$) | Death rate ($\mu$) | Infection rate ($\beta$) | Incubation period ($1/\vartheta$) | Recovery rate ($\pi$) |
|---|---|---|---|---|
| 0.013 | 0.013 | 0.06 | 14 | 0.98 |

*source* own

**Table 9.2** Maximum absolute errors for Example 1 with $\Lambda = 0.013$, $\mu = 0.013$, $\beta = 0.06$, $1/\vartheta = 14$ and $\pi = 0.98$

| $\alpha$ | $M = 100$ | $M = 200$ | $M = 400$ | $M = 800$ | $M = 1600$ |
|---|---|---|---|---|---|
| 0.1 | 6.1152e−02 | 3.1069e−02 | 1.5659e−02 | 7.8606e−03 | 3.9381e−03 |
| 0.2 | 5.9707e−02 | 3.0276e−02 | 1.5245e−02 | 7.6491e−03 | 3.8312e−03 |
| 0.3 | 5.7925e−02 | 2.9330e−02 | 1.4758e−02 | 7.4021e−03 | 3.7069e−03 |
| 0.4 | 5.5896e−02 | 2.8272e−02 | 1.4218e−02 | 7.1294e−03 | 3.5698e−03 |
| 0.5 | 5.3692e−02 | 2.7136e−02 | 1.3641e−02 | 6.8388e−03 | 3.4240e−03 |
| 0.6 | 5.1370e−02 | 2.5948e−02 | 1.3040e−02 | 6.5369e−03 | 3.2726e−03 |
| 0.7 | 4.8978e−02 | 2.4731e−02 | 1.2427e−02 | 6.2289e−03 | 3.1183e−03 |
| 0.8 | 4.6552e−02 | 2.3503e−02 | 1.1809e−02 | 5.9190e−03 | 2.9631e−03 |
| 0.9 | 4.4121e−02 | 2.2277e−02 | 1.1194e−02 | 5.6107e−03 | 2.8088e−03 |
| 1.0 | 4.1705e−02 | 2.1064e−02 | 1.0586e−02 | 5.3066e−03 | 2.6567e−03 |

*Source* own

**Table 9.3** Convergence rates for Example 1 with $\Lambda = 0.013$, $\mu = 0.013$, $\beta = 0.06$, $1/\vartheta = 14$ and $\pi = 0.98$

| $\alpha$ | $M = 200$ | $M = 400$ | $M = 800$ | $M = 1600$ |
|---|---|---|---|---|
| 0.1 | 0.98 | 0.99 | 0.99 | 1.00 |
| 0.2 | 0.98 | 0.99 | 0.99 | 1.00 |
| 0.3 | 0.98 | 0.99 | 1.00 | 1.00 |
| 0.4 | 0.98 | 0.99 | 1.00 | 1.00 |
| 0.5 | 0.98 | 0.99 | 1.00 | 1.00 |
| 0.6 | 0.99 | 0.99 | 1.00 | 1.00 |
| 0.7 | 0.99 | 0.99 | 1.00 | 1.00 |
| 0.8 | 0.99 | 0.99 | 1.00 | 1.00 |
| 0.9 | 0.99 | 0.99 | 1.00 | 1.00 |
| 1.0 | 0.99 | 0.99 | 1.00 | 1.00 |

*Source* own

**Table 9.4** Parameter values for Example 1

| Birth rate ($\lambda$) | Death rate ($\mu$) | Infection rate ($\beta$) | Incubation period ($1/\vartheta$) | Recovery rate ($\pi$) |
|---|---|---|---|---|
| 0.023 | 0.023 | 0.10 | 7 | 0.70 |

*Source* own

**Table 9.5** Maximum absolute errors for Example 1 with $\Lambda = 0.023$, $\mu = 0.023$, $\beta = 0.10$, $1/\vartheta = 7$ and $\pi = 0.70$

| $\alpha$ | $M = 100$ | $M = 200$ | $M = 400$ | $M = 800$ | $M = 1600$ |
|---|---|---|---|---|---|
| 0.1 | 1.5661e−01 | 8.1330e−02 | 4.1509e−02 | 2.0978e-02 | 1.0547e−02 |
| 0.2 | 1.3930e−01 | 7.2084e−02 | 3.6713e−02 | 1.8533e-02 | 9.3120e−03 |
| 0.3 | 1.2548e−01 | 6.4751e−02 | 3.2926e−02 | 1.6607e-02 | 8.3404e−03 |
| 0.4 | 1.1445e−01 | 5.8938e−02 | 2.9934e−02 | 1.5088e-02 | 7.5750e−03 |
| 0.5 | 1.0574e−01 | 5.4365e−02 | 2.7586e−02 | 1.3898e-02 | 6.9759e−03 |
| 0.6 | 9.8970e−02 | 5.0832e−02 | 2.5777e−02 | 1.2983e-02 | 6.5152e−03 |
| 0.7 | 9.3914e−02 | 4.8203e−02 | 2.4435e−02 | 1.2304e-02 | 6.1741e−03 |
| 0.8 | 9.0418e−02 | 4.6397e−02 | 2.3516e−02 | 1.1840e-02 | 5.9410e−03 |
| 0.9 | 8.8420e−02 | 4.5378e−02 | 2.3001e−02 | 1.1581e-02 | 5.8109e−03 |
| 1.0 | 8.7951e−02 | 4.5161e−02 | 2.2896e−02 | 1.1530e-02 | 5.7857e−03 |

*Source* own

**Table 9.6** Convergence rates for Example 1 with $\Lambda = 0.023$, $\mu = 0.023$, $\beta = 0.10$, $1/\vartheta = 7$ and $\pi = 0.70$

| $\alpha$ | $M = 200$ | $M = 400$ | $M = 800$ | $M = 1600$ |
|---|---|---|---|---|
| 0.1 | 0.95 | 0.97 | 0.98 | 0.97 |
| 0.2 | 0.95 | 0.97 | 0.99 | 0.99 |
| 0.3 | 0.95 | 0.98 | 0.99 | 0.99 |
| 0.4 | 0.96 | 0.98 | 0.99 | 0.99 |
| 0.5 | 0.96 | 0.98 | 0.99 | 0.99 |
| 0.6 | 0.96 | 0.98 | 0.99 | 0.98 |
| 0.7 | 0.97 | 0.98 | 0.99 | 0.99 |
| 0.8 | 0.97 | 0.98 | 0.99 | 0.99 |
| 0.9 | 0.97 | 0.98 | 0.99 | 0.99 |
| 1.0 | 0.97 | 0.98 | 0.99 | 0.99 |

*Source* own

## Results and Discussion

In this section, we consider two scenarios of the spread of COVID-19 in Namibia. In Example 1, we consider a case when proper quarantining, and lockdown regulations are followed and in Example 1, we consider a case where the government does not effectively implement and enforce the quarantining protocols to those individuals who travelled from outside the country, for example, truck drivers and essential goods suppliers. Apart from the opening of the Namibian borders to essential goods and services providers to and from Namibia, the proposition of Example 1 was also necessitated by the envisaged opening of Namibian borders to tourists by July 2020. These two reasons are believed to be key risk factors to importation of COVID-19 infections into the country. The projections herein are made from July 2020 up to

July 2021 under the condition that no effective and affordable vaccine or cure for COVID-19 is made available until then.

To simulate our model under the two considered scenarios, the generic initial conditions to the model are that the susceptible population is the country's entire population, i.e. ($S_0 = 2,200,000$ people), exposed individuals ($E_0 = 1680$ people), infected individuals ($I_0 = 202$ people) and recovered individuals ($R_0 = 27$ people).

***Remark 1*** In this scenario, the following model parameters were used, birth rate ($\Lambda = 0.013$), death rate ($\mu = 0.013$), infection rate ($\beta = 0.06$), incubation period ($1/\vartheta = 14$) and recovery rate ($\pi = 0.98$).

*Example 1* In this scenario, we assume the violation of established quarantine protocols, inconsistencies in management of positive cases, no proper contact tracing and early isolations. With these assumptions in place, the following model parameters were used, birth rate ($\Lambda = 0.023$), death rate ($\mu = 0.023$), infection rate ($\beta = 0.10$), incubation period ($1/\vartheta = 7$) and recovery rate ($\pi = 0.70$).

General results as observed in Figs. 9.2 and 9.4 indicate that the fractional calculus approach proposed in this chapter is highly predictive for when $0 < \alpha \leq 0.5$ and conservative when $0.5 < \alpha < 1$. These observations substantiate the use of fractional order derivatives over full order derivatives. From the numerical point of view, when $0.5 < \alpha < 1$ the order of the fractional derivative operator is more closer or equal to unit, in which case, we have a classical SEIR model. Whereas, when $0 < \alpha \leq 0.5$, we have super-diffusive fractal dynamics characterised by persistent non-Gaussian increments in the underlying process with memory (Figs. 9.2 and 9.4).

In order to show how convergent, stable and robust, the proposed method is in solving the fractional SEIR model presented in (9.1), we present the numerical stability and convergence results in Tables 9.2 and 9.3 for Example 1, as well as Tables 9.5 and 9.6 for Example 1. These results were computed using the double mesh principle described in Sect. Analysis of the Numerical Method. As one can see from the results presented in Tables 9.2 and 9.5, the proposed method is very robust and unconditionally stable for a range of values of $\alpha$ $(0 < \alpha \leq 1)$. The results in Tables 9.3 and 9.6 further indicate that indeed the numerical method is convergent with order ($\mathcal{O}(1)$). Therefore, from the numerical point of view, the fact that the method is unconditionally stable, $\alpha$ can be chosen to be small or large without affecting the order of accuracy the proposed method.

Furthermore, from the application point of view, results from the best case scenario presented in Example 1 indicate that given current (as of 11 July 2020) statistics, if the government continue to exercise proper quarantining of individuals coming from affected countries, isolating the infected individuals, timely tracing contacts of the infected individuals, as well as heavily investing in public health awareness campaigns to sensitise the general public on the danger and impact of COVID-19 to the country, then, the spread of the virus can be reasonably contained at around 1800 positive cases by October 2020. It is further projected in Fig. 9.3a that secondary cases may raise from 10 to around 60 cases from July 2020 and will subside by October 2020 and that the mortality rate will remain substantially below 3% of the infected cases under the same reporting period (see Fig. 9.3b).

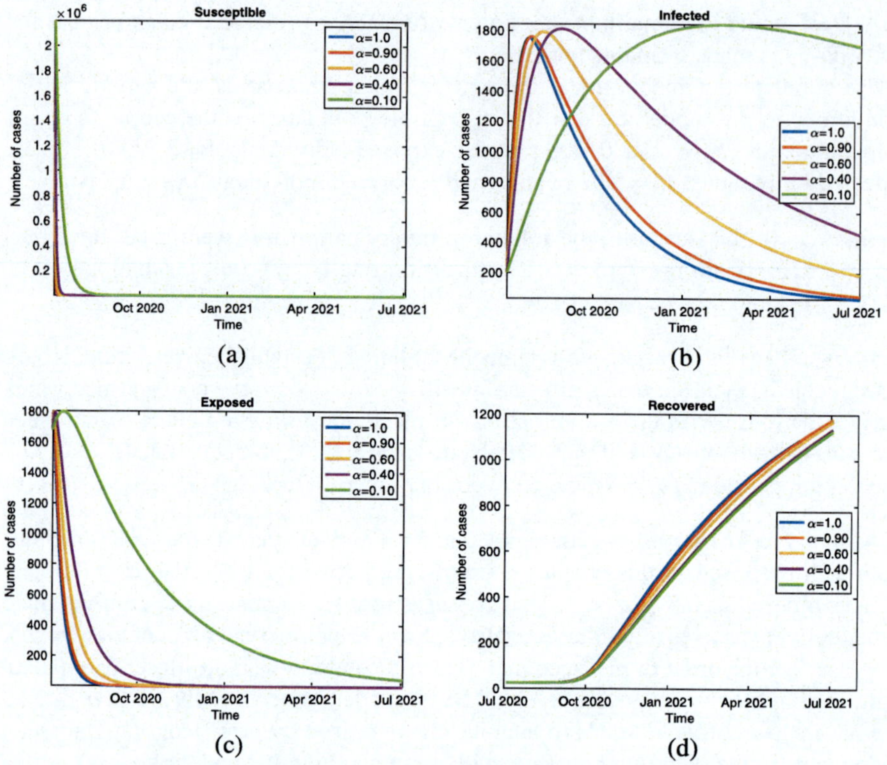


**Fig. 9.2** Numerical solution to Eq. (9.1) under Example 1: **a** susceptible population, **b** infected population, **c** exposed population and **d** recovered population at different values of $\alpha$. Model parameter values are given in Table 9.1. *Source* own

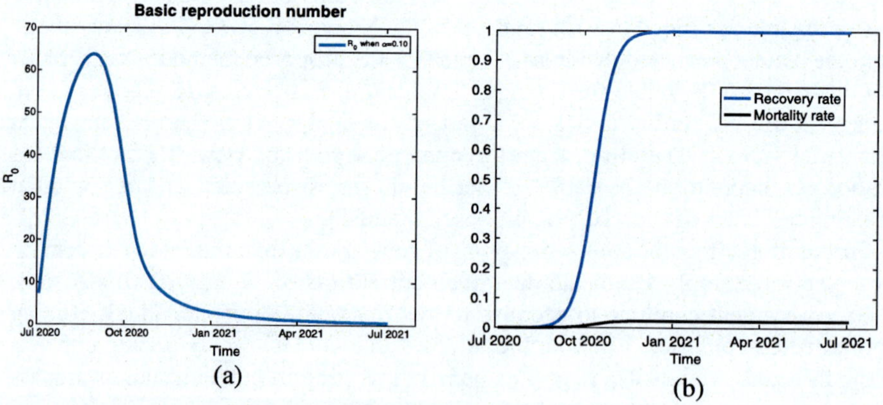


**Fig. 9.3** **a** $R_o$ and **b** projected recovery and mortality rate under Example 1 at $\alpha = 0.10$ for model parameters under Table 9.1. *Source* own

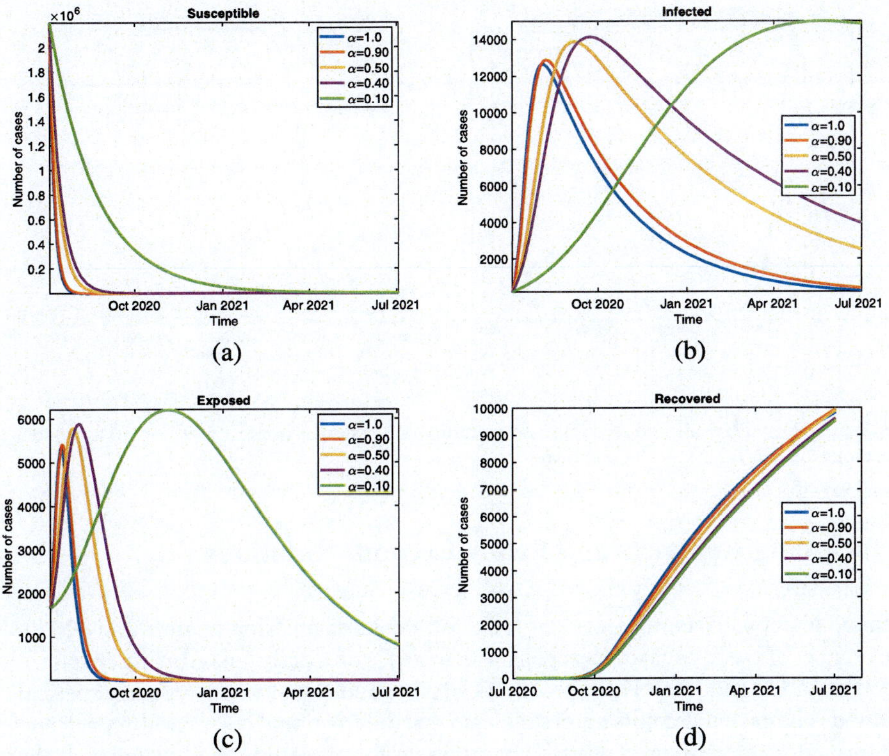


**Fig. 9.4** Numerical solution to Eq. (9.1) under Example 1: **a** susceptible population, **b** infected population, **c** exposed population and **d** recovered population at different values of $\alpha$. Model parameter values are given in Table 9.4. *Source* own

Moreover, if the Namibian government reinforce its efforts in kerbing further spread of the virus, secondary cases are projected to reach zero by July 2021, see Fig. 9.3a and also that by the end of 2020, a very significant proportion of the existing positive cases would have recovered and further spread substantially contained, see Fig. 9.2d. The overall mortality rate as projected in Figs. 9.3b and 9.5b for both scenarios will remain way below a 3% level for the period under investigation.

In the worse-case scenario under Example 1, given the current reported trend of widespread community transmission across the country, our results indicate that if government does not take drastic steps to kerb further community transmissions in major regions of the country, namely Khomas, Erongo as well as a majority of northern regions where a few cases of local transmissions have been reported, and Namibia can potential record over 14, 000 cases of positive COVID-19 infections in a very short period of time, see Fig. 9.4. In the absence of an effective cure or vaccine for COVID-19, our projections will remain valid.

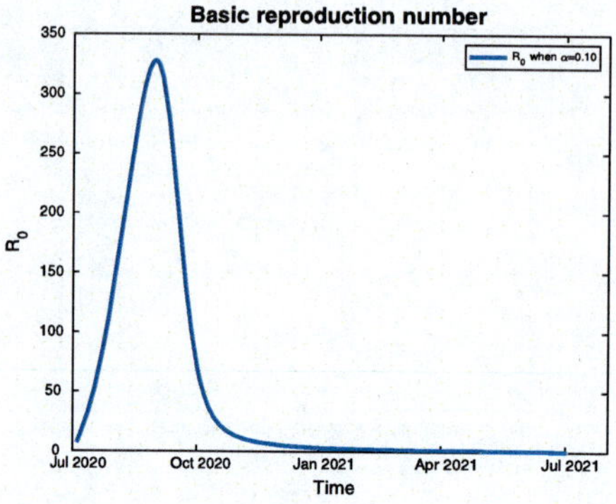


(a)

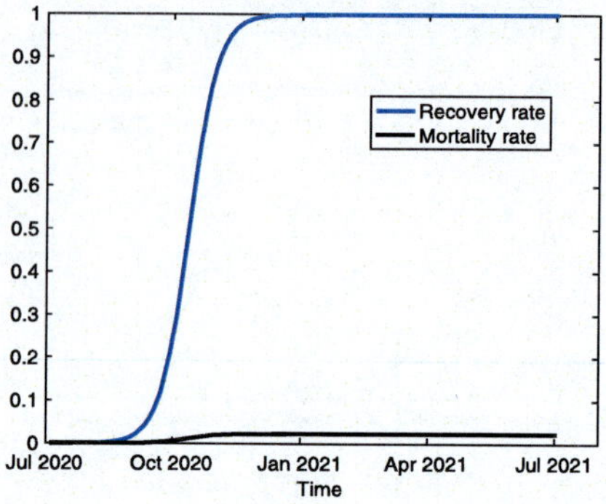


(b)

**Fig. 9.5** **a** $R_o$ and **b** projected recovery and mortality rate under Example 1 at $\alpha = 0.10$ for model parameters under Table 9.4. *Source* own

## Concluding Remarks and Policy Recommendations

Though Namibia did not record any cases of local transmission up until May 2020, the number of positive cases has been on an increase on a daily basis. In this chapter, a fractional SEIR model and its robust first-order unconditionally convergent numerical method is proposed. Results herein indicate that the fractional calculus approach and the numerical method used are well appropriate for modelling the dynamics of the spread of COVID-19. From the numerical point of view, the results indicate that the considered method is robust and unconditionally converges with order 1. From the application point of view, we considered two scenarios characterised by different parameterisation of the fractional SEIR model. The first scenario is regarded as the best case scenario, which is characterised by proper quarantining protocols, effective contact tracing and isolation of positive individuals and their contacts. In the second scenario, we looked at a worst-case scenario characterised by ineffective quarantining and isolation procedures, non-compliance of the general public to the set guidelines for kerbing the spread of the virus and enforcement of the set guidelines by the relevant assigned authorities. The two considered scenarios present similar structural features in terms of profiling the three compartments of the considered SEIR model. However, the spread of the virus is amplified in the worst-case scenario with longer delays in the recovery of the infected individuals.

In an effort to help isolate and kerb further spread of the virus, this current work draws some policy recommendations:

1. It is recommended that government revert back to stage 1 of the declaration of a national health emergency on COVID-19, at least for a period of 14 days in those regions with a high community transmission.
2. Enforce working from home for non-essentially entities.

3. Alcohol outlets should only open on a takeaway basis until a point where the country has attained a full control of the virus.
4. Furthermore, it is recommended that schools and high education institutions should take their teaching and learning to 100% e-learning at least for the remainder for the current academic year 2020.
5. Those schools which are unable to offer their classes online should be allowed to cancel the current academic year.

**Acknowledgements** The authors would like to thank the two anonymous reviewers whose comments and suggestions helped improve this chapter.

## References

1. Djilali, S., & Ghanbari, B. (2020). Coronavirus pandemic: A predictive analysis of the peak outbreak epidemic in South Africa. *Turkey, and Brazil, Chaos, Solitons & Fractals*, *138*, 109971.
2. Mishra, A. M., Purohit, S. D., Owolabi, K. M., & Sharma, Y. D. (2020). A nonlinear epidemiological model considering asymptotic and quarantine classes for SARS CoV-2 Virus. *Chaos, Solitons & Fractals*, *138*, 109953.
3. Hu, Y., Sun, J., Dai, Z., Deng, H., Li, X., Huang, Q., et al. (2020). Prevalence and severity of corona virus disease 2019 (COVID-19): A systematic review and meta-analysis. *Journal of Clinical Virology*, *127*, 104371.
4. Postnikov, E. B. (2020). Estimation of COVID-19 dynamics "on a back-of-envelope": Does the simplest SIR model provide quantitative parameters and predictions? Chaos. *Solitons & Fractals*, *135*, 109841.
5. Manotosh, M., Soovoojeet, J., Swapan, K. N., Anupam, K., Sayani, A., & Kar, T. K. (2020). A model based study on the dynamics of COVID-19: Prediction and control. *Chaos, Solitons & Fractals*, *136*, 109889.
6. Kassa, S. M., Njagarah, J. B. H., & Terefe, Y. A. (2020). Analysis of the mitigation strategies for COVID-19: From mathematical modelling perspective. *Chaos, Solitons & Fractals*, *138*, 109968.
7. Fanelli, D., & Piazza, F. (2020). Analysis and forecast of COVID-19 spreading in China. *Italy and France, Chaos, Solitons & Fractals*, *134*, 109761.
8. Ullah, S., & Khan, M. A. (2020). Modeling the impact of non-pharmaceutical interventions on the dynamics of novel coronavirus with optimal control analysis with a case study, Chaos. *Solitons & Fractals*, *139*, 110075. https://doi.org/10.1016/j.chaos.2020.110075.
9. Barmparis, G. D., & Tsironis, G. P. (2020). Estimating the infection horizon of COVID-19 in eight countries with a data-driven approach. *Chaos, Solitons & Fractals*, *135*, 109842.
10. Willis, M. J., Díaz, V. H. G., Prado-Rubio, O. A., & von Stosch, M. (2020). Insights into the dynamics and control of COVID-19 infection rates, Chaos. *Solitons & Fractals*, *138*, 109937.
11. Amina, R., Shah, K., Asifa, M., Khana, I., & Ullaha, F. (2020). An efficient algorithm for numerical solution of fractional integro-differential equations via Haar wavelet. *Journal of Computational and Applied Mathematics*, *381*, 113028.
12. Alzaid, S. S., & Alkahtani, B. S. T. (2019). Modified numerical methods for fractional differential equations. *Alexandria Engineering Journal*, *58*, 1439–1447.
13. Al-Zhour, Z., Al-Mutairi, N., Alrawajeh, F., & Alkhasawneh, R. (2019). Series solutions for the Laguerre and Lane-Emden fractional differential equations in the sense of conformable fractional derivative. *Alexandria Engineering Journal*, *58*, 1413–1420.
14. Das, S. (2008). *Functional Fractional Calculus for System Identification and Controls*. Berlin, Heidelberg: Springer-Verlag.

15. Alkahtani, B. S. T., & Alzaid, S. S. (2020). A novel mathematics model of covid-19 with fractional derivative. *Stability and Numerical Analysis, Chaos, Solitons & Fractals, 138*, 110006.
16. Atangana, A. (2020). Modelling the spread of COVID-19 with new fractal-fractional operators: Can the lockdown save mankind before vaccination. *Chaos, Solitons & Fractals, 136*, 3109860.
17. Nuugulu, S.M., Gideon, F. & Patidar, K. C. (2020). A robust $\theta$-method for a time-space-fractional black-scholes equation, *Springer Proceedings, Article in Press*.
18. Nuugulu, S. M., Gideon, F., & Patidar, K. C. (2020). *An efficient finite difference approximation for a time-fractional black-scholes PDE arising via fractal market hpothesis*. Under Review: Alexandria Engineering Journal.
19. Owolabi, K. M. (2019). Mathematical modelling and analysis of love dynamics: A fractional approach. *Physica A, 525*, 849–865.
20. Prakash, A., & Kaur, H. (2019). Analysis and numerical simulation of fractional order Cahn-Allen model with Atangana-Baleanu derivative. *Chaos, Solitons & Fractals, 124*, 134–142.
21. Nuugulu, S. M., Gideon, F., & Patidar, K. C. (2020). *An efficient numerical method for pricing double barrier options on an underlying asset following a fractal stochastic process*. Under Review: Applied and Computational Mathematics.
22. Nuugulu, S. M., Gideon, F., & Patidar, K. C. (2020). *A robust numerical scheme for a time fractional Black-Scholes partial differential equation describing stock exchange dynamics, Chaos*. Solitons & Fractals: Article In Press.
23. Julius, E. T., Nuugulu, S. M., & Julius L. H. (2020). Estimating the economic impact of Covid-19: a case study of Namibia, *MPRA* (preprint) https://mpra.ub.uni-muenchen.de/99641/.
24. Khan, M. A., & Atangana, A. (2020). Modeling the dynamics of novel coronavirus (2019-nCov) with fractional derivative. *Alexandria Engineering Journal, 59*, 2379–2389. https://doi.org/10.1016/j.aej.2020.02.033.
25. Zhang, Y., Yu, X., Sun, H., Tick, G. R., Wei, W., & Jin, B. (2020). Applicability of time fractional derivative models for simulating the dynamics and mitigation scenarios of COVID-19. *Chaos, Solitons & Fractals, 138*, 109959.
26. Khan, M. A., Atangana, A., Alzahrani, E., & Fatmawati, E. (2020). The dynamics of COVID-19 with quarantined and isolation. *Advances in Difference Equations, 1*, 425. https://doi.org/10.1186/s13662-020-02882-9.
27. Diethelm, K., & Ford, N. J. (2002). Analysis of fractional differential equations. *Journal of Mathematical Analysis and Applications, 265*, 229–248.
28. Li, M. Y., & Wang, L. (2020). Global stability in some seir epidemic models. *The IMA Volumes in Mathematics and its Applications***126**. (2002) (Springer, New York)

# Chapter 10
# Impact of COVID-19 in India and Its Metro Cities: A Statistical Approach

**Radha Gupta, Kokila Ramesh, N. Nethravathi, and B. Yamuna**

**Abstract** The infectious coronavirus disease is spreading at an alarming rate, not only in India but also globally too. The impact of coronavirus disease (COVID- 19) outbreak needs to be analyzed statistically and modelled to know its behaviour so as to predict the same for future. An exhaustive statistical analysis of the data available for the spread of this infection, specifically on the number of positive cases, active cases, death cases and recovered cases, and connection between them could probably suggest some key factors. This has been achieved in this paper by analyzing these four dominant cases. This helped to know the relationship between the current and the past cases. Hence, in this paper, an approach of statistical analysis of COVID-19 data specific to metropolitan cities of India is done. A regression model has been developed for prediction of active cases with 10 lag days in four metropolitan cities of India. The data used for developing the model is considered from 26th April to 31st July (97 days), tested for the month of August. Further, an Artificial Neural Network (ANN) model using back propagation algorithm for active cases for all India and Bangalore has been developed to see the comparison between the two models. This is different from the other existing ANN models as it uses the lag relationships to predict the future scenario. In this case, data is divided into training, validation and testing sets. Model is developed on the training sets and is checked on the validation set, tested on the remaining, and then, it is implemented for prediction.

**Keywords** COVID-19 · Data analysis · ANN modelling · Prediction

---

R. Gupta (✉) · N. Nethravathi · B. Yamuna
Department of Mathematics, Dayananda Sagar College of Engineering, Bangalore, Karnataka, India
e-mail: radha.gaurav.gupta@gmail.com

N. Nethravathi
e-mail: nandukowshik@gmail.com

B. Yamuna
e-mail: yamunabmsc@gmail.com

K. Ramesh
Department of Mathematics, FET, Jain (Deemed-to-be-University), Bangalore, Karnataka, India
e-mail: r.kokila@jainuniversity.ac.in

 
N. H. Shah and M. Mittal (eds.), *Mathematical Analysis for Transmission of COVID-19*, Mathematical Engineering, https://doi.org/10.1007/978-981-33-6264-2_10

**Mathematics Subject Classification** 60G25 · 62J05 · 62M10

## Introduction

The newly discovered coronavirus disease (COVID-19) is infectious disease caused with common symptoms such as fever, tiredness, body aches, nasal congestion, runny nose, sore throat or diarrhoea and dry cough. The outbreak of coronavirus disease (COVID- 19) has happened in China in December 2019, the first case being found in Wuhan. In March, it was declared as pandemic by the World Health Organization (WHO) exposing the world to public health emergencies.

Several preventive measures have been taken by governments' of various countries across the globe to prevent the spread of COVID infection. These preventive measures include social distancing, wearing the masks, face shields, frequent sanitization, quarantine in case of travel or home isolation in case of suspected symptoms. In India, first confirmed case happened to be in Kerala on 30-Jan-2020. The man, who was studying in WUHAN University and had travelled to India, tested positive for the virus. India's first death was confirmed in Karnataka (Kalaburgi) on 12-Mar-2020. The man who returned from Saudi Arabia and had a history of Hypertension, Diabetes and Asthma succumbed to the disease.

Some of the worst affected cities in India are Delhi, Mumbai, Chennai and Bangalore. Bangalore's first case got confirmed on 08-Mar-2020, when a software engineer who returned from Austin, US along with his wife and daughter tested positive. The first death case was of a 70 year old woman from Chikkabalapur in Bangalore, on 24-Mar-2020. She had travelled to Mecca and arrived in India on 14 March.

Chennai's first COVID case was confirmed on 07-Mar-2020, when a man travelled to Chennai from Oman. He was admitted to Hospital on 05-Mar-2020 with complaints of fever and cough, and finally, the reports on 7 March showed that he was positive. The first death cases of Chennai confirmed on 5-Apr-2020, where a 71 year old man from Ramanathpuram and 60 year old man from Washermanpet died at Government hospital, Chennai.

Mumbai's first confirmed case was on 11-Mar-2020. A couple from Andheri had tested positive post their return from a Dubai and Abu Dhabi trip. The first death case was confirmed on 17-Mar-2020, where a 71 year old man with a history of high blood pressure returned from Dubai. He developed pneumonia and inflammation of heart muscles and increased heart rate leading to death.

A 45 years old person from East Delhi, with a history of travel from Italy, was the first confirmed case in Delhi on 02-Mar-2020. Delhi's first death was of a 68 years old woman who got the virus from her son who returns from Switzerland. It was confirmed on 14-Mar-2020.

Researchers are trying to contribute their bit in every possible way that can lead to some solution. While conventional methods are precise and deterministic, artificial intelligence (AI) techniques could give high-quality predictive models. In this study, authors use a publically available dataset from which contains information on positive cases, recovered cases, active cases and death cases in four metropolitan cities of India over 97 days (from 26th April to 31st July 2020). In the present study, a detailed statistical analysis has been performed on the data procured. It is observed that the data has a strong relationship with past days, an autoregressive model for active cases with 10 lag days data as input and an artificial neural network (ANN) model to extract the non-linearity between the data if it exists has been developed and verified both the models in the training as well as testing periods. Needless to say that the methods leading to reliable prediction of spreading of COVID-19 would be of big help in taking preventive measures to minimize its spread, deaths, active cases and maximizing recovery cases.

## Literature

Ahmed [1] did an exhaustive review to understand the epidemiological evidences, clinical manifestations, investigations and treatment given to COVID cases who are admitted in various hospitals of Wuhan city and other parts of China.

Luo et al. [2] made an effort to expand screening capacity, reviewed advances and challenges in the rapid detection of COVID-19 by targeting nucleic acids, antigens or antibodies. They summarized some of the effective treatments and vaccines against COVID-19. They also discussed about possible reduction of viral progression post ongoing clinical trials of interventions.

Poletto et al. [3] published a comment, highlighting some of the important discoveries as a result of predictive modelling to diverse data sources. These results had an impact on clinical and policy decisions.

Through link [4], one will find expert, curated information on SARS-CoV-2 (the novel coronavirus) and COVID-19 (the disease), that will help the research and health community to work together. All these resources are free to access and include clear guidelines for clinicians and patients.

Shah et al. [5] proposed a generalized SEIR model of COVID-19 to study the behaviour of its transmission under different control strategies. This model considered all possible cases, where transmission happens from one human to another and formulated its reproduction number to analyze the accuracy of transmission dynamics of the coronavirus outbreak. Further, they applied optimal control theory to demonstrate the impact of various intervention strategies, people in quarantine and isolation of infected individuals, immunity boosters and hospitalization.

An epidemic model describing its spread in a population was formulated by Arino and Portet [6]. This model considered an Erlang distribution of times of sojourn in incubating, symptomatically and asymptomatically infectious compartments.

Fong et al. [7] proposed a methodology that embraces three virtues, (1) augmentation of existing data, (2) selecting a panel to pick the best forecasting model from many existing models and (3) tweaking the parameters of an individual forecasting model so that the accuracy of data mining is highest possible.

Shah et al. [8] further made an attempt to assess the impact of inter-state, foreign travel and public health interventions imposed by the US Government in response to the COVID-19 pandemic. They developed a disjoint mutually exclusive compartmental to study the transmission dynamics of the coronavirus. Formulation of system of non-linear differential equations, computation of basic reproduction number $R_0$ and stability of the model at the equilibrium points was discussed in detail.

A visionary perspective on data usage and management for infectious diseases is provided by Wong et al. [9]. They highlighted that there is ample opportunity for researchers to make use of artificial intelligence methods to enable reliable and data-oriented disease monitoring in this information age. It is concluded that together with reliable data management platforms AI methods will enable effective analysis of infectious disease. It will also provide surveillance data to support risk and resource analysis for government agencies, healthcare service providers and medical professionals in the future.

Dey [10] developed a time series model for number of total infected cases in India, considering data from 3rd to 7th March 2020. They had developed two models during the initial days of COVID which were discarded because they lost their statistical validity. But later on they developed another model as a third degree polynomial that has remained stable since 8 Apr , with $R^2 > 0.998$ consistently. This model is used for forecasting total number of confirmed COVID cases after cautionary discussion of triggers that would invalidate the model.

Hu et al. [11] also proposed the artificial intelligence (AI)-inspired methods for real-time forecasting of COVID-19 for estimating the size, lengths and ending time of COVID-19 across China. They developed a modified stacked auto-encoder for modelling the transmission dynamics of the epidemics and applied this model for forecasting the real-time confirmed cases of COVID-19 across China. The data collected for this study varied from 11 January to 27 February, 2020 from WHO.

Car et al. [12] transformed a time series dataset into a regression dataset and used it in training a multilayer perceptron (MLP) artificial neural network (ANN). By training this dataset, they tried to achieve a worldwide model of the maximal number of patients across all locations in each time unit. Hyper parameters were varied using a grid search algorithm, and a total of 48,384 ANNs were trained. Their study models showed high robustness of the deceased patient model, good robustness for confirmed and low robustness for recovered patient model.

For our present study, we collected data from [13–15].

## Data

There are various sources that are tracking the coronavirus data. They are updated at different times and are gathered in different ways, so the data might differ from source to source. As on 21st August 2020, WHO website quotes 21,294,845 confirmed cases, 761,779 confirmed deaths, total of 216 countries/territories affected with this respiratory disease. According to revised guidelines on public health surveillance for

COVID-19 by WHO (on 13-08-2020), emphasis should be on information on the importance of the collection of metadata for analysis and interpretation of surveillance data. The data has been collected from [13–15], for the purpose of studying the trend in four urban cities Bangalore, Delhi, Chennai and Mumbai along with India. Four major parameters, namely confirmed, recovered, active and death cases have been considered.

The basic statistics of the data for the period of 113 days (26 April to 17 August) for four metropolitan cities of India and India as a whole which is given in Table 10.1. The data has been plotted to see the pattern if visible in figure and also for the visual appreciation of the distribution in Fig. 10.1.

It is clearly visible from the basic statistics mentioned in Table 10.1 that data considered for the present study is non-Gaussian in nature as in all the cases, the skewness is not nearly or equal to 0 except for Indian case, but even in this case, the Kurtosis is not near 3. The most important property of COVID-19 data is its non-Gaussian nature. Hence, even though the mean and the standard deviation are valuable descriptors, when questioned about the severity of the spread, the assumption of normality (Gaussianity) will not be applicable. This can further be seen in the form of the probability distributions. The data of the four cases are normalized using the relation $D_i = (d_i - m_i)/s_i$ where $D_i$ is the normalized data, $d_i$ is the actual data, $m_i$ is the average of the given sample length, and $s_i$ is the standard deviation of the

**Table 10.1** Basic statistics of COVID-19 data for all different types such as positive, active, recovered and death cases of all India and its four important metropolitan cities for the period of 113 days from 26 April to 17 August (*Source* Own)

| S. no. | | 1 | 2 | 3 | 4 | 5 |
|---|---|---|---|---|---|---|
| Place | | India | Bengaluru | Chennai | Mumbai | Delhi |
| Average | Positive cases | 484,809 | 9151 | 39,060 | 56,590 | 55,464 |
| | Active cases | 184,427 | 6753 | 11,571 | 21,021 | 14,817 |
| | Recovered cases | 287,561 | 2219 | 26,867 | 32,605 | 37,976 |
| | Death cases | 12,821 | 180 | 626 | 2876 | 1651 |
| Standard deviation | Positive cases | 456,229.05 | 15,259.24 | 33,246.22 | 34,419.53 | 47,213.22 |
| | Active cases | 145,474.97 | 11,104.84 | 7018.75 | 7142.94 | 9065.65 |
| | Recovered cases | 300,728.24 | 3928.90 | 27,734.52 | 28,105.36 | 40,847.39 |
| | Death cases | 10,593.51 | 294.95 | 660.42 | 2148.90 | 1462.78 |
| Skewness | Positive cases | 0.07 | 1.72 | −1.31 | −1.32 | −1.45 |
| | Active cases | −0.02 | 1.72 | −1.14 | −0.15 | −1.35 |
| | Recovered cases | 0.15 | 3.68 | −0.81 | −1.17 | −0.95 |
| | Death cases | −0.80 | 1.73 | −0.40 | −1.59 | −1.61 |
| Kurtosis | Positive cases | 1.05 | 1.72 | 0.44 | 0.05 | 0.41 |
| | Active cases | 0.95 | 1.77 | 0.03 | −0.99 | 0.16 |
| | Recovered cases | 1.10 | 2.11 | 0.79 | 0.47 | 0.79 |
| | Death cases | 0.67 | 2.36 | 0.91 | 0.24 | 0.27 |

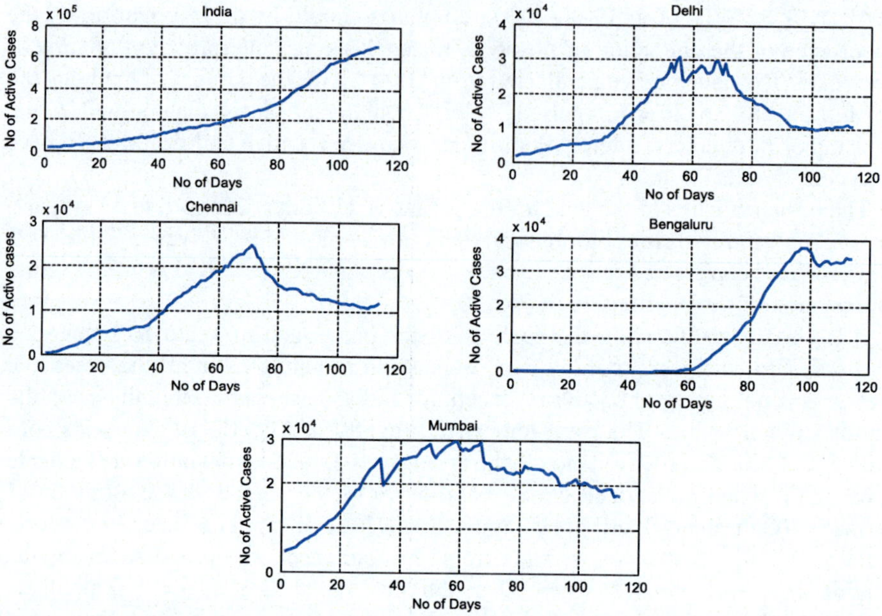


**Fig. 10.1** Observed data plot of active cases in India and four metropolitan cities, namely Bengaluru, Chennai, Mumbai and Delhi for the period of 113 days (*Source* Own)

given sample length for each *i* for the period of 113 days from 26 April to 17 August. Even after normalizing the data, the skewness and kurtosis remain the same. The data distribution of the normalized data has been plotted as a histogram to see patterns in the distribution so as to choose the model appropriately and is shown in Fig. 10.2.

A common assumption that is popular with any time series data is to consider it to be a stationary random process. This helps in defining the long-term average and long-term deviation which remain as reference values in modelling and forecasting exercises. For a stationary process, basic statistical parameters such as mean and standard deviation of the long period remain time independent. However, if they vary widely over a period of time, then the stationary assumption will not be valid. In Fig. 10.3, the non-stationarity of long-term average and long-term deviation of COVID-19 active cases data for all India, and its metropolitan cities is shown by changing the sample length. In all cases, the number of samples does not lead to a constant value of the average. The data considered for modelling purpose in the present study focuses on active cases of all India and the four metropolitan cities mentioned in Table 10.1.

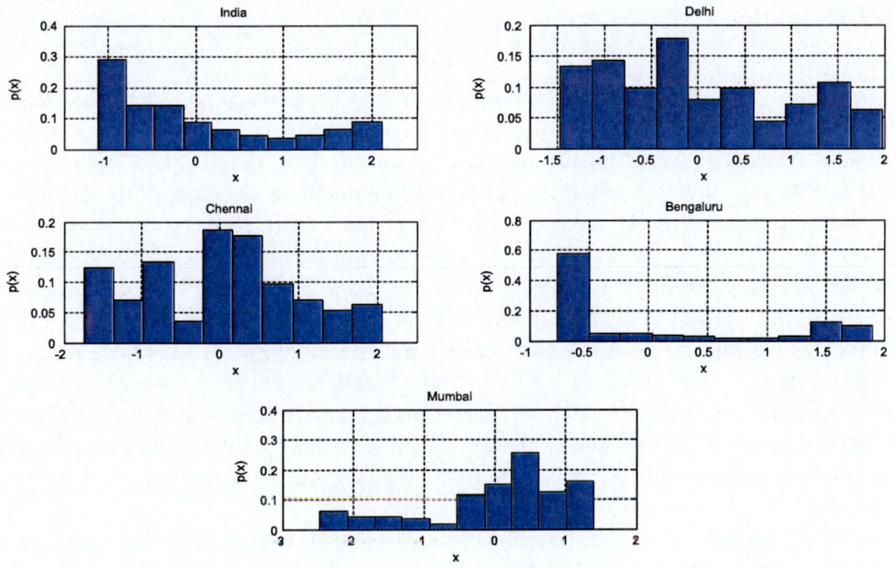


**Fig. 10.2** Histogram plot of COVID-19 active cases of all India and four metropolitan cities mentioned in the diagram (*Source* Own)

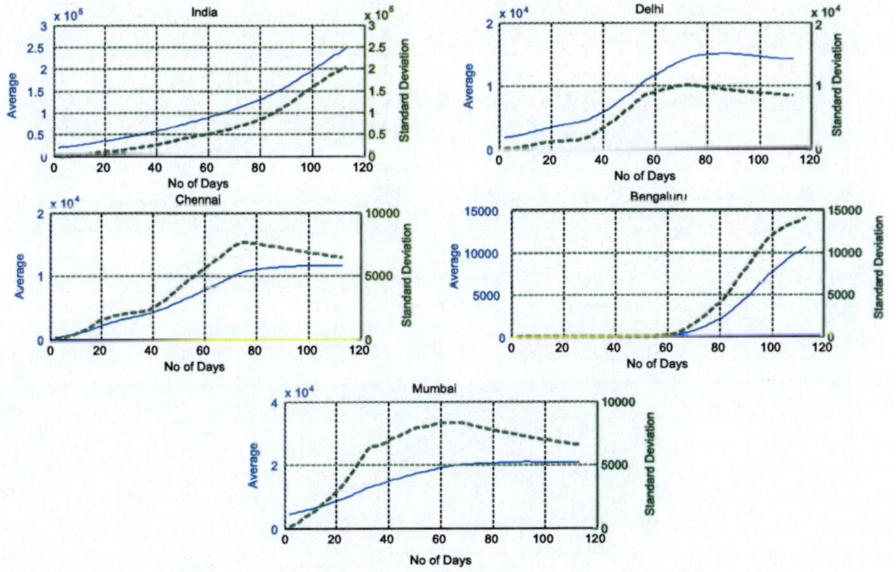


**Fig. 10.3** Running mean and the standard deviation of COVID-19 data of the active cases for the data from 26 April to 17 August of 2020 for all India and four metropolitan cities of India (*Source* Own)

## Methodology

To model any data, one has to understand the hidden structure or pattern in the data. In order to understand the data well, specifically for the active cases, data has been normalized using its own data and standard deviation. Also, detailed analysis of the same will help in understanding the pattern. The autocorrelation function indicates that a strong connection is there in data lags at least until 10 days lag as shown in Fig. 10.4. The data has been divided into training period (26 April to 31 July) with the sample size 97 and testing period (1 August to 31 August) with the sample size 31. In the training period, data will be trained for a particular model with the appropriate parameters with the number of parameters being less than 50% of the sample size; otherwise, it will be a polynomial fit for the entire length. Model will be validated using the measures such as the root mean square error and the coefficient of determination. If all the measures stay well within the confidence bands, then it will be implemented and checked again in the testing period mentioned.

Model 1: (Auto-Regressive Model)

Based on the autocorrelation function plotted in Fig. 10.4, an auto-regressive (AR) model considering the past 10 lag days in the regression equation as the variable is constructed for the present active cases. This is due to a strong correlation which exists in 10 lag days, beyond which it starts reducing. For the sample size considered for the

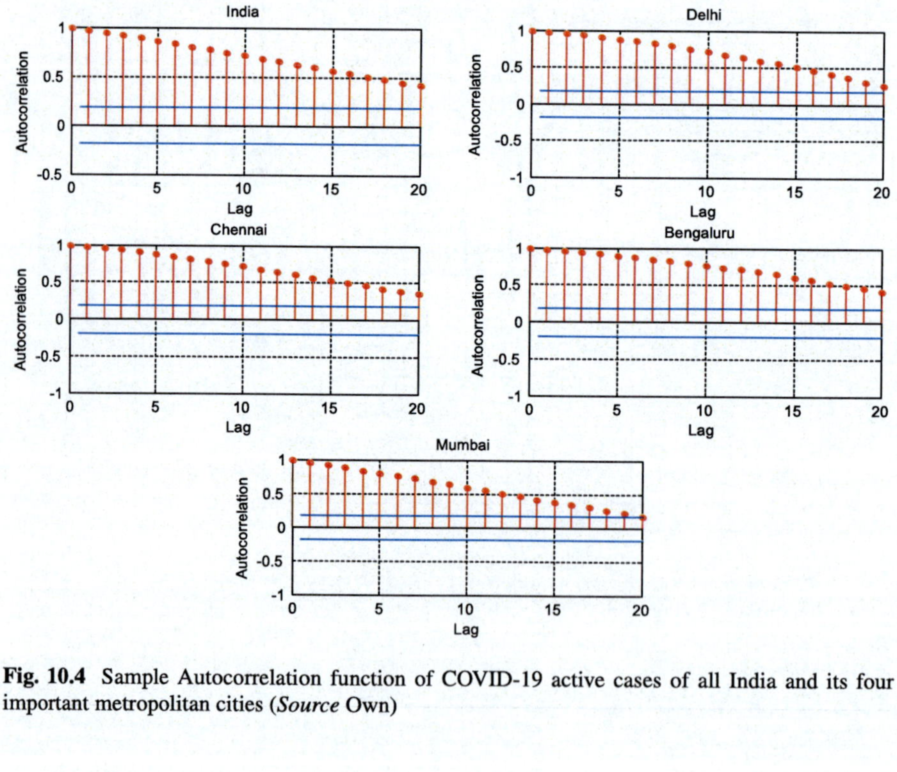


**Fig. 10.4** Sample Autocorrelation function of COVID-19 active cases of all India and its four important metropolitan cities (*Source* Own)

modelling purpose, a significant correlation is 0.6. Hence, until 10 lag information has been incorporated in modelling the active cases of all India and the four metropolitan cities considered for the present study. As mentioned in the previous section, the data has been trained using this regression model for the sample size of 97 days (from 26 April to 31 July) using the following equation:

$$A_t = B0 + B1A_{t-1} + B2A_{t-2} + B3A_{t-3} + B4A_{t-4} + B5A_{t-5} + B6A_{t-6} + B7A_{t-7} + B8A_{t-8} + B9A_{t-9} + B10A_{t-10} \quad (10.1)$$

Here, $A$ represents active cases and $t$ in days. The parameters in equation for all India and four metropolitan cities during training the model have been given in Table 10.2.

Comparison between the actual data, i.e. number of active cases and the AR model fit is shown in Fig. 10.5. Also the basic statistics such as average, standard deviation have been matched with the model, and the measures such as correlation coefficient (CC) have been found and listed in Table 10.3. In Fig. 10.5, it can be clearly seen that the model exactly matches with the observed data, and hence, the same be tested in the testing period data for using it in forecasting.

Model 2: Artificial Neural Network (ANN)

The non-Gaussianness of the data cannot be ignored at this moment as there can be some non-linearity hidden in the data; it is seen in the data distribution plotted as histograms. Hence, the artificial neural network (ANN) model has been used here with 6 days lag as the inputs in the input layer with one hidden layer and one output in the output layer. Network has been trained in the training period using back propagation algorithm using sigmoid function. There are totally 49 weights used while training the network. Network used for modelling is shown in Fig. 10.6. In the diagram, I represents the input layer, H represents the hidden layer, and O represents the output layer.

The network has been experimented only for two regions active cases data, i.e. for all India and Bengaluru active cases to compare with the regression model used in the previous section. The comparison between the actual data and the network model for these two cases is shown in Fig. 10.7.

The moment comparison is shown in Table 10.4 for ANN model. It can be clearly observed that the comparison between the actual data and the network fit is really appreciable even though only 6 lag information was used in the network for training the entire length data. Both model 1, i.e. Auto-regressive model and ANN model are performing good in the training period; check has to be in the testing period. The model which performs better in the testing period can be considered for future forecasting.

**Table 10.2** Parameters of the regression Eq. (10.1) for all India and four metropolitan cities of India (*Source* Own)

| Place | B0 | B1 | B2 | B3 | B4 | B5 | B6 | B7 | B8 | B9 | B10 |
|---|---|---|---|---|---|---|---|---|---|---|---|
| India | 0.0248 | 1.1170 | 0.0765 | −0.1507 | 0.0041 | 0.0203 | −0.0860 | 0.3402 | −0.4715 | 0.1052 | 0.0648 |
| Delhi | 0.0029 | 0.8634 | 0.0651 | 0.2501 | −0.1855 | 0.0184 | −0.0266 | 0.2158 | −0.0411 | −0.0268 | −0.1653 |
| Chennai | 0.0069 | 1.3064 | −0.1226 | −0.0258 | −0.1571 | 0.1233 | −0.1875 | 0.0108 | 0.1465 | −0.2535 | 0.1467 |
| Bengaluru | −0.0027 | 1.7168 | −0.7592 | 0.3016 | −0.3415 | 0.0848 | 0.0030 | 0.6354 | −1.2750 | 0.8317 | −0.2066 |
| Mumbai | 0.02287 | 0.9299 | 0.04812 | −0.0023 | −0.0754 | 0.0611 | −0.1051 | 0.0853 | 0.2495 | −0.0926 | −0.1512 |

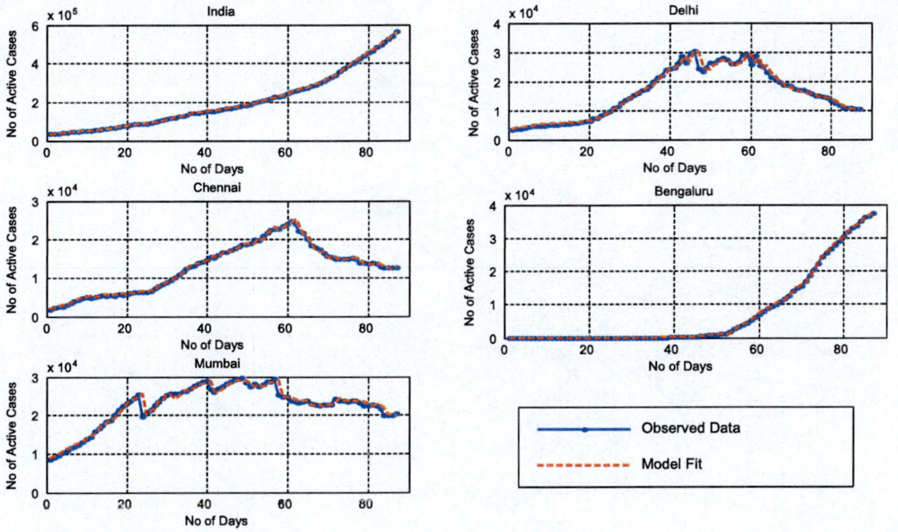


**Fig. 10.5** Comparison between the observed data and the mode fit of active cases of all India and four metropolitan cities (*Source* Own)

**Table 10.3** Comparison between the observed data and the model parameters for the training period (*Source* Own)

| Parameters | | India | Delhi | Chennai | Bengaluru | Mumbai |
|---|---|---|---|---|---|---|
| Average | Actual | 202,624 | 16,238 | 12,802 | 7522 | 22,728 |
| | Model fit | 202,624 | 16,238 | 12,802 | 7522 | 22,728 |
| Standard deviation | Actual | 142,731.8576 | 8479.2639 | 6332.0540 | 11,483.2192 | 5307.9156 |
| | Model fit | 142,715.1130 | 8395.2372 | 6318.1238 | 11,481.7111 | 5218.2834 |
| Correlation coefficient | Between actual and fit | 0.99 | 0.98 | 0.98 | 0.99 | 0.98 |

## Results and Discussion

It is observed that both AR model and ANN model have performed nearly same in the modelling or training period. In this section, both the models will be compared in the testing period; both the models have its own advantages and their disadvantages; one would be interested in the model which performs better in the testing period of 31 days (1 Aug to 31 Aug). AR model has been tested for all the five regions considered in the present study, whereas ANN model is performed only for two regions (All India and Bengaluru). In Table 10.5, the comparison between the AR model and ANN model for the two regions is shown:

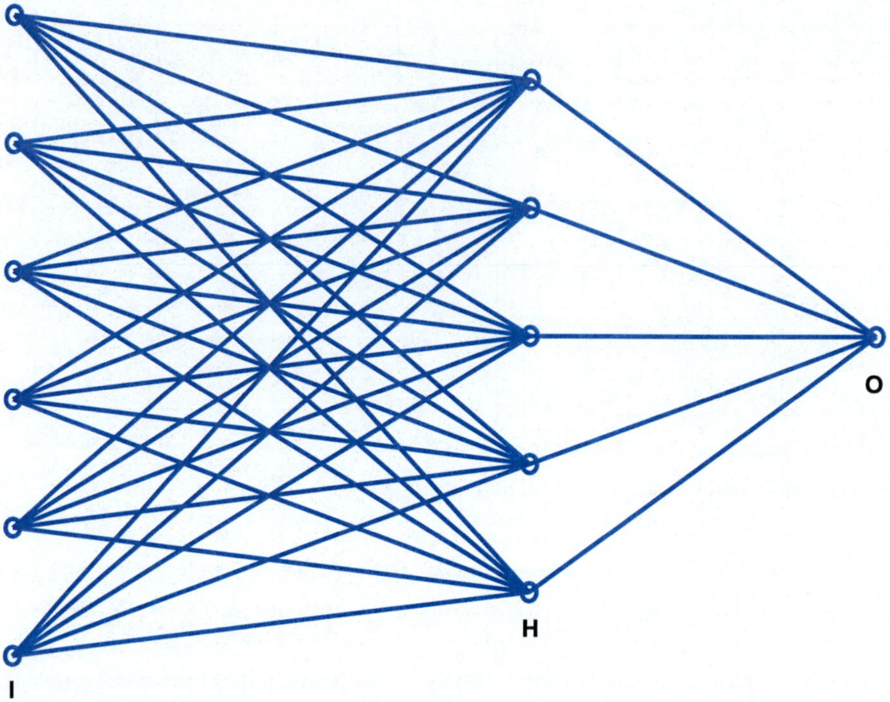


**Fig. 10.6** Network used with six inputs in the input layer (I), one hidden layer with six neurons (H) and one output in output layer (O) (*Source* Own)

Day-wise forecast comparison has been listed in Table 10.5 for number of active cases comparison with the observed data so as to check each day whether it is nearly matching with the actual data or not. This may not be the correct measure to show the comparison. Hence, comparison in terms of root mean square error (RMSE) between observed data and the forecast by AR and ANN models, correlation coefficient (CC) between actual data and forecast and the performance parameter (PP) between the actual data and forecast has been given in Table 10.6 which is considered to be the best parameters for the comparison.

It is clearly visible that training period and testing periods will not have the same relations as in terms of the measures which are used to compare the model in the forecast period as well as the training periods. Hence, one has to definitely check the measures in both the periods. But in the present cases, specifically in two cases, i.e. in India and Bengaluru, both methods perform similar as the result shows with minor variations in the measures, so any model can be used to forecast the present situation of the active cases. Similarly, in the remaining cases, only AR model has been tried and tested for the parameters which are well within the significance bands of their nature. Hence, the models can be accepted for modelling as well as forecasting

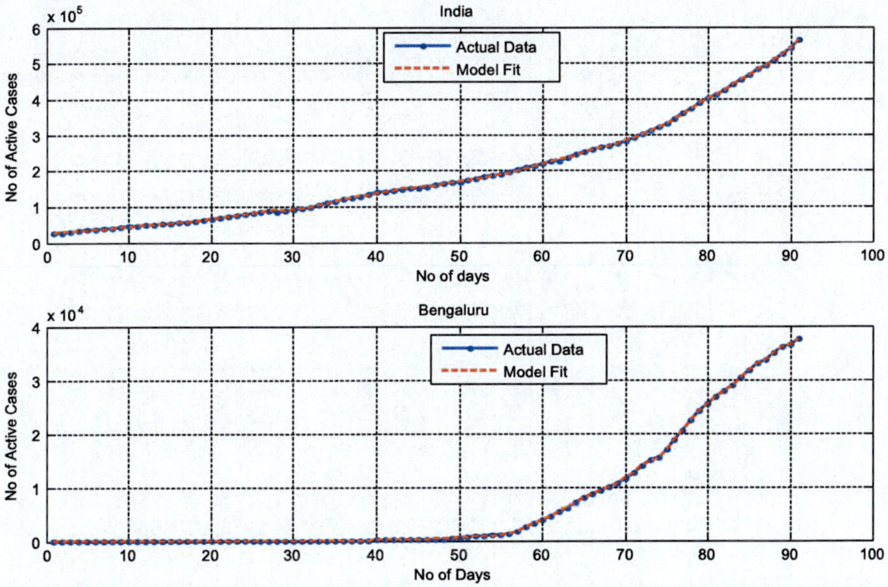


**Fig. 10.7** Comparison between the actual data and the model fit using ANN model

**Table 10.4** Comparison between the observed data and the model parameters for the training period (*Source* Own)

| Parameters | | India | Bengaluru |
|---|---|---|---|
| Average | Actual | 195,063 | 7526 |
| | Model fit | 195,023 | 7520 |
| Standard deviation | Actual | 143,959.27 | 11,709.33 |
| | Model fit | 143,907.42 | 11,709.42 |
| Correlation coefficient | Between actual and fit | 0.99 | 0.99 |
| Performance parameter | Between actual and fit | 0.99 | 0.99 |

purposes of COVID-19 data. Also, due to limitation of time and also due to restriction of the work, model has been tried only on the active cases.

## Conclusion

In the present study, AR and ANN models have been tried on the present situation of COVID-19 pandemic, specifically on the active cases. Before applying these models, an exhaustive statistical analysis has been performed on the data due to understand the nature and pattern of the data. Stationarity test has been performed by plotting

**Table 10.5** Day-wise comparison of the forecast for the period of 31 days (1 Aug to 31 Aug) using AR and ANN model with observed data (*Source* Own)

| Day | India | | | Bengaluru | | | Chennai | | Mumbai | | Delhi | |
|---|---|---|---|---|---|---|---|---|---|---|---|---|
| | Actual data | AR forecast | ANN forecast | Actual data | AR forecast | ANN forecast | Actual data | AR forecast | Actual data | AR forecast | Actual data | AR forecast |
| 1 Aug | 567,730 | 581,773 | 570,306 | 37,760 | 38,499 | 37,129 | 12,439 | 12,662 | 20,731 | 20,692 | 10,596 | 10,476 |
| 2 Aug | 579,357 | 586,262 | 576,822 | 37,513 | 37,847 | 37,912 | 12,193 | 12,438 | 21,394 | 21,032 | 10,356 | 10,316 |
| 3 Aug | 586,298 | 593,057 | 583,934 | 36,290 | 37,232 | 37,702 | 11,986 | 12,063 | 20,528 | 21,697 | 10,207 | 10,225 |
| 4 Aug | 586,244 | 602,699 | 598,355 | 34,021 | 35,780 | 35,939 | 11,859 | 11,832 | 20,309 | 20,512 | 9897 | 10,014 |
| 5 Aug | 595,501 | 602,137 | 598,516 | 32,757 | 32,143 | 28,487 | 11,814 | 11,852 | 20,679 | 19,945 | 10,072 | 10,055 |
| 6 Aug | 607,384 | 608,994 | 595,085 | 32,314 | 31,118 | 32,782 | 11,723 | 11,753 | 20,546 | 20,513 | 10,348 | 9866 |
| 7 Aug | 619,088 | 625,355 | 610,965 | 33,308 | 32,013 | 33,141 | 11,609 | 11,712 | 20,124 | 20,802 | 10,409 | 10,156 |
| 8 Aug | 628,747 | 630,759 | 624,741 | 33,726 | 33,056 | 33,134 | 11,737 | 11,581 | 19,914 | 20,389 | 10,667 | 10,368 |
| 9 Aug | 634,945 | 646,127 | 637,378 | 33,815 | 33,631 | 33,892 | 11,657 | 11,755 | 19,700 | 20,279 | 10,729 | 10,541 |
| 10 Aug | 639,929 | 650,812 | 637,362 | 32,985 | 33,234 | 34,600 | 11,331 | 11,713 | 19,172 | 20,141 | 10,346 | 10,761 |
| 11 Aug | 643,948 | 652,994 | 641,410 | 33,070 | 31,369 | 33,093 | 11,133 | 11,291 | 18,887 | 19,303 | 10,868 | 10,794 |
| 12 Aug | 653,622 | 660,845 | 656,412 | 33,489 | 33,236 | 33,822 | 10,956 | 11,055 | 19,047 | 18,974 | 10,946 | 10,485 |
| 13 Aug | 661,595 | 670,546 | 662,099 | 33,148 | 33,344 | 32,779 | 10,871 | 10,900 | 19,314 | 19,391 | 10,975 | 10,966 |
| 14 Aug | 667,950 | 678,252 | 669,630 | 33,432 | 33,373 | 33,048 | 11,212 | 10,812 | 19,337 | 19,626 | 11,366 | 11,048 |
| 15 Aug | 677,714 | 683,705 | 674,768 | 34,858 | 33,039 | 34,700 | 11,324 | 11,279 | 17,591 | 19,504 | 11,489 | 11,323 |
| 16 Aug | 676,900 | 693,145 | 679,620 | 34,584 | 35,470 | 34,267 | 11,501 | 11,484 | 17,825 | 17,896 | 10,823 | 11,593 |
| 17 Aug | 673,213 | 692,458 | 675,657 | 34,408 | 33,657 | 33,548 | 12,006 | 11,657 | 17,704 | 18,034 | 10,852 | 11,693 |
| 18 Aug | 676,549 | 686,458 | 670,446 | 33,081 | 34,968 | 33,974 | 12,106 | 12,250 | 17,693 | 17,830 | 11,068 | 11,104 |
| 19 Aug | 686,395 | 691,832 | 683,454 | 33,280 | 31,726 | 33,598 | 12,259 | 12,395 | 17,914 | 17,942 | 11,137 | 11,126 |
| 20 Aug | 692,030 | 702,187 | 691,463 | 34,186 | 32,729 | 33,575 | 12,290 | 12,467 | 18,172 | 18,189 | 11,271 | 11,127 |

(continued)

**Table 10.5** (continued)

| Day | India | | | Bengaluru | | | Chennai | | Mumbai | | Delhi | |
|---|---|---|---|---|---|---|---|---|---|---|---|---|
| | Actual data | AR forecast | ANN forecast | Actual data | AR forecast | ANN forecast | Actual data | AR forecast | Actual data | AR forecast | Actual data | AR forecast |
| 21 Aug | 697,330 | 707,541 | 689,907 | 34,532 | 34,720 | 34,556 | 12,711 | 12,415 | 18,299 | 18,744 | 11,426 | 11,376 |
| 22 Aug | 707,668 | 714,209 | 699,734 | 34,224 | 35,297 | 34,534 | 12,965 | 13,017 | 18,301 | 18,651 | 11,594 | 11,524 |
| 23 Aug | 710,771 | 722,043 | 712,653 | 34,877 | 32,913 | 33,606 | 13,226 | 13,109 | 18,567 | 18,195 | 11,778 | 11,659 |
| 24 Aug | 704,348 | 726,006 | 701,456 | 34,735 | 35,440 | 33,797 | 13,258 | 13,457 | 18,267 | 18,644 | 11,626 | 11,674 |
| 25 Aug | 706,851 | 720,222 | 700,865 | 35,430 | 33,426 | 34,828 | 13,374 | 13,415 | 17,938 | 18,572 | 11,998 | 11,818 |
| 26 Aug | 725,991 | 723,062 | 718,567 | 36,053 | 36,789 | 36,510 | 13,520 | 13,477 | 18,979 | 18,223 | 12,520 | 11,756 |
| 27 Aug | 742,023 | 742,754 | 739,876 | 35,989 | 36,295 | 35,916 | 13,453 | 13,635 | 19,463 | 19,239 | 13,208 | 12,197 |
| 28 Aug | 752,424 | 761,294 | 754,576 | 36,521 | 35,529 | 36,533 | 13,533 | 13,491 | 19,407 | 19,835 | 13,550 | 12,614 |
| 29 Aug | 765,302 | 773,116 | 771,234 | 37,315 | 36,507 | 36,998 | 13,656 | 13,616 | 19,971 | 19,757 | 14,040 | 13,353 |
| 30 Aug | 781,975 | 783,666 | 783,456 | 37,703 | 38,201 | 37,359 | 13,475 | 13,692 | 20,321 | 20,183 | 14,793 | 13,761 |
| 31 Aug | 785,996 | 799,731 | 786,543 | 37,116 | 37,251 | 37,153 | 13,227 | 13,481 | 20,511 | 20,619 | 14,626 | 14,299 |

**Table 10.6** Comparison in terms of RMSE between observed and forecast values for a period of 31 days, from 1 August to 31 August (*Source* Own)

| Place | Forecast model | Root mean square error (RMSE) | CC | PP |
|---|---|---|---|---|
| India | AR | 5.3321e+003 | 0.99 | 0.99 |
| | ANN | 4.8385e+003 | 0.99 | 0.99 |
| Bengaluru | AR | 1.0637e+003 | 0.87 | 0.62 |
| | ANN | 1.0345e+003 | 0.88 | 0.64 |
| Chennai | AR | 173.7424 | 0.98 | 0.95 |
| Mumbai | AR | 524.1394 | 0.88 | 0.76 |
| Delhi | AR | 420.9962 | 0.95 | 0.89 |

running mean and standard deviation in the same plot to see if they converge to a particular value and concluded that it is non-stationary. Hence, AR model with 10 days lag having strong correlation with the data is modelled, and the same model has been tested in the testing period or forecast period also. In order to understand the non-linear structure if it exists in the data, ANN model is constructed, and both the models are compared so as to identify the best model for further forecasting of the other cases such as death cases and recovered cases. Both models have outperformed in terms of the parameters used for measuring the same. Hence, any model can be used for the remaining cases. For future study of the same link between active cases, recovered cases and death cases has to be found, and if possible, then a combined ANN model with three outputs have to be developed which will forecast all three cases at a time. The future work will be concentrated on this combined study.

## References

1. Ahmed, S. S. (2020). The coronavirus disease 2019 (COVID-19): A review. *JAMMR, 32*(4), 1–9. https://doi.org/10.9734/jammr/2020/v32i430393
2. Luo, Z., Ang, M. J. Y., Chan, S. Y., Yi, Z., Goh, Y. Y., Yan, S., et al. (2020). Combating the coronavirus pandemic: early detection, medical treatment, and a concerted effort by the global community. *Research* https://doi.org/10.34133/2020/6925296.
3. Poletto, C., Scarpino, S. V., & Volz, E. M. (2020). Applications of predictive modelling early in the COVID-19 epidemic. *Lancet Digital Health, 2*(10), Published Online August 10, https://doi.org/10.1016/S2589-7500(20)30196-5
4. Elsevier, Novel Coronavirus Information Center. (2020). https://www.elsevier.com/connect/coronavirus-information-center.
5. Shah, N. H., Suthar, A. H., & Jayswal, E. N. (2020). Control strategies to curtail transmission of COVID-19. *Hindawi International Journal of Mathematics and Mathematical Sciences*, Article ID 2649514. https://doi.org/10.1155/2020/2649514.
6. Arino, J., & Portet, S. (2020). A simple model for COVID-19. *Infectious Disease Modelling, 5*, 309–315. www.keaipublishing.com/idm, Production & Hosting by Elsevier B V on behalf of KeAi Communications Co. Ltd, an open access journal under CC-BY-NC-ND 4.0 International license.

7. Fong, S. J., Li, G., Dey, N., Crespo, R. G., & Herrera-Viedma, E. (2020). Finding an accurate early forecasting model from small dataset: a case of 2019-nCoV novel coronavirus outbreak. *Int Journal of Interactive Multimedia and Artificial Intelligence, 6*(1), 132–140.
8. Shah, N. H., Sheoran, N., Jayswal, E., Shukla D,, Shukla, N., Shukla, J. & Shah Y. (2020) Modelling COVID-19 transmission in the united states through interstate and foreign travels and evaluating impact of governmental public health interventions. medRxiv preprint the copyright holder for this preprint this version posted. https://doi.org/10.1101/2020.05.23.20110999
9. Wong, Z. S., Zhou, J., & Zhang, Q. (2019). Artificial intelligence for infectious disease big data analytics. *Infection, Disease & Health, 24*(1), 44–48.
10. Smarajit D. E. Y. (2020). Modeling Covid19 In India (Mar 3-May 7, 2020): How flat is flat, and other hard facts. medRxiv preprint, https://doi.org/10.1101/2020.05.11.20097865.
11. Hu, Z., Ge, Q., Jin, L., & Xiong, M. (2020). Artificial intelligence forecasting of covid-19 in China. http://arxiv.org/abs/2002.07112.
12. C, Z., Šegota, S. B., Anđelić, N., Lorencin, I., & Mrzljak, V. (2020). Modeling the spread of COVID-19 infection using a multilayer perceptron. *Computational and Mathematical Methods in Medicine.* https://doi.org/10.1155/2020/5714714.
13. https://www.accuweather.com/en/in/national/covid-19.
14. https://www.bing.com/covid/local/delhi_india?vert=graph.
15. Ixigo train application—Coronavirus Live Tracker.

# Chapter 11
# A Fractional-Order SEQAIR Model to Control the Transmission of nCOVID 19

**Jitendra Panchal and Falguni Acharya**

**Abstract** The ensuing paper expounds a new mathematical model for a pandemic instigated by novel coronavirus (COVID-19) with influence of quarantine on transmission of COVID-19, using Caputo fractional-order derivative for various fractional order. Basic reproduction number for the SEQAIR model has been calculated in the study and additionally proving the existence and uniqueness of the solution using the fixed-point theorem. Furthermore, numerical solution is revealed using the Adams–Bashforth–Moulton method, and its application for real-world data is deliberated.

**Keywords** nCOVID-19 · Fractional-order derivative · SEQAIR model · Fixed-point theorem · Numerical simulation

**MSC** 34A08 · 35R10 · 81T80

## Introduction

Environmental scenarios and role of human activities in it have been extensively discussed, studied and analyzed by numerous ecologists, scientists and naturalist, since the time it came into understanding. It is observed that a slew of human actions such as air and water pollution, cutting down forests, overconsumption of fuel, etc. have radically impacted the ecology and degraded the natural resources. The generation today is hence experiencing unfamiliar events and challenges which are quite difficult to control. One of the recent outbreaks humanity is facing is the prevalence of Severe Acute Respiratory Syndrome Coronavirus 2 (SARS-Cov-2), which was declared as a pandemic by WHO on March 11, 2020. The virus introduced itself

---

J. Panchal (✉) · F. Acharya
Department of Applied Sciences, Parul Institute of Engineering and Technology, Parul University, Vadodara, Gujarat, India
e-mail: jitendra.jit.panchal@gmail.com; jitendrakumar.panchal@paruluniversity.ac.in

F. Acharya
e-mail: falguni.acharya@paruluniversity.ac.in

 
N. H. Shah and M. Mittal (eds.), *Mathematical Analysis for Transmission of COVID-19*, Mathematical Engineering, https://doi.org/10.1007/978-981-33-6264-2_11

on December 31, 2019, in Wuhan city of China and begun to spread, marking its foot prints on 213 countries universally leaving 25,925,495 people infected, 861,667 dead, and 18,209,744 recovered [1, 2]. The COVID-19 has exhibited diverse symptoms such as dry cough, fever, losing sense of smell and taste, shortness of breath, etc. [3], eventually changing itself rapidly leading to infected people without the former symptoms. Apparently, in myriad cases, the symptoms are diagnosed only after 14 days, which makes the supervisory steps vague. The primary medium for spread is respiratory dews of the diseased person. Due to the common symptoms shared by coronavirus, flu, cold and viral fever, it becomes hard to scrutinize them and work out accordingly for prevention and cure. The nature of virus being adaptable to survive in the environment for quite a few days, and it can be transmitted faster and risk of contamination upsurges [4, 5], especially among senior citizens, children, diseased patient and low immune individual.

Moreover, the disease has no vaccine or cure yet, consequently the ministry of various countries was compelled to impose lockdown to avoid contact of individuals, in order to restrict the further spread of infection from the carriers. Specifically, in a country such as India whose huge population rather support the wide and quick spread of the COVID-19; therefore, the decision of lockdown by Indian government was the only scope to rein the proliferation of the virus. Clearly, witnessing the mushrooming of infected individuals, day after day and understanding the fact of limited resources and huge population, Indian government was too impelled to implement the idea of lockdown to combat the ferocious virus.

The unpredictable vagaries and dynamism observed in the study of the disease have proved challenging for intellectuals in the field of biology and mathematics (see, for example, [5–9] and [10]). In recent years, numerous papers have been published on the subject of Caputo–Fabrizio fractional derivative (see, for example, [11, 8, 12]) and mathematical models are used for simulating the transmission of coronavirus (see, for example, [13–15]). A generalized SEIR model by Peng aptly integrates the impact of unseen, exposed, and infectious cases of COVID-19 [16]. However, Khan and Atangana discussed the interaction between various cases of transmission through the mathematical fractional model [17]. Aside, a SEIARW model based on the transmission routes through a market to individual and community for specific age groups was developed by Zhao, susceptible to people from Wuhan city [17]. Similarly, to make a primary estimate of the coronavirus outbreak in Mainland China, Zhong created a mathematical model through epidemiological data and analyses from characteristics of historical epidemics [18].

As a matter of fact, the fractional-order derivative is the generalization of the integer-order derivative and has recently produced improved results in modeling the real phenomena; in the light of that, the study inspects the transmission model of COVID-19 using the Caputo fractional-order derivative. In addition, the research includes mathematical simulation predicting the influence of one of the plausible effective ways to restrict the rampant COVID-19 worldwide, that is isolation, with the aid of available statistical data. Quarantine as expected to be one of the predominant practices to restrain the prevalence of coronavirus, the paper investigates its role in the frequency of variation in the spread of infection among the community.

In [19], Panchal and Acharya formulated and proved the impact of quarantine on the transmission of COVID-19 using an integer-order mathematical model.

In [20], authors have proved existence and uniqueness of SEIARW model using fixed-point theory. In this paper, authors have proved SEQAIR model showing influence of quarantine on transmission of COVID-19, using fixed-point technique [20]. Moreover, extend the work done in [19] by modifying an integer-order model to the fractional-order model for different fractional orders and proved the existence of the solution using fixed-point technique with numerical solutions. In Sect. Preliminary Results and Definitions, important definitions and concepts of fractional calculus are mentioned. The fractional-order model for nCOVID-19 transmission is discussed in Sect. Model Formulation. The existence and uniqueness of the solution for the system have been proved in Sect. Stability Results. In Sect. Existence and Uniqueness of Solution, the result is also analyzed using numerical method.

## Preliminary Results and Definitions

This section discusses the prerequisites and elementary notions of fractional differential equation comprising three types of fractional derivatives and fractional integrals, one of which is Caputo.

**Definition 2.1** References [21, 22] For an integrable function $f$, the Caputo derivative of fractional order $\alpha \in (0, 1)$ is given by

$$^{C}D^{\alpha} f(t) = \frac{1}{\Gamma(m-\alpha)} \int_{0}^{t} \frac{f^{(m)}(v)}{(t-v)^{\alpha-m+1}} \mathrm{d}v, \quad m = [\alpha] + 1.$$

Also, the corresponding fractional integral of order $\alpha$ with $\mathrm{Re}(\alpha) > 0$ is given by

$$^{C}I^{\alpha} f(t) = \frac{1}{\Gamma(\alpha)} \int_{0}^{t} (t-v)^{\alpha-1} f(v) \mathrm{d}v.$$

**Definition 2.2** References [23, 24] For $f \in H^{1}(c, d)$ and $d > c$. The Caputo–Fabrizio derivative of fractional order $\alpha \in (0, 1)$ for $f$ is given by

$$^{CF}D^{\alpha} f(t) = \frac{M(\alpha)}{(1-\alpha)} \int_{c}^{t} \exp\left(\frac{-\alpha}{1-\alpha}(t-v)\right) f'(v) \mathrm{d}v,$$

where $t \geq 0$, $M(\alpha)$ is a normalization function that depends on $\alpha$ and $M(0) = M(1) = 1$. If $f \notin H^{1}(c, d)$ and $0 < \alpha < 1$, this derivative for $f \in L^{1}(-\infty, d)$ is given by

$$ {}^{CF}D^{\alpha} f(t) = \frac{\alpha M(\alpha)}{(1-\alpha)} \int_{-\infty}^{d} (f(t) - f(v)) \exp\left(\frac{-\alpha}{1-\alpha}(t-v)\right) f'(v)dv. $$

Also, the corresponding C–F fractional integral is presented by

$$ {}^{\mathrm{CF}}I^{\alpha} f(t) = \frac{2(1-\alpha)}{(2-\alpha)M(\alpha)} f(t) + \frac{2\alpha}{(2-\alpha)M(\alpha)} \int_{0}^{t} f(v)dv. $$

Below is the definition of Laplace transform, a significant tool for solving differential equations, for two different types of fractional derivative

**Definition 2.3** References [21, 22] The Laplace transform of Caputo fractional differential operator of order $\alpha$ is given by

$$ L\left[{}^{C}D^{\alpha} f(t)\right](s) = s^{\alpha} Lf(t) - \sum_{i=0}^{m-1} s^{\alpha-i-1} f^{(i)}(0), m-1 < \alpha \leq m \in N. $$

which can also be obtained in the form

$$ L\left[{}^{C}D^{\alpha} f(t)\right](s) = \frac{s^{m} L[f(t)] - s^{m-1} f(0) - \cdots - f^{(m-1)}}{s^{m-\alpha}} $$

# Model Formulation

In India, migrating inter-states for earning livelihood is quite common. However, due to the crisis levied by the lockdown made the life of daily wage workers miserable, since many industries, factory outlets and businesses were closed down. As a result, they choose to commute back to their home town, inspite of food and shelter provided by government, which in turn became one of the major reasons for the prevalence of virus at various places. Hence, the virus was still pervading. However, with due implementation of lockdown and voluntary quarantine the mass was successful in controlling the spread up to certain level. Therefore, without losing generality, the below model is formulated considering all the plausible ways of transmissions among the community, including travel to various infected cities or countries, especially China.

Assuming that the population is under lockdown/shutdown condition and to analyze the human to human transmission dynamics of coronavirus, we constructed a compartmental model. The transmission dynamics of COVID-19 is described graphically in Fig. 11.1.

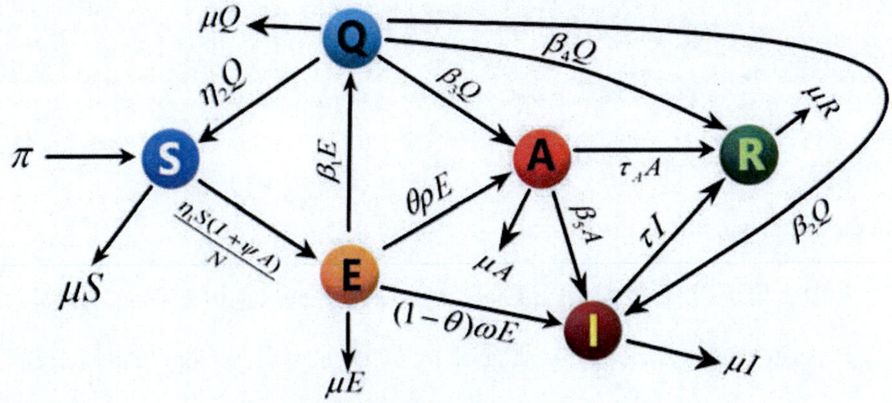


**Fig. 11.1** Transmission dynamics of COVID-19 [19]

$N(t)$ denotes the total population which is classified further into six different classes, the susceptible $S(t)$, the exposed $E(t)$, the quarantine $Q(t)$, the asymptomatic infected $A(t)$, the infected $I(t)$ and the recovered or the removed people $R(t)$.

The birth and natural death rate is denoted by $\pi$ and $\mu$, respectively. $\eta_1 SI$ is the rate at which susceptible people $S$ get infected through the infected people $I$, where $\eta_1$ denotes the coefficient of disease transmission. As discussed in [19], the asymptomatic infected people can spread the virus to healthy people and the transmission is given by $\eta_1 \psi SA$, where $\psi$ is the transmissibility multiple of $A$ to $I$ and $\psi = [0, 1]$. When $\psi = 0$, then there will be no infection, and when $\psi = 1$, then the infection will be the same as $I$. $\theta$ represents the proportion of asymptomatic infection. The parameters $\omega$ and $\rho$, respectively, represent the transmission rate for joining the class $I$ and $A$. $\beta_1$ is the rate of exposed at which join class $Q$. $\eta_2$ is the rate at which quarantine person $Q$ transfer to class $S$. $\beta_2$ is the rate at which quarantine individuals become infected $I$. $\beta_4$ represents the rate at which quarantine people join class $R$. $\beta_5$ is the transfer rate of the people from asymptomatic class $A$ becomes infected and joining the class $I$. $\beta_3$ is the rate at which people from class $Q$ transfer to class $A$. The rates $\tau$ and $\tau_A$ represents the removal or recovery rate of class $I$ and class $A$ joining the class $R$.

The dynamical system of the nCOVID-19 model [19] with quarantine effect is given by

$$
\begin{aligned}
\frac{\mathrm{d}S}{\mathrm{d}t} &= \pi + \eta_2 Q - \frac{\eta_1 S(I + \psi A)}{N} - \mu S \\
\frac{\mathrm{d}E}{\mathrm{d}t} &= \frac{\eta_1 S(I + \psi A)}{N} - (\beta_1 + \theta\rho + (1-\theta)\omega)E - \mu E \\
\frac{\mathrm{d}Q}{\mathrm{d}t} &= \beta_1 E - (\eta_2 + \beta_2 + \beta_3 + \beta_4)Q - \mu Q
\end{aligned}
$$

$$\frac{dA}{dt} = \theta\rho E + \beta_3 Q - (\beta_5 + \lambda_A)A - \mu A$$
$$\frac{dI}{dt} = (1-\theta)\omega E + \beta_2 Q + \beta_5 A - \lambda I - \mu I$$
$$\frac{dR}{dt} = \lambda_A A + \lambda I + \beta_4 Q - \mu R \tag{11.1}$$

with the initial conditions

$$S(0) \geq 0,\ E(0) \geq 0,\ Q(0) \geq 0,\ A(0) \geq 0,\ I(0) \geq 0,\ R(0) \geq 0 \tag{11.2}$$

The total dynamics of the people obtained by adding all six equations of the model (11.1) is,

$$\frac{dN}{dt} = \pi - \mu N$$

The system (11.1) is moderated by substituting the Caputo fractional time derivative. In this moderate system, the dimension of the system will not remain the same for the right and left sides. We use $\chi$ an auxiliary parameter to resolve this problem. According to the explanation, the fractional-order model for $t > 0$ and $\alpha \in (0, 1)$ with the same initial conditions is given by

$$\chi^{\alpha-1} {}^C D_t^\alpha S(t) = \pi + \eta_2 Q(t) - \frac{\eta_1 S(t)(I(t) + \psi A(t))}{N} - \mu S(t)$$
$$\chi^{\alpha-1} {}^C D_t^\alpha E(t) = \frac{\eta_1 S(t)(I(t) + \psi A(t))}{N} - (\beta_1 + \theta\rho + (1-\theta)\omega)E(t) - \mu E(t)$$
$$\chi^{\alpha-1} {}^C D_t^\alpha Q(t) = \beta_1 E(t) - (\eta_2 + \beta_2 + \beta_3 + \beta_4)Q(t) - \mu Q(t)$$
$$\chi^{\alpha-1} {}^C D_t^\alpha A(t) = \theta\rho E(t) + \beta_3 Q(t) - (\beta_5 + \lambda_A)A(t) - \mu A(t)$$
$$\chi^{\alpha-1} {}^C D_t^\alpha I(t) = (1-\theta)\omega E(t) + \beta_2 Q(t) + \beta_5 A(t) - \lambda I(t) - \mu I(t)$$
$$\chi^{\alpha-1} {}^C D_t^\alpha R(t) = \lambda_A A(t) + \lambda I(t) + \beta_4 Q(t) - \mu R(t) \tag{11.3}$$

The feasible region for the model (11.2) is given by

$$\Omega = \left\{ (S(t), E(t), Q(t), A(t), I(t), R(t)) \in \mathbb{R}_+^6 : N(t) \leq \frac{\pi}{\mu} \right\} \tag{11.4}$$

we prove that the closed set $\Omega$ is the feasible region of the system (11.3).

**Lemma 3.1** *The closed set $\Omega$ is a positive invariant concerning the fractional system* (11.3).

***Proof*** We add all the terms in the system (11.3) to obtain the fractional derivative of the overall population.

i.e., $\chi^{\alpha-1C}D_t^{\alpha}N(t) = \pi - \mu N(t)$.

where $N(t) = S(t)+E(t)+Q(t)+A(t)+I(t)+R(t)$. To obtain the population size, we use Laplace transform as follows,

$$N(t) = N(0)E_{\alpha}(-\mu\chi^{1-\alpha}t^{\alpha}) + \int_0^t \pi\chi^{1-\alpha}\theta^{\alpha-1}E_{\alpha,\alpha}(-\mu\chi^{1-\alpha}\theta^{\alpha})\mathrm{d}\theta,$$

where $N(0)$ is the initial population size, after simplifying we get

$$N(t) = N(0)E_{\alpha}(-\mu\chi^{1-\alpha}t^{\alpha}) + \int_0^t \pi\chi^{1-\alpha}\theta^{\alpha-1}\sum_{i=0}^{\infty}\frac{(-1)^i\mu^i\chi^{i(1-\alpha)}\theta^{i\alpha}}{\Gamma(i\alpha+\alpha)}\mathrm{d}\theta,$$

A fractional-order model for nCOVID-19 transmission using the Caputo derivative is given as

$$N(t) = \frac{\pi}{\mu} + E_{\alpha}(-\mu\chi^{1-\alpha}t^{\alpha})\left(N(0) - \frac{\pi}{\mu}\right),$$

Thus, if $N(0) \leq \frac{\pi}{\mu}$, then for $t > 0$, $N(t) \leq \frac{\pi}{\mu}$.

Consequently, the closed set $\Omega$ is positive invariant concerning fractional-order model (11.3).

## Stability Results

Let us define the disease-free equilibrium $E_0$ and the basic reproduction number $\Re_0$ for the stability of model (11.3)

$$E_0 = (S_0, 0, 0, 0, 0, 0) = \left(\frac{\pi}{\mu}, 0, 0, 0, 0, 0\right).$$

is the disease-free equilibrium for the system (11.3).

To compute, the basic reproduction number for the given system (11.3) refers [17], where the required computation of the matrices $F$ and $V$ is given by,

$$F = \begin{pmatrix} 0 & 0 & \eta_1\psi & \eta_1 \\ 0 & 0 & 0 & 0 \\ 0 & 0 & 0 & 0 \\ 0 & 0 & 0 & 0 \end{pmatrix},$$

$$V = \begin{pmatrix} \beta_1 + \theta\rho + (1-\theta)\omega + \mu & 0 & 0 & 0 \\ -\beta_1 & \eta_2 + \beta_2 + \beta_3 + \beta_4 + \mu & 0 & 0 \\ -\theta\rho & -\beta_3 & \beta_5 + \lambda_A + \mu & 0 \\ (1-\theta)\omega & -\beta_2 & -\beta_5 & \lambda + \mu \end{pmatrix}$$

The basic reproduction number of the system (11.3) is given as

$$\Re_0 = \frac{\begin{array}{c}\eta_1\psi(\lambda+\mu)(\theta\rho(\eta_2+\beta_2+\beta_3+\beta_4)+\beta_1\beta_3) \\ +\eta_1\beta_5((\eta_2+\beta_2+\beta_3+\beta_4)(\theta\rho+(1-\theta)\omega)+\beta_1(\beta_2+\beta_3))\end{array}}{(\lambda+\mu)(\beta_5+\lambda_A+\mu)(\beta_1+\theta\rho+(1-\theta)\omega+\mu)(\eta_2+\beta_2+\beta_3+\beta_4)}$$

where the spectral radius is $\gamma(\mathrm{FV}^{-1})$. Further, $\Re_0$ can be also represented as $\Re_0 = \Re_1 + \Re_2$, where

$$\Re_1 = \frac{\eta_1\psi(\lambda+\mu)(\theta\rho(\eta_2+\beta_2+\beta_3+\beta_4)+\beta_1\beta_3)}{(\lambda+\mu)(\beta_5+\lambda_A+\mu)(\beta_1+\theta\rho+(1-\theta)\omega+\mu)(\eta_2+\beta_2+\beta_3+\beta_4)}$$

and

$$\Re_2 = \frac{\eta_1\beta_5((\eta_2+\beta_2+\beta_3+\beta_4)(\theta\rho+(1-\theta)\omega)+\beta_1(\beta_2+\beta_3))}{(\lambda+\mu)(\beta_5+\lambda_A+\mu)(\beta_1+\theta\rho+(1-\theta)\omega+\mu)(\eta_2+\beta_2+\beta_3+\beta_4)}$$

**Theorem 4.1** *The disease-free equilibrium $E_0$ for the model* (11.3) *is locally asymptotically stable if $\Re_0 < 1$.*

***Proof*** To obtain disease-free equilibrium at a point $E_0$, the Jacobian matrix below,

$$J = \begin{pmatrix} -\mu & 0 & \eta_2 & -\eta_1\psi & -\eta_1 & 0 \\ 0 & -(\beta_1+\theta\rho+(1-\theta)\omega+\mu) & 0 & -\eta_1\psi & -\eta_1 & 0 \\ 0 & \beta_1 & -(\eta_2+\beta_2+\beta_3+\beta_4+\mu) & 0 & 0 & 0 \\ 0 & \theta\rho & \beta_3 & -(\beta_5+\lambda_A+\mu) & 0 & 0 \\ 0 & (1-\theta)\omega & \beta_2 & \beta_5 & -(\lambda+\mu) & 0 \\ 0 & 0 & \beta_4 & \lambda_A & \lambda & -\mu \end{pmatrix}$$

The Jacobian matrix has two negative eigenvalues as $-\mu$ (twice) and the remaining eigenvalues can be obtained by the following characteristics equation:

$$\lambda^4 + a_1\lambda^3 + a_2\lambda^2 + a_3\lambda + a_4 = 0$$

where

$$\begin{aligned} a_1 &= \lambda + \lambda_A + \theta\rho + (1-\theta)\omega + \eta_2 + \beta_1 + \beta_2 + \beta_3 + \beta_4 + \beta_5 + 4\mu \\ a_2 &= (\beta_1 + \theta\rho + (1-\theta)\omega + \mu)(\eta_2 + \beta_2 + \beta_3 + \beta_4 + \mu) + (\beta_5 + \lambda_A + \mu) \\ &\quad (\beta_1 + \theta\rho + (1-\theta)\omega + \mu) + \eta_1\psi\theta\rho + (\lambda+\mu)(\beta_1 + \theta\rho + (1-\theta)\omega + \mu) \\ &\quad + \eta_1(1-\theta)\omega + (\lambda+\mu)(\beta_5 + \lambda_A + \mu) + (\lambda+\mu)(\eta_2 + \beta_2 + \beta_3 + \beta_4 + \mu) \end{aligned}$$

$$+ (\eta_2 + \beta_2 + \beta_3 + \beta_4 + \mu)(\beta_5 + \lambda_A + \mu)$$
$$a_3 = (\lambda + \mu)(\beta_1 + \theta\rho + (1-\theta)\omega + \mu)[(\lambda + \mu)(1 - \Re_1) + (\lambda_A + \mu)(1 - \Re_2)]$$
$$+ (\eta_2 + \beta_2 + \beta_3 + \beta_4 + \mu)[(\beta_5 + \lambda_A + \mu) - \eta_1\psi\theta\rho]$$
$$+ (\beta_5 + \lambda_A + \mu)(\eta_2 + \beta_2 + \beta_3 + \beta_4 + \mu)(\beta_1 + \theta\rho + (1-\theta)\omega + \mu)$$
$$a_4 = (\beta_1 + \theta\rho + (1-\theta)\omega + \mu)(\eta_2 + \beta_2 + \beta_3 + \beta_4 + \mu)(\beta_5 + \lambda_A + \mu)(\lambda + \mu)(1 - \Re_0)$$

The first four terms in the coefficient $a_2$ are less than $\Re_0$ and the coefficient $a_4$ is positive when $\Re_0 < 1$, and hence, all the coefficients are positive. Moreover, the coefficients satisfy the Routh–Hurwitz criteria for the fourth-order polynomial. Thus, the model (11.3) at the disease-free equilibrium is locally asymptotically stable if $\Re_0 < 1$.

## *Endemic Equilibria*

We use the following equations to determine the equilibrium points for the fractional-order model (11.3),

$$^{C}D^{\alpha}S(t) = {}^{C}D^{\alpha}E(t) = {}^{C}D^{\alpha}Q(t) = {}^{C}D^{\alpha}A(t) = {}^{C}D^{\alpha}I(t) = {}^{C}D^{\alpha}R(t) = 0$$

The algebraic solution of the equations provides equilibrium points of the system. The disease-free equilibrium point, given as $E_1^0 = \left(\frac{\pi}{\mu}, 0, 0, 0, 0, 0\right)$.

And if $\Re_0 > 1$, then the system (11.3) has a positive endemic equilibrium $E_1^* = (S^*, E^*, Q^*, A^*, I^*, R^*)$

where

$$S^* = \frac{\pi}{\lambda^* + \mu}, \quad E^* = \frac{\lambda^* S^*}{\beta_1 + \theta\rho + (1-\theta)\omega + \mu}, \quad Q^* = \frac{\beta_1 E^*}{\eta_2 + \beta_2 + \beta_3 + \beta_4 + \mu}$$
$$A^* = \frac{\theta\rho E^* + \beta_3 Q^*}{\beta_5 + \lambda_A + \mu}, \quad I^* = \frac{(1-\theta)\omega E^* + \beta_2 Q^* + \beta_5 A^*}{\lambda + \mu}, \quad R^* = \frac{\lambda_A A^* + \lambda I^*}{\mu}$$

and

$$\lambda^* = \frac{\eta_1(\psi A^* + I^*)}{S^* + E^* + Q^* + A^* + I^* + R^*}$$

which satisfies the equation $P(\lambda^*) = m_1(\lambda^*)^2 + m_2\lambda^* = 0$ and $m_1, m_2$ are given by

$$m_1 = (\beta_1 + \theta\rho + (1-\theta)\omega + \mu)(\eta_2 + \beta_2 + \beta_3 + \beta_4 + \mu)(\beta_5 + \lambda_A + \mu)(\lambda + \mu)$$
$$m_2 = (\beta_1 + \theta\rho + (1-\theta)\omega + \mu)(\eta_2 + \beta_2 + \beta_3 + \beta_4 + \mu)(\beta_5 + \lambda_A + \mu)(\lambda + \mu)(1 - \Re_0).$$

Obviously, $m_1 \geq 0$ and $m_2 \geq 0$ whenever $\Re_0 < 1$, so that $\lambda^* = \frac{-m_2}{m_1} \leq 0$. The endemic equilibrium does not hold for $\Re_0 < 1$.

## Existence and Uniqueness of Solution

In this section, we used fixed-point technique given in [20] to prove the uniqueness of the solution for the system (11.3) as follows

$$\chi^{\alpha-1C}D_t^{\alpha}S(t) = G_1(t, S(t))$$
$$\chi^{\alpha-1C}D_t^{\alpha}E(t) = G_2(t, E(t))$$
$$\chi^{\alpha-1C}D_t^{\alpha}Q(t) = G_3(t, Q(t))$$
$$\chi^{\alpha-1C}D_t^{\alpha}A(t) = G_4(t, A(t))$$
$$\chi^{\alpha-1C}D_t^{\alpha}I(t) = G_5(t, I(t))$$
$$\chi^{\alpha-1C}D_t^{\alpha}R(t) = G_6(t, R(t))$$

By lemma (3.1), the above system can be written as follows

$$S(t) - S(0) = \frac{\chi^{1-\alpha}}{\Gamma\alpha}\int_0^t G_1(\tau, S)(t-\tau)^{\alpha-1}\mathrm{d}\tau$$

$$E(t) - E(0) = \frac{\chi^{1-\alpha}}{\Gamma\alpha}\int_0^t G_2(\tau, E)(t-\tau)^{\alpha-1}\mathrm{d}\tau$$

$$Q(t) - Q(0) = \frac{\chi^{1-\alpha}}{\Gamma\alpha}\int_0^t G_3(\tau, Q)(t-\tau)^{\alpha-1}\mathrm{d}\tau$$

$$A(t) - A(0) = \frac{\chi^{1-\alpha}}{\Gamma\alpha}\int_0^t G_4(\tau, A)(t-\tau)^{\alpha-1}\mathrm{d}\tau$$

$$I(t) - I(0) = \frac{\chi^{1-\alpha}}{\Gamma\alpha}\int_0^t G_5(\tau, I)(t-\tau)^{\alpha-1}\mathrm{d}\tau$$

$$R(t) - R(0) = \frac{\chi^{1-\alpha}}{\Gamma\alpha}\int_0^t G_6(\tau, R)(t-\tau)^{\alpha-1}\mathrm{d}\tau \tag{11.5}$$

In the ensuing theorem, the kernels $G_i,\ i = 1, 2, 3, 4, 5, 6$ satisfy the Lipschitz condition and contraction.

**Theorem 5.1** *The kernel $G_1$ satisfies the Lipschitz condition and contraction if the inequality given below holds*

$$0 \leq \eta_1(M_4 + \psi M_5) + \mu < 1$$

.

***Proof*** For $S$ and $S_1$, we have

$$\|G_1(t, S) - G_1(t, S_1)\| \leq [\eta_1(M_4 + \psi M_5) + \mu]||S - S_1||$$

Suppose that $d_1 = \eta_1(M_4 + \psi M_5) + \mu$, where $||S|| \leq M_1$, $||E|| \leq M_2$, $||Q|| \leq M_3$, $||A|| \leq M_4$, $||I|| \leq M_5$ and $||R|| \leq M_6$, is a bounded function, So

$$\|G_1(t, S) - G_1(t, S_1)\| \leq d_1 \|S(t) - S_1(t)\|. \tag{11.6}$$

Thus, for $G_1$, the Lipchitz condition is obtained, and if $0 \leq \eta_1(M_4 + \psi M_5) + \mu < 1$, then $G_1$ is a contraction.

Similarly, the Lipschitz condition for $G_i,\ i = 2, 3, 4, 5, 6$ given as follows

$$\begin{aligned}
&\|G_2(t, E) - G_2(t, E_1)\| \leq d_2 \|E(t) - E_1(t)\| \\
&\|G_3(t, Q) - G_3(t, Q_1)\| \leq d_3 \|Q(t) - Q_1(t)\| \\
&\|G_4(t, A) - G_4(t, A_1)\| \leq d_4 \|A(t) - A_1(t)\| \\
&\|G_5(t, I) - G_5(t, I_1)\| \leq d_5 \|I(t) - I_1(t)\| \\
&\|G_6(t, R) - G_6(t, R_1)\| \leq d_6 \|R(t) - R_1(t)\|
\end{aligned}$$

where $d_2 = \beta_1 + \theta\rho + (1-\theta)\omega + \mu$, $d_3 = \eta_2 + \beta_2 + \beta_3 + \beta_4 + \mu$, $d_4 = \beta_5 + \lambda_A + \mu$, $d_5 = \lambda + \mu$ and $d_6 = \mu$ are bounded functions, if $0 \leq d_i < 1, i = 2, 3, 4, 5, 6$, then $G_i, i = 2, 3, 4, 5, 6$ are contraction. According to the system (11.5), consider the following recursive forms

$$Z_{1n}(t) = S_n(t) - S_{n-1}(0) = \frac{\chi^{1-\alpha}}{\Gamma\alpha} \int_0^t [G_1(\tau, S_{n-1}) - G_1(\tau, S_{n-2})](t-\tau)^{\alpha-1} d\tau$$

$$Z_{2n}(t) = E_n(t) - E_{n-1}(0) = \frac{\chi^{1-\alpha}}{\Gamma\alpha} \int_0^t [G_2(\tau, E_{n-1}) - G_2(\tau, E_{n-2})](t-\tau)^{\alpha-1} d\tau$$

$$Z_{3n}(t) = Q_n(t) - Q_{n-1}(0) = \frac{\chi^{1-\alpha}}{\Gamma\alpha} \int_0^t [G_3(\tau, Q_{n-1}) - G_3(\tau, Q_{n-2})](t-\tau)^{\alpha-1} d\tau$$

$$Z_{4n}(t) = A_n(t) - A_{n-1}(0) = \frac{\chi^{1-\alpha}}{\Gamma\alpha} \int_0^t [G_4(\tau, A_{n-1}) - G_4(\tau, A_{n-2})](t-\tau)^{\alpha-1} d\tau$$

$$Z_{5n}(t) = I_n(t) - I_{n-1}(0) = \frac{\chi^{1-\alpha}}{\Gamma\alpha} \int_0^t [G_5(\tau, I_{n-1}) - G_5(\tau, I_{n-2})](t-\tau)^{\alpha-1} d\tau$$

$$Z_{6n}(t) = R_n(t) - R_{n-1}(0) = \frac{\chi^{1-\alpha}}{\Gamma\alpha}\int_0^t [G_6(\tau, R_{n-1}) - G_6(\tau, R_{n-2})](t-\tau)^{\alpha-1}\mathrm{d}\tau$$

with the initial conditions $S_0(t) = S(0)$, $E_0(t) = E(0)$, $Q_0(t) = Q(0)$, $A_0(t) = A(0), I_0(t) = I(0)$ and $R_0(t) = R(0)$. We take the norm of the first equation in the above system, then

$$\|Z_{1n}(t)\| = \|S_n(t) - S_{n-1}(0)\| = \left\| \frac{\chi^{1-\alpha}}{\Gamma\alpha}\int_0^t [G_1(\tau, S_{n-1}) - G_1(\tau, S_{n-2})](t-\tau)^{\alpha-1}\mathrm{d}\tau \right\|$$

$$\leq \frac{\chi^{1-\alpha}}{\Gamma\alpha}\int_0^t \left\|[G_1(\tau, S_{n-1}) - G_1(\tau, S_{n-2})](t-\tau)^{\alpha-1}\right\|\mathrm{d}\tau.$$

with Lipchitz condition (11.6), we have

$$\|Z_{1n}(t)\| \leq \frac{\chi^{1-\alpha}}{\Gamma\alpha} d_1 \int_0^t \left\|Z_{1(n-1)}(\tau)\right\|\mathrm{d}\tau. \tag{11.7}$$

As a similar way, we obtained

$$\|Z_{2n}(t)\| \leq \frac{\chi^{1-\alpha}}{\Gamma\alpha} d_2 \int_0^t \left\|Z_{2(n-1)}(\tau)\right\|\mathrm{d}\tau$$

$$\|Z_{3n}(t)\| \leq \frac{\chi^{1-\alpha}}{\Gamma\alpha} d_3 \int_0^t \left\|Z_{3(n-1)}(\tau)\right\|\mathrm{d}\tau$$

$$\|Z_{4n}(t)\| \leq \frac{\chi^{1-\alpha}}{\Gamma\alpha} d_4 \int_0^t \left\|Z_{4(n-1)}(\tau)\right\|\mathrm{d}\tau$$

$$\|Z_{5n}(t)\| \leq \frac{\chi^{1-\alpha}}{\Gamma\alpha} d_5 \int_0^t \left\|Z_{5(n-1)}(\tau)\right\|\mathrm{d}\tau$$

$$\|Z_{6n}(t)\| \leq \frac{\chi^{1-\alpha}}{\Gamma\alpha} d_6 \int_0^t \left\|Z_{6(n-1)}(\tau)\right\|\mathrm{d}\tau \tag{11.8}$$

Thus, we can write that

$$S_n(t) = \sum_{j=1}^{n} Z_{1j}(t),\ E_n(t) = \sum_{j=1}^{n} Z_{2j}(t),\ Q_n(t) = \sum_{j=1}^{n} Z_{3j}(t),$$

$$A_n(t) = \sum_{j=1}^{n} Z_{4j}(t),\ I_n(t) = \sum_{j=1}^{n} Z_{5j}(t),\ R_n(t) = \sum_{j=1}^{n} Z_{6j}(t),$$

In the next theorem, we prove the existence of a solution.

**Theorem 5.2** *A system of solutions given by the fractional COVID-19 model* (11.3) *exists if there exists* $t_1$ *such that*

$$\frac{\chi^{1-\alpha}}{\Gamma\alpha} t_1 d_i < 1.$$

***Proof*** From recursive technique, and Eqs. (11.7) and (11.8), we conclude that

$$\|Z_{1n}(t)\| \le \|S_n(0)\| \left[\frac{\chi^{1-\alpha}}{\Gamma\alpha} d_1 t\right]^n, \|Z_{2n}(t)\| \le \|E_n(0)\| \left[\frac{\chi^{1-\alpha}}{\Gamma\alpha} d_2 t\right]^n,$$
$$\|Z_{3n}(t)\| \le \|Q_n(0)\| \left[\frac{\chi^{1-\alpha}}{\Gamma\alpha} d_3 t\right]^n, \|Z_{4n}(t)\| \le \|A_n(0)\| \left[\frac{\chi^{1-\alpha}}{\Gamma\alpha} d_4 t\right]^n,$$
$$\|Z_{5n}(t)\| \le \|I_n(0)\| \left[\frac{\chi^{1-\alpha}}{\Gamma\alpha} d_5 t\right]^n, \|Z_{6n}(t)\| \le \|R_n(0)\| \left[\frac{\chi^{1-\alpha}}{\Gamma\alpha} d_6 t\right]^n$$

Thus, the system has a continuous solution. Now we show that the above functions construct a solution for the model (11.3), and we assume that

$$S(t) - S(0) = S_n(t) - W_{1n}(t),\ E(t) - E(0) = E_n(t) - W_{2n}(t),$$
$$Q(t) - Q(0) = Q_n(t) - W_{3n}(t),\ A(t) - A(0) = A_n(t) - W_{4n}(t),$$
$$I(t) - I(0) = I_n(t) - W_{5n}(t),\ R(t) - R(0) = R_n(t) - W_{6n}(t)$$

So,

$$\|W_{1n}(t)\| \le \frac{\chi^{1-\alpha}}{\Gamma\alpha} \int_0^t \|G_1(\tau, S) - G_1(\tau, S_{n-1})\| d\tau \le \frac{\chi^{1-\alpha}}{\Gamma\alpha} d_1 \|S - S_{n-1}\| t.$$

By repeating the method, we obtain

$$\|W_{1n}(t)\| \le \left[\frac{\chi^{1-\alpha}}{\Gamma\alpha} t\right]^{n+1} d_1^{n+1} h.$$

At $t_1$, we get

$$\|W_{1n}(t)\| \le \left[\frac{\chi^{1-\alpha}}{\Gamma\alpha} t_1\right]^{n+1} d_1^{n+1} h.$$

As $n$ approaches to $\infty$, implies $\|W_{1n}(t)\| \rightarrow 0$. Similarly, we can obtain $\|W_{in}(t)\| \rightarrow 0, i = 2, 3, 4, 5, 6$. Hence, the theorem is proved.

To show the uniqueness of the solution, we suppose that the system has another solution such as $S_1(t)$, $E_1(t)$, $Q_1(t)$, $A_1(t)$, $I_1(t)$ and $R_1(t)$, then we have

$$S(t) - S_1(t) = \frac{\chi^{1-\alpha}}{\Gamma\alpha} \int_0^t (G_1(\tau, S) - G_1(\tau, S_1))\mathrm{d}\tau.$$

We take norm from this equation

$$\|S(t) - S_1(t)\| \leq \frac{\chi^{1-\alpha}}{\Gamma\alpha} \int_0^t \|(G_1(\tau, S) - G_1(\tau, S_1))\|\mathrm{d}\tau.$$

It follows from Lipschitz condition (11.6) that

$$\|S(t) - S_1(t)\| \leq \frac{\chi^{1-\alpha}}{\Gamma\alpha} d_1 t \|S(t) - S_1(t)\|.$$

Thus,

$$\|S(t) - S_1(t)\| \left(1 - \frac{\chi^{1-\alpha}}{\Gamma\alpha} d_1 t\right) \leq 0. \tag{11.9}$$

**Theorem 5.3** *The solution of nCOVID-19 model* (11.3) *is unique if below condition holds*

$$\left(1 - \frac{\chi^{1-\alpha}}{\Gamma\alpha} d_1 t\right) > 0.$$

***Proof*** Suppose that condition (11.9) holds

$$\|S(t) - S_1(t)\| \left(1 - \frac{\chi^{1-\alpha}}{\Gamma\alpha} d_1 t\right) \leq 0.$$

Then $\|S(t) - S_1(t)\| = 0$. Therefore, we get $S(t) = S_1(t)$. Likewise, the same equality can be shown for $E$, $Q$, $A$, $I$ and $R$.

## Numerical Results

In this section, we present the numerical results for the nCOVID-19 model (11.3). We used the Adams–Bashforth–Moulton scheme [25]. Set $h = \frac{T}{N}, t_n = nh, \ n =$

$0, 1, 2, \ldots, N \in \mathbb{Z}^+$, we can write the system (11.3) as follows

$$S_{n+1} = S_0 + \frac{h^\eta \lambda_1^{1-\eta}}{\Gamma(\eta+2)}\left[\pi + \eta_2 Q_{n+1}^p - \eta_1 S_{n+1}^p \frac{(I_{n+1}^p + \psi A_{n+1}^p)}{N} - \mu S_{n+1}^p\right] + \frac{h^\eta \lambda_1^{1-\eta}}{\Gamma(\eta+2)} \sum_{i=0}^{n} a_{i,n+1}\left[\pi + \eta_2 Q_i - \eta_1 S_i \frac{(I_i + \psi A_i)}{N} - \mu S_i\right]$$

$$E_{n+1} = E_0 + \frac{h^\eta \lambda_1^{1-\eta}}{\Gamma(\eta+2)}\left[\eta_1 S_{n+1}^p \frac{(I_{n+1}^p + \psi A_{n+1}^p)}{N} - (\beta_1 + \theta\rho + (1-\theta)\omega) E_{n+1}^p - \mu E_{n+1}^p\right] + \frac{h^\eta \lambda_1^{1-\eta}}{\Gamma(\eta+2)} \sum_{i=0}^{n} a_{i,n+1}\left[\eta_1 S_i \frac{(I_i + \psi A_i)}{N} - (\beta_1 + \theta\rho + (1-\theta)\omega) E_i - \mu E_i\right]$$

$$Q_{n+1} = Q_0 + \frac{h^\eta \lambda_1^{1-\eta}}{\Gamma(\eta+2)}[\beta_1 E_{n+1}^p - (\eta_2 + \beta_2 + \beta_3 + \beta_4) Q_{n+1}^p - \mu Q_{n+1}^p] + \frac{h^\eta \lambda_1^{1-\eta}}{\Gamma(\eta+2)} \sum_{i=0}^{n} a_{i,n+1}[\beta_1 E_i - (\eta_2 + \beta_2 + \beta_3 + \beta_4) Q_i - \mu Q_i]$$

$$A_{n+1} = A_0 + \frac{h^\eta \lambda_1^{1-\eta}}{\Gamma(\eta+2)}[\theta\rho E_{n+1}^p + \beta_3 Q_{n+1}^p - (\beta_5 + \lambda_A) A_{n+1}^p - \mu A_{n+1}^p] + \frac{h^\eta \lambda_1^{1-\eta}}{\Gamma(\eta+2)} \sum_{i=0}^{n} a_{i,n+1}[\theta\rho E_i + \beta_3 Q_i - (\beta_5 + \lambda_A) A_i - \mu A_i]$$

$$I_{n+1} = I_0 + \frac{h^\eta \lambda_1^{1-\eta}}{\Gamma(\eta+2)}[(1-\theta)\omega E_{n+1}^p + \beta_2 Q_{n+1}^p + \beta_5 A_{n+1}^p - \lambda I_{n+1}^p - \mu I_{n+1}^p] + \frac{h^\eta \lambda_1^{1-\eta}}{\Gamma(\eta+2)} \sum_{i=0}^{n} a_{i,n+1}[(1-\theta)\omega E_i + \beta_2 Q_i + \beta_5 A_i - \lambda I_i - \mu I_i]$$

$$R_{n+1}^p = R_0 + \frac{\lambda_1^{1-\eta}}{\Gamma\eta} \sum_{i=0}^{n} \varphi_{i,n+1}[\lambda_A A_i + \lambda I_i + \beta_4 Q_i - \mu R_i]$$

where,

$$S_{n+1}^p = S_0 + \frac{\lambda_1^{1-\eta}}{\Gamma\eta} \sum_{i=0}^{n} \varphi_{i,n+1}\left[\pi + \eta_2 Q_i - \eta_1 S_i \frac{(I_i + \psi A_i)}{N} - \mu S_i\right]$$

$$E_{n+1}^p = E_0 + \frac{\lambda_1^{1-\eta}}{\Gamma\eta} \sum_{i=0}^{n} \varphi_{i,n+1}\left[\eta_1 S_i \frac{(I_i + \psi A_i)}{N} - (\beta_1 + \theta\rho + (1-\theta)\omega) E_i - \mu E_i\right]$$

$$Q_{n+1}^p = Q_0 + \frac{\lambda_1^{1-\eta}}{\Gamma\eta} \sum_{i=0}^{n} \varphi_{i,n+1}[\beta_1 E_i - (\eta_2 + \beta_2 + \beta_3 + \beta_4) Q_i - \mu Q_i]$$

$$A_{n+1}^{p} = A_0 + \frac{\lambda_1^{1-\eta}}{\Gamma\eta}\sum_{i=0}^{n}\varphi_{i,n+1}\left[\theta\rho E_i + \beta_3 Q_i - (\beta_5 + \lambda_A)A_i - \mu A_i\right]$$

$$I_{n+1}^{p} = I_0 + \frac{\lambda_1^{1-\eta}}{\Gamma\eta}\sum_{i=0}^{n}\varphi_{i,n+1}\left[(1-\theta)\omega E_i + \beta_2 Q_i + \beta_5 A_i - \lambda I_i - \mu I_i\right]$$

$$R_{n+1}^{p} = R_0 + \frac{\lambda_1^{1-\eta}}{\Gamma\eta}\sum_{i=0}^{n}\varphi_{i,n+1}\left[\lambda_A A_i + \lambda I_i + \beta_4 Q_i - \mu R_i\right]$$

In which

$$a_{i,n+1} = \begin{cases} n^{\eta_j+1} - (n-\eta_j)(n+1)^{\eta_j};\ i = 0, \\ (n-i+2)^{\eta_j+1} + (n-i)^{\eta_j+1} - 2(n-i+1)^{\eta_j+1};\ 1 \le i \le n, \\ 1;\ i = n+1 \end{cases}$$

and

$$\varphi_{i,n+1} = \frac{h^{\eta_j}}{\eta_j}\left((n-i+1)^{\eta_j} - (n-i)^{\eta_j}\right);\ 0 \le i \le n, \text{ and } j = 1, 2, 3.$$

## *Simulation*

The parametric values used for simulation given in Table 11.1.

The following graphical simulation is represented for fractional order $\alpha = 0.5, 0.55, 0.6, 0.65, 0.7, 0.75, 0.8, 0.85, 0.9, 0.95, 0.9999 \simeq 1$.

In Fig. 11.2, we can observe that the numbers of susceptible individuals decrease with respect to time.

In Fig. 11.3, we can observe that initially, the numbers of exposed persons increase, and due to decrease in the number of susceptible with respect to time, the number of exposed persons will decrease over the time.

In Fig. 11.4, we can observe that the number of infected persons is increasing with respect to time due to exposed persons.

In Fig. 11.5, we can observe that as the numbers of infected peoples are increasing, the numbers of quarantine individuals also increase over the time.

In Fig. 11.6, we can observe that the numbers of asymptomatic individuals increase with respect to time as many peoples come into contact with infected persons.

In Fig. 11.7, we can observe that the numbers of recovered or removed(dead) individual over the time are increasing as there is no vaccination.

**Table 11.1** Parameters used in the model (11.3) [1]

| Parameters | Description | Value |
|---|---|---|
| $\pi$ | Birth rate | 0.0242 |
| $\mu$ | Natural mortality rate | 0.0325 |
| $\eta_1$ | Contact rate | 0.5 |
| $\beta_1$ | Rate at which exposed individual transfer to quarantine class | 0.0359 |
| $\theta$ | The proportion of asymptomatic infection | 0.023 |
| $\omega$ | Rate at which exposed individual become infected | 0.9606 |
| $\rho$ | Rate at which exposed individual become asymptomatic infected | 0.35 |
| $\eta_2$ | Rate at which quarantine individual become susceptible | 0.37 |
| $\beta_2$ | Rate at which quarantine individual become infected | 0.002 |
| $\beta_3$ | Rate at which quarantine individual become asymptomatic infected | 0.0002 |
| $\beta_4$ | Rate at which quarantine individual joining class $R$ | 0.3722 |
| $\beta_5$ | Rate at which asymptomatic individual become infected | 0.5312 |
| $\lambda$ | Removal or recovery rate of $I$ | 0.32829392 |
| $\lambda_A$ | Removal or recovery rate of $A$ | 0.02672 |

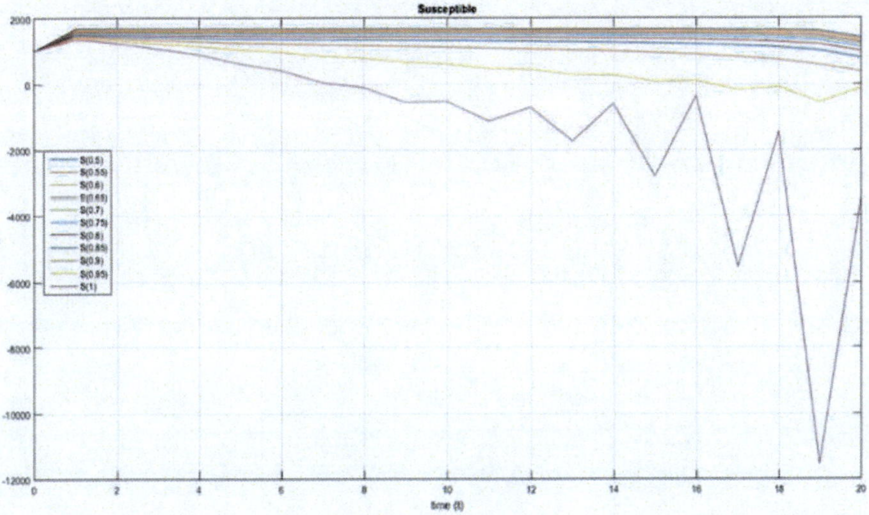


**Fig. 11.2** Numbers of susceptible individuals S(t) versus time (t)

## Conclusion

Authors have formulated and proved the influence of quarantine on the transmission of COVID-19 using the fractional-order model with Caputo derivative for various

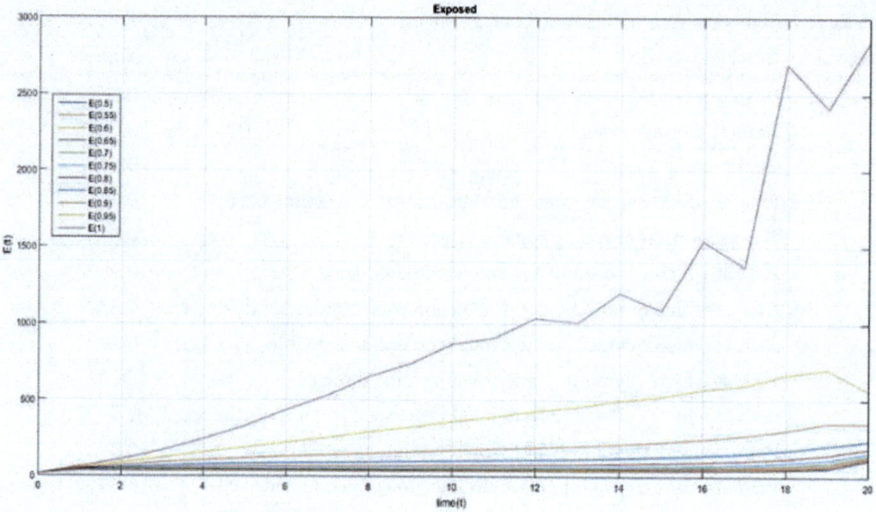


**Fig. 11.3** Numbers of exposed individuals E(t) versus time (t)

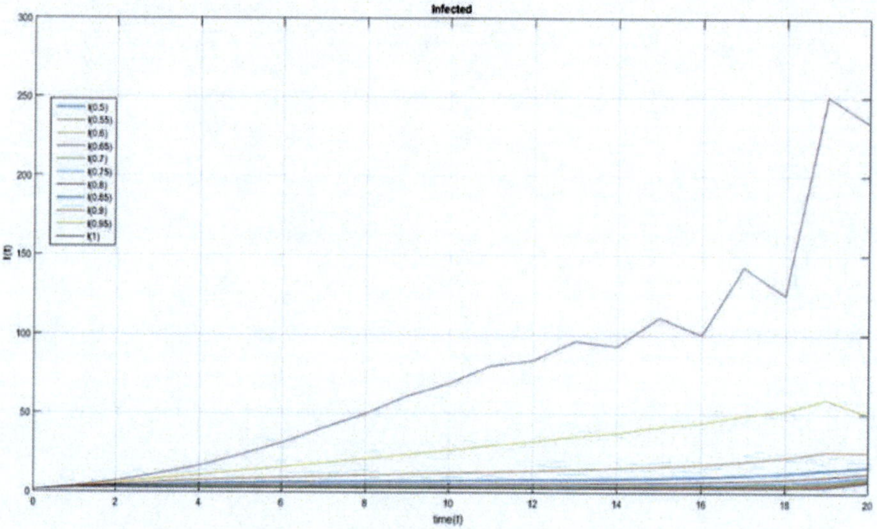


**Fig. 11.4** Numbers of infected individuals I(t) versus time (t)

fractional orders. In addition, the existence of the solution is verified using fixed-point technique, together with justifying the result by means of numerical method and its graphical representation to demonstrate the flow of all compartments at the following values of alpha.

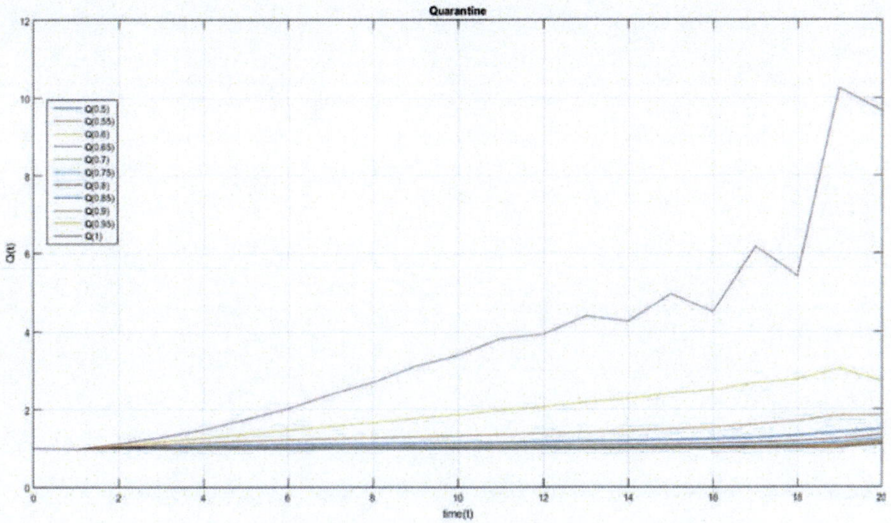


Fig. 11.5 Numbers of quarantine individuals Q(t) versus time (t)

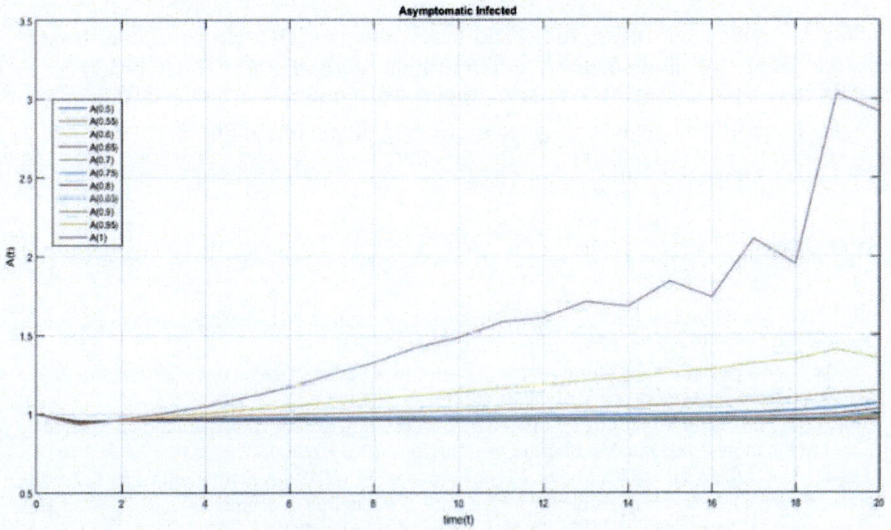


Fig. 11.6 Numbers of asymptomatic individuals A(t) versus time (t)

Visibly, the study exhibits the significant role of self-isolation and lockdown in restraining the spread of the nCOVID-19. The dynamics of the coronavirus under the situation of shutdown is revealed clearly using numerical simulation of mathematical model. Also, the existence and uniqueness of the

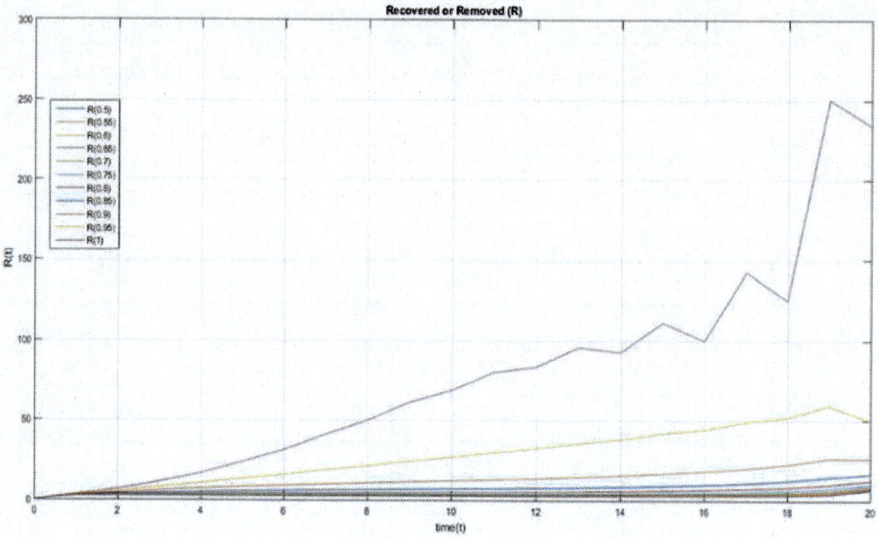


**Fig. 11.7** Numbers of recovered or removed individuals R(t) versus time (t)

solution is evidenced using the fixed-point theory. The above model represents the nature of flow in each compartment with the fractional order $\alpha = 0.5, 0.55, 0.6, 0.65, 0.7, 0.75, 0.8, 0.85, 0.9, 0.95, 0.9999 \simeq 1$ and stability results show that the coronavirus is locally asymptomatically stable if $\Re_0 < 1$.

## References


1. https://www.who.int/emergencies/diseases/novel-coronavirus-2019/events-as-they-Happen.
2. https://www.worldometers.info/coronavirus/country/india/.
3. "Is the World Ready for the corona virus?" Editorial. The New York Times. 29 January 2020. Archived from the original on 30 January 2020.
4. Atangana, A. (2020). Modelling the spread of COVID-19 with new fractal-fractional operators: Can the lockdown save mankind before vaccination? 136, 109860.
5. Dighe, A., Jombart, T., van Kerkhove, M., & Ferguson, N. (2019). A mathematical model of the transmission of middle East respiratory syndrome coronavirus in dromedary camels (Camelus dromedarius). *International Journal of Infectious Diseases, 79*(1), 1–150.
6. Li, C., & Tao, C. (2009). On the fractional Adams method. *Computers & Mathematics with Applications, 58*(8), 1573–1588.
7. Chen, Y., Cheng, J., Jiang, Y., & Liu, K., (2020). A time delay dynamic system with external source for the local outbreak of 2019-nCoV. *Applicable Analysis*, 1–12.
8. Baleanu, D., Jajarmi, A., Mohammadi H., & Rezapour, Sh. (2020). A new study on the mathematical modelling of human liver with Caputo-Fabrizio fractional derivative. *Chaos, Solitons & Fractals*, 134.
9. Akbari Kojabad, E., & Rezapour, Sh. (2017). *Approximate solutions of a sum-type fractional integro-differential equation by using Chebyshev and Legendre polynomials* (p. 351). Eq: Adv. Diff.

10. Ucar, E., Ozdemir, N., & Altun, E. (2019). Fractional order model of immune cells influenced by cancer cells. *Mathematical Modelling of Natural Phenomena, 14*(3), 308.
11. Haq, F., Shah, K., Rahman, G., & Shahzad, M. (2017). Numerical analysis of fractional order model of HIV-1 infection of CD4 + T-cells. *Computational Methods for Differential Equations, 5*(1), 1–11.
12. Singh, H., Dhar, J., Bhatti, H. S., & Chandok, S. (2016). An epidemic model of childhood disease dynamics with maturation delay and latent period of infection. *Modeling Earth Systems and Environment, 2,* 79.
13. Koca, I. (2018). Analysis of rubella disease model with non-local and non-singular fractional derivatives. *Journal of Theories and Applications, 8*(1), 17–25.
14. Losada, J., & Nieto, J. J. (2015). Properties of the new fractional derivative without singular kernel. *Progress in Fractional Differentiation and Applications, 1*(2), 87–92.
15. Tchuenche, J. M., Dube, N., Bhunu, C. P., Smith, R. J., & Bauch, C. T. (2011). The impact of media coverage on the transmission dynamics of human influenza. *BMC Public Health, 11*(Suppl 1), S5.
16. Khan, A. K., & Atangana, A. (2020). Modelling the dynamics of novel corona virus (2019-nCov) with fractional derivative. *Alexandria Engineering Journal,* 1–11.
17. Lauer, S. A., Grantz, K. H., Bi, Q., Jones, F. K., Zheng, Q., Meredith, H. R., Azman, A. S., Reich, N. G., & Lessler, J. (2020). The Incubation Period of corona virus disease 2019 (COVID-19) from publicly reported confirmed cases: Estimation and application. *Annals of Internal Medicine, 172,* 9.
18. Dokuyucu, M. A., Celik, E., Bulut, H., & Baskonus, H. M. (2018). Cancer treatment model with the Caputo-Fabrizio fractional derivative. *European Physical Journal—Plus, 133,* 92.
19. Caputo, M., & Fabrizio, M. (2015). A new definition of fractional derivative without singular kernel. *Progress in Fractional Differentiation and Applications, 1*(2), 73–85.
20. Talaee, M., Shabibi, M., Gilani, A., & Rezapour, Sh. (2020). On the existence of solutions for a pointwise defined multi-singular integro-differential equation with integral boundary condition. *Advances in Difference Equations, 2020,* 41.
21. Owolabi, K. M., & Atangana, A. (2019). Mathematical analysis and computational experiments for an epidemic system with nonlocal and nonsingular derivative. *Chaos, Solitons & Fractals, 1*(126), 41–49.
22. Panchal, J., & Acharya, F. (2020). Mathematical Model on Impact of Quarantine to control the transmission of Corona Virus Disease 2019 (COVID-19). *International Journal of Advanced Science and Technology, 29*(7s), 4235–4244.
23. Rothan HA, Byrareddy SN (2020). The epidemiology and pathogenesis of corona virus disease (COVID-19) outbreak. *Journal of Autoimmunity.* https://doi.org/10.1016/j.jaut.2020.102433
24. Upadhyay, R. K., & Roy, P. (2014). Spread of a disease and its effect on population dynamics in an eco-epidemiological system. *Communications in Nonlinear Science and Numerical Simulation, 19*(12), 4170–4184.
25. Rezapour, S., Mohammadi, H., & Samei, M. E. (2020). SEIR epidemic model for COVID-19 transmission by Caputo derivative of fractional order. *Advances in Difference Equations, 490.*
26. Samko, S. G., Kilbas A. A., & Marichev, O. I. (1993). *Fractional integrals and derivatives: Theory and applications.* CRC Press.
27. Qureshi, S., & Memon, Z. N. (2020). Monotonically decreasing behavior of measles epidemic well captured by Atangana-Baleanu-Caputo fractional operator under real measles data of Pakistan. *Chaos. Solitons and Fractals, 131,* 109478.
28. Rida, S. Z., Arafa, A. A. M., & Gaber, Y. A. (2016). Solution of the fractional epidemic model by L-ADM. *Journal of Fractional Calculus and Applications, 7*(1), 189–195.
29. Tang, B., Wang, X., Li, Q., Bragazzi, N. L., Tang, S., Xiao, Y. et al. (2020). Estimation of the transmission risk of the 2019-nCoV and its implication for public health interventions. *Journal of Clinical Medicine, 9*(2), 462.
30. Tuan, N. H., Mohammadi, H., & Rezapour, S. (2020). A mathematical model for COVID-19 transmission by using the Caputo fractional derivative. *Chaos. Solitons & Fractals, 140,* 110107.

31. Zhou, Y., Ma Z., and Brauer, F. (2004). A discrete epidemic model for SARS transmission and control in China. *Mathematical and Computer Modelling, 40*(13), 1491–1506.
32. Zhao, Z., Zhu, Y. Z., Xu, J. W., Hu, Q. Q., Lei, Z., Rui, J., Liu, X., Wang, Y., Luo, L., Yu, S. S., & Li, J. (2020). A mathematical model for estimating the age-specific transmissibility of a novel corona virus; medRxiv.
33. Zhong, L., Mu, L., Li, J., Wang, J., Yin, Z., & Liu, D. (2020). Early Prediction of the 2019 Novel corona virus Outbreak in the Mainland China based on simple mathematical model. *IEEE Access*, 51761–51769.

# Chapter 12
# Analysis of Novel Corona Virus (COVID-19) Pandemic with Fractional-Order Caputo–Fabrizio Operator and Impact of Vaccination

**A. George Maria Selvam, R. Janagaraj, and R. Dhineshbabu**

**Abstract** Within a very short period, the corona infection virus (COVID-19) has created a global emergency situation by spreading worldwide. This virus has dissimilar effects in different geographical regions. In the beginning of the spread, the number of new cases of active corona virus has shown exponential growth across the globe. At present, for such infection, there is no vaccination or anti-viral medicine specific to the recent corona virus infection. Mathematical formulation of infection models is exceptionally successful to comprehend epidemiological models of ailments, just as it causes us to take vital proportions of general wellbeing interruptions to control the disease transmission and the spread. This work based on a new mathematical model analyses the dynamic behaviour of novel corona virus (COVID-19) using Caputo–Fabrizio fractional derivative. A new modified SEIRQ compartment model is developed to discuss various dynamics. The COVID-19 transmission is studied by varying reproduction number. The basic number of reproduction $R_0$ is determined by applying the next generation matrix. The equilibrium points for disease-free and endemic states are computed with the help of basic reproduction number $R_0$ to check the stability property. The Picard approximation and Banach's fixed point theorem based on iterative Laplace transform are useful in establishing the existence and stability behaviour of the fractional-order system. Finally, numerical computations of the COVID-19 fractional-order system are presented to analyse the dynamical behaviour of the solutions of the model. Also, a fractional-order SEIRQ COVID-19 model with vaccinated people has also been formulated and its dynamics with impact on the propagation behaviour is studied.

A. George Maria Selvam (✉)
Department of Mathematics, Sacred Heart College (Autonomous), Tirupattur, Tamil Nadu 635601, India
e-mail: agmshc@gmail.com

R. Janagaraj
Department of Mathematics, Faculty of Engineering, Karpagam Academy of Higher Education, Coimbatore, Tamil Nadu 641021, India

R. Dhineshbabu
Department of Mathematics, Sri Venkateswara College of Engineering and Technology (Autonomous), Chittoor, Andhra Pradesh 517127, India

 
N. H. Shah and M. Mittal (eds.), *Mathematical Analysis for Transmission of COVID-19*, Mathematical Engineering, https://doi.org/10.1007/978-981-33-6264-2_12

**Keywords** Mathematical modelling · Caputo–Fabrizio derivative · Corona virus · Basic reproduction number · Existence and stability · Fixed point theory · Numerical simulation

**2010 Mathematics Subject Classifications** 34A08 · 34A25 · 34D20 · 35A20 · 35A22

## Introduction

Towards the early December 2019, in Wuhan, China, the first COVID-19 patient was identified to have infection with corona virus. In the next few weeks, the illness spreads generally in China terrain and then to different countries, which increased worldwide panic. The virus was named "SARS-CoV-2", and the malady it causes was named "corona virus ailment 2019 (abridged "COVID-19"). According to official statement from the Indian government, there were 3,043,436 people diagnosed with the disease and 56,846 deaths until 22 August 2020 [1]. Several tactics are implemented by the governments of Indian and other countries around the globe to prevent the spread of the virus, such as traffic stoppage, citywide lockdown, social distancing, community control and promotion of health education awareness.

Not like the severe acute respiratory syndrome (SARS) and different irresistible illnesses, important factor of COVID-19 is that there are asymptomatic populations (who have gentle side effects). These asymptomatic populations are ignorant of their infectious capacity, thereby infecting further ignorant people. In this situation, the transmission rate can rise drastically. Just 80% and 20% of COVID-19 patients have fever and dry cough respectively, as indicated by the ongoing reports from WHO [2]. In the event that we use body temperature level as a way to identify affected cases of COVID-19, it is difficult to detect more than 10% of infected individuals.

The Indian government has been vigilant after identifying the first case of an infected individual in Kerala in late January 2020. Indian government is taking precautionary estimates well in time by reporting across the nation lockdown on 22 March 2020 and its fourth stage, Lockdown 4.0, began on 18 May 2020. Specific measures [3] like restriction of public meetings, airport passenger screening, increased sample testing, suspension of transportation involving buses, trains and planes, dedicated COVID-19 hospitals, increased quarantine facilities, etc. have been taken by the state during the Lockdown time.

Researchers have conducted several studies to understand the nature of this pandemic [4–7]. Several researchers have examined the prediction of COVID-19 in the sense of India using epidemiological [8–11] and mathematical [12–15] models however have restricted investigations of individual states [16]. Taking a gander at the demographic and topographical assorted variety in India, a state-wise investigation of COVID-19 pestilence is a very important necessity.

As an absolutely novel epidemic, COVID-19 considerably affects general well-being and the worldwide economy. Computational analysis plays a significant role in forecasting and controlling the currently underway pandemic COVID-19. In the epidemic, one of the most critical issues is that we must not only establish computational models to forecast disease evolution and predict the efficacy of different action initiatives but also their economic effects. The models have a wide influence on diagnostic decision and policy making. Usually, there are three types of epidemiological mathematical model, such as computational models, involving statistical, machine learning and dynamical methods [17]. Because of the inadequate data on this epidemic, it is important to give more consideration to the dynamical model at this point. Epidemiological models consolidate deterministic or stochastic strategies dependent on various populations in subgroups [18].

This chapter discusses a new mathematical model to analyse the dynamic behaviour of novel corona virus (COVID-19) pandemic by using Caputo–Fabrizio fractional derivative (C-FFD). The following section presents fundamental definitions and basic theories of fractional calculus. A mathematical model of 1st order SEIRQ COVID-19 differential equations formulates in Sect. Mathematical Formulation of COVID-19 Model. Section Existence Criteria of CF Model (3.2) by Picard Approximation obtains the existence results for the modified transmission COVID-19 fractional-order Caputo–Fabrizio (CF) model by using Picard approximation technique. The equilibrium points of disease-free and endemic states are calculated for the transmission of COVID-19 model (12.2) and analysed its stability in Sect. Equilibrium Points and Its Stability. A solution to the fractional COVID-19 system described by (12.2) is achieved by applying Laplace transform (LT) and utilising the fixed point theorem to demonstrate the stability analysis via iterative technique in Sect. Stability Analysis via Iterative Scheme.

## Preliminaries

This section introduces some auxiliary fundamental definitions and primitive concepts of fractional calculus.

**Definition 2.1** References [19, 20] Suppose that $\rho \in (t-1, t]$, such that $t = \lceil \rho \rceil + 1$. For a function $U \in AC^{\rho}_{\mathbb{R}}([0, \infty))$, the Caputo type fractional derivative operator is illustrated by

$$^{C}D_0^{\rho} U(\iota) = \frac{1}{\Gamma(t-\rho)} \int_0^{\iota} (\iota - \lambda)^{t-\rho-1} U^{(t)}(\lambda) d\lambda,$$

provided the integral value is finite.

Recently, in the year 2015, a new fractional-order operator has been considered without singular kernel and was introduced by two Italian mathematicians, namely Caputo and Fabrizio.

**Definition 2.2** References [21] Let $m > n,\ U \in H^1(m, n)$ and $\rho \in (0, 1)$. Then the C-FFD operator is expressed by

$$^{\mathrm{CF}}D_\iota^\rho U(\iota) = \frac{\psi(\rho)}{(1-\rho)} \int_a^\iota \exp\left[-\frac{\rho(\iota-\lambda)}{1-\iota}\right] U'(\lambda)\mathrm{d}\lambda, \quad \iota \geq 0,$$

where $\Psi(\rho)$ is a normalisation function which depends on $\rho$ and satisfies $\psi(0) = \psi(1) = 1$.

Later Losada and Nieto analysed the above C-FFD operator in 2015 and is given in the definition as follows.

**Definition 2.3** References [22] For a function $U$, the fractional-order Caputo–Fabrizio integral operator of order $\rho \in (0, 1)$ is defined as

$$^{\mathrm{CF}}J_0^\rho U(\iota) = \frac{2(1-\rho)}{(2-\rho)\psi(\rho)} U(\iota) + \frac{2\rho}{(2-\rho)\psi(\rho)} \int_0^\iota U(\lambda)\mathrm{d}\lambda, \quad \iota \geq 0,$$

when $U$ is a constant function, then $^{\mathrm{CF}}D_\iota^\rho U(\iota) = 0$.

**Definition 2.4** References [23] The LT for the C-FFD of order $\rho \in (0, 1)$ for $\sigma \in \mathbb{N}$ is presented by

$$L\left(^{\mathrm{CF}}D_\iota^{\sigma+\rho} U(\iota)\right)(\zeta) = \frac{1}{1-\rho} L\left(U^{(\sigma+1)}(\iota)\right) L\left(\exp\left(-\frac{\rho}{1-\rho}\iota\right)\right)$$
$$= \frac{\zeta^{\sigma+1} L(U(\iota)) - \zeta^\sigma U(0) - \zeta^{\sigma-1} U'(0) - \cdots - U^{(\sigma)}(0)}{\zeta + \rho(1-\zeta)}.$$

Specifically,

$$L\left(^{\mathrm{CF}}D_\iota^{\rho+\sigma} U(\iota)\right)(\zeta) = \begin{cases} \frac{\zeta L(U(\iota)) - U(0)}{\zeta + \rho(1-\zeta)}, & \sigma = 0, \\ \frac{\zeta^2 L(U(\iota)) - \zeta U(0) - U'(0)}{\zeta + \rho(1-\zeta)}, & \sigma = 1. \end{cases}$$

## Mathematical Formulation of COVID-19 Model

The aim of infection prevention is to create a predictive strategy to effectively recognise the main factors influencing COVID-19 transmission in India. Epidemiological models such as SEIR present a best essential framework for understanding the spreading disease macroscopic picture. The fundamental idea is to separate the large populations considered into individual chambers of the categories between susceptible people ($S$), exposed people ($E$), infected people ($I$), recovered/removed people

($R$) and COVID-19 in the reservoir (quarantine) people ($Q$). A person in the model is moving from one chamber to another focused on intrinsic parameters that control the dynamics of the spread of disease within the population. A mathematical formulation of 1st order SEIRQ COVID-19 differential equations of the model is given as follows

$$
\begin{aligned}
D_\iota S(\iota) &= \varpi - \eta S(\iota)I(\iota) - \varepsilon S(\iota)Q(\iota) - \kappa S(\iota),\\
D_\iota E(\iota) &= \eta S(\iota)I(\iota) + \varepsilon S(\iota)Q(\iota) - (1-\xi)\vartheta E(\iota) - \kappa E(\iota),\\
D_\iota I(\iota) &= (1-\xi)\vartheta E(\iota) - \nu I(\iota) - \kappa I(\iota),\\
D_\iota R(\iota) &= \nu I(\iota) - \kappa R(\iota),\\
D_\iota Q(\iota) &= \chi I(\iota) - \tau Q(\iota),
\end{aligned}
\tag{12.1}
$$

and $S_\iota(0) = S_0$, $E_\iota(0) = E_0$, $I_\iota(0) = I_0$, $R_\iota(0) = R_0$, $Q_\iota(0) = Q_0$ are the initial conditions of the model (12.1). Also where $\varpi = n \times N$, $n$ is the birth rate and $N$ is the total number of populations, $\eta$ is the contact rate from $I$ to $S$, $\varepsilon$ is the disease transmission coefficient rate from $Q$ to $S$, $\kappa$ is the natural death rate of population, $\xi$ is the proportion of asymptomatic infected population, $\vartheta$ is the period of incubation, $\nu$ is the infectious period of $I$, $\chi$ is the contribution of the infection to $Q$ by $I$, $\tau$ is removing rate of infection from $Q$.

Fractional calculus (FC) is the analysis of non-integer-order derivatives and integrals that gives an excellent technique for understanding the memory and genetic properties of dynamic systems [19, 24, 25]. It has gained tremendous popularity over the past few decades and has become an effective tool for better modelling real-world problems involving epidemiology [26, 27], ecology [28, 29], electrical circuits [30], mathematical biology [31], medicine [32]. Furthermore, a number of biological and engineering problems are currently used to model and solve fractional differential equations (FDEs) [33–35]. Fractional derivatives (FDs) provide numerous approaches for non-integer order equations, which can help us, use the FDE that can best explain the model's dynamics.

Recently, several mathematicians have been paying attention to the concept of the FDs. There are some limitations to the well-known concepts of Caputo and Riemann–Liouville (R-L) FDs. The derivative of a constant by R-L is not zero. R-L and Caputo FDs concepts require a single kernel and this deficiency influences the real-life situations. In 2015, a new FD with non-singular kernel was introduced by Caputo and Fabrizio [21] and Losada [22] studied its properties. Therefore, it is more fitting to research Caputo–Fabrizio concept with non-singular kernel. The modified transmission COVID-19 fractional-order CF model for $\iota \geq 0$ and $\rho \in (0, 1)$ is of the form

$$
\begin{aligned}
{}^{\mathrm{CF}}D_\iota^\rho S(\iota) &= \varpi - \eta S(\iota)I(\iota) - \varepsilon S(\iota)Q(\iota) - \kappa S(\iota),\\
{}^{\mathrm{CF}}D_\iota^\rho E(\iota) &= \eta S(\iota)I(\iota) + \varepsilon S(\iota)Q(\iota) - (1-\xi)\vartheta E(\iota) - \kappa E(\iota),\\
{}^{\mathrm{CF}}D_\iota^\rho I(\iota) &= (1-\xi)\vartheta E(\iota) - \nu I(\iota) - \kappa I(\iota),
\end{aligned}
$$

$$\begin{aligned}
{}^{\mathrm{CF}}D_{\iota}^{\rho}R(\iota) &= \nu I(\iota) - \kappa R(\iota),\\
{}^{\mathrm{CF}}D_{\iota}^{\rho}Q(\iota) &= \chi I(\iota) - \tau Q(\iota),
\end{aligned} \tag{12.2}$$

## Existence Criteria of CF Model (12.2) by Picard Approximation

In this section, we shall discuss the existence of a unique solution for the modified transmission COVID-19 fractional-order CF model with Picard approximation technique. For this purpose, considering the fractional-order CF integral, given by Losada and Nieto [22], on both sides of model (12.2), we get

$$\begin{aligned}
S(\iota) &= S(0) + {}^{\mathrm{CF}}J_{\iota}^{\rho}[\varpi - \eta S(\iota)I(\iota) - \varepsilon S(\iota)Q(\iota) - \kappa S(\iota)],\\
E(\iota) &= E(0) + {}^{\mathrm{CF}}J_{\iota}^{\rho}[\eta S(\iota)I(\iota) + \varepsilon S(\iota)Q(\iota) - (1-\xi)\vartheta E(\iota) - \kappa E(\iota)],\\
I(\iota) &= I(0) + {}^{\mathrm{CF}}J_{\iota}^{\rho}[(1-\xi)\vartheta E(\iota) - \nu I(\iota) - \kappa I(\iota)],\\
R(\iota) &= R(0) + {}^{CF}J_{\iota}^{\rho}[\nu I(\iota) - \kappa R(\iota)],\\
Q(\iota) &= Q(0) + {}^{\mathrm{CF}}J_{\iota}^{\rho}[\chi I(\iota) - \tau Q(\iota)].
\end{aligned}$$

Definition 2.3 leads to

$$\begin{aligned}
S(\iota) = S(0) &+ \frac{2(1-\rho)}{(2-\rho)\psi(\rho)}[\varpi - \eta S(\iota)I(\iota) - \varepsilon S(\iota)Q(\iota) - \kappa S(\iota)]\\
&+ \frac{2\rho}{(2-\rho)\psi(\rho)}\int_0^{\iota}[\varpi - \eta S(\lambda)I(\lambda) - \varepsilon S(\lambda)Q(\lambda) - \kappa S(\lambda)]\mathrm{d}\lambda,\\
E(\iota) = E(0) &+ \frac{2(1-\rho)}{(2-\rho)\psi(\rho)}[\eta S(\iota)I(\iota) + \varepsilon S(\iota)Q(\iota) - (1-\xi)\vartheta E(\iota) - \kappa E(\iota)]\\
&+ \frac{2\rho}{(2-\rho)\psi(\rho)}\int_0^{\iota}[\eta S(\lambda)I(\lambda) + \varepsilon S(\lambda)Q(\lambda) - (1-\xi)\vartheta E(\lambda) - \kappa E(\lambda)]\mathrm{d}\lambda,\\
I(\iota) = I(0) &+ \frac{2(1-\rho)}{(2-\rho)\psi(\rho)}[(1-\xi)\vartheta E(\iota) - \nu I(\iota) - \kappa I(\iota)]\\
&+ \frac{2\rho}{(2-\rho)\psi(\rho)}\int_0^{\iota}[(1-\xi)\vartheta E(\lambda) - \nu I(\lambda) - \kappa I(\lambda)]\mathrm{d}\lambda,
\end{aligned}$$

$$R(\iota) = R(0) + \frac{2(1-\rho)}{(2-\rho)\psi(\rho)}[\nu I(\iota) - \kappa R(\iota)]$$

$$+\frac{2\rho}{(2-\rho)\psi(\rho)}\int_0^{\iota}[\nu I(\lambda)-\kappa R(\lambda)]\mathrm{d}\lambda,$$

$$Q(\iota)=Q(0)+\frac{2(1-\rho)}{(2-\rho)\psi(\rho)}[\chi I(\iota)-\tau R(\iota)]$$

$$+\frac{2\rho}{(2-\rho)\psi(\rho)}\int_0^{\iota}[\chi I(\lambda)-\tau Q(\lambda)]\mathrm{d}\lambda.$$

Equivalently

$$\begin{aligned}
S(\iota)&=S(0)+\frac{2(1-\rho)}{(2-\rho)\psi(\rho)}[A_1(\iota,S)]+\frac{2\rho}{(2-\rho)\psi(\rho)}\int_0^{\iota}[A_1(\lambda,S)]\mathrm{d}\lambda,\\
E(\iota)&=E(0)+\frac{2(1-\rho)}{(2-\rho)\psi(\rho)}[A_2(\iota,E)]+\frac{2\rho}{(2-\rho)\psi(\rho)}\int_0^{\iota}[A_2(\lambda,E)]\mathrm{d}\lambda,\\
I(\iota)&=I(0)+\frac{2(1-\rho)}{(2-\rho)\psi(\rho)}[A_3(\iota,I)]+\frac{2\rho}{(2-\rho)\psi(\rho)}\int_0^{\iota}[A_3(\lambda,I)]\mathrm{d}\lambda,\\
R(\iota)&=R(0)+\frac{2(1-\rho)}{(2-\rho)\psi(\rho)}[A_4(\iota,R)]+\frac{2\rho}{(2-\rho)\psi(\rho)}\int_0^{\iota}[A_4(\lambda,R)]\mathrm{d}\lambda,\\
Q(\iota)&=Q(0)+\frac{2(1-\rho)}{(2-\rho)\psi(\rho)}[A_5(\iota,Q)]+\frac{2\rho}{(2-\rho)\psi(\rho)}\int_0^{\iota}[A_5(\lambda,Q)]\mathrm{d}\lambda.
\end{aligned}\tag{12.3}$$

Here

$$\begin{aligned}
A_1(\iota,S)&=\varpi-\eta S(\iota)I(\iota)-\varepsilon S(\iota)Q(\iota)-\kappa S(\iota),\\
A_2(\iota,E)&=\eta S(\iota)I(\iota)+\varepsilon S(\iota)Q(\iota)-(1-\xi)\vartheta E(\iota)-\kappa E(\iota),\\
A_3(\iota,I)&=(1-\xi)\vartheta E(\iota)-\nu I(\iota)-\kappa I(\iota),\\
A_4(\iota,R)&=\nu I(\iota)-\kappa R(\iota),\\
A_5(\iota,Q)&=\chi I(\iota)-\tau Q(\iota).
\end{aligned}$$

Now considering the functions $S(\iota)$ and $S_*(\iota)$, we obtain

$$\begin{aligned}
&\|A_1(\iota,S(\iota))-A_1(\iota,S_*(\iota))\|\\
&=\|\varpi-\eta S(\iota)I(\iota)-\varepsilon S(\iota)Q(\iota)-\kappa S(\iota)-\varpi+\eta S_*(\iota)I(\iota)+\varepsilon S_*(\iota)Q(\iota)+\kappa S_*(\iota)\|\\
&=\|-\eta I(\iota)[S(\iota)-S_*(\iota)]-\varepsilon Q(\iota)[S(\iota)-S_*(\iota)]-\kappa[S(\iota)-S_*(\iota)]\|
\end{aligned}$$

$$\leq (\eta\|I(\iota)\| + \varepsilon\|Q(\iota)\| + \kappa)\|S(\iota) - S_*(\iota)\|$$
$$\leq \delta_1\|S(\iota) - S_*(\iota)\|.$$

Here $\delta_1 = \eta z_1 + \varepsilon z_2 + \kappa$ and $z_1 = \|I(\iota)\|$, $z_2 = \|Q(\iota)\|$, which are bounded. Thus $A_1$ satisfies the Lipchitz condition. Moreover, $A_1$ is a contraction when $0 < \delta_1 \leq 1$. Similarly,

$$\|A_2(\iota, E(\iota)) - A_2(\iota, E_*(\iota))\| \leq \delta_2\|E(\iota) - E_*(\iota)\|,$$
$$\|A_3(\iota, I(\iota)) - A_3(\iota, I_*(\iota))\| \leq \delta_3\|I(\iota) - I_*(\iota)\|,$$
$$\|A_4(\iota, R(\iota)) - A_4(\iota, R_*(\iota))\| \leq \delta_4\|R(\iota) - R_*(\iota)\|,$$
$$\|A_5(\iota, Q(\iota)) - A_5(\iota, Q_*(\iota))\| \leq \delta_5\|Q(\iota) - Q_*(\iota)\|.$$

Hence, the Lipchitz condition is satisfied for $A_2, A_3, A_4, A_5$.
Now, consider the recursive formula:

$$S_j(\iota) = \frac{2(1-\rho)}{(2-\rho)\psi(\rho)}[A_1(\iota, S_{j-1})] + \frac{2\rho}{(2-\rho)\psi(\rho)}\int_0^{\iota}[A_1(\lambda, S_{j-1})]d\lambda,$$

$$E_j(\iota) = \frac{2(1-\rho)}{(2-\rho)\psi(\rho)}[A_2(\iota, E_{j-1})] + \frac{2\rho}{(2-\rho)\psi(\rho)}\int_0^{\iota}[A_2(\lambda, E_{j-1})]d\lambda,$$

$$I_j(\iota) = \frac{2(1-\rho)}{(2-\rho)\psi(\rho)}[A_3(\iota, I_{j-1})] + \frac{2\rho}{(2-\rho)\psi(\rho)}\int_0^{\iota}[A_3(\lambda, I_{j-1})]d\lambda,$$

$$R_j(\iota) = \frac{2(1-\rho)}{(2-\rho)\psi(\rho)}[A_4(\iota, R_{j-1})] + \frac{2\rho}{(2-\rho)\psi(\rho)}\int_0^{\iota}[A_4(\lambda, R_{j-1})]d\lambda,$$

$$Q_j(\iota) = \frac{2(1-\rho)}{(2-\rho)\psi(\rho)}[A_5(\iota, Q_{j-1})] + \frac{2\rho}{(2-\rho)\psi(\rho)}\int_0^{\iota}[A_5(\lambda, Q_{j-1})]d\lambda. \tag{12.4}$$

From (12.4)

$$B_{1j} = S_j(\iota) - S_{j-1}(\iota)$$
$$= \frac{2(1-\rho)}{(2-\rho)\psi(\rho)}[A_1(\iota, S_{j-1}) - A_1(\iota, S_{j-2})]$$
$$+ \frac{2\rho}{(2-\rho)\psi(\rho)}\int_0^{\iota}[A_1(\lambda, S_{j-1}) - A_1(\lambda, S_{j-2})]d\lambda, \quad B_{2j} = E_j(\iota) - E_{j-1}(\iota)$$
$$= \frac{2(1-\rho)}{(2-\rho)\psi(\rho)}[A_2(\iota, E_{j-1}) - A_2(\iota, E_{j-2})]$$

$$+\frac{2\rho}{(2-\rho)\psi(\rho)}\int_0^{\iota}\big[A_2(\lambda, E_{j-1})-A_2(\lambda, E_{j-2})\big]\mathrm{d}\lambda,\ B_{3j}=I_j(\iota)-I_{j-1}(\iota)$$
$$=\frac{2(1-\rho)}{(2-\rho)\psi(\rho)}\big[A_3(\iota, I_{j-1})-A_3(\iota, I_{j-2})\big]$$
$$+\frac{2\rho}{(2-\rho)\psi(\rho)}\int_0^{\iota}\big[A_3(\lambda, I_{j-1})-A_3(\lambda, I_{j-2})\big]\mathrm{d}\lambda,\ B_{4j}=R_j(\iota)-R_{j-1}(\iota)$$
$$=\frac{2(1-\rho)}{(2-\rho)\psi(\rho)}\big[A_4(\iota, R_{j-1})-A_4(\iota, R_{j-2})\big]$$
$$+\frac{2\rho}{(2-\rho)\psi(\rho)}\int_0^{\iota}\big[A_4(\lambda, R_{j-1})-A_4(\lambda, R_{j-2})\big]\mathrm{d}\lambda,\ B_{5j}=Q_j(\iota)-Q_{j-1}(\iota)$$
$$=\frac{2(1-\rho)}{(2-\rho)\psi(\rho)}\big[A_5(\iota, Q_{j-1})-A_5(\iota, Q_{j-2})\big]$$
$$+\frac{2\rho}{(2-\rho)\psi(\rho)}\int_0^{\iota}\big[A_5(\lambda, Q_{j-1})-A_5(\lambda, Q_{j-2})\big]\mathrm{d}\lambda.$$

From the above equations, one can conclude that

$$S_j(\iota)=\sum_{h=o}^{j}B_{1h}(\iota),\ E_j(\iota)=\sum_{h=o}^{j}B_{2h}(\iota),\ I_j(\iota)=\sum_{h=o}^{j}B_{3h}(\iota),$$
$$R_j(\iota)=\sum_{h=o}^{j}B_{4h}(\iota),\ Q_j(\iota)=\sum_{h=o}^{j}B_{5h}(\iota). \tag{12.5}$$

Applying triangular inequality to the definition of $B_{1j}'s$ leads to

$$\|B_{1j}(\iota)\|=\|S_j(\iota)-S_{j-1}(\iota)\|$$
$$=\left\|\frac{2(1-\rho)}{(2-\rho)\psi(\rho)}\big[A_1(\iota, S_{j-1})-A_1(\iota, S_{j-2})\big]\right.$$
$$\left.+\frac{2\rho}{(2-\rho)\psi(\rho)}\int_0^{\iota}\big[A_1(\lambda, S_{j-1})-A_1(\lambda, S_{j-2})\big]\mathrm{d}\lambda\right\|,$$

$$\|B_{1j}(\iota)\|\le\frac{2(1-\rho)}{(2-\rho)\psi(\rho)}\big\|\big[A_1(\iota, S_{j-1})-A_1(\iota, S_{j-2})\big]\big\|$$
$$+\frac{2\rho}{(2-\rho)\psi(\rho)}\left\|\int_0^{\iota}\big[A_1(\lambda, S_{j-1})-A_1(\lambda, S_{j-2})\big]\mathrm{d}\lambda\right\|.$$

Since $A_1$ is fulfilled the Lipchitz condition, hence

$$\left\|S_j(\iota) - S_{j-1}(\iota)\right\| \leq \frac{2(1-\rho)}{(2-\rho)\psi(\rho)}\delta_1 \left\|S_{j-1} - S_{j-2}\right\| + \frac{2\rho}{(2-\rho)\psi(\rho)}\delta_1 \int_0^\iota \left\|S_{j-1} - S_{j-2}\right\| d\lambda.$$

Thus,

$$\left\|B_{1j}(\iota)\right\| \leq \frac{2(1-\rho)}{(2-\rho)\psi(\rho)}\delta_1 \left\|B_{1j-1}(\iota)\right\| + \frac{2\rho}{(2-\rho)\psi(\rho)}\delta_1 \int_0^\iota \left\|B_{1j-1}(\lambda)\right\| d\lambda. \tag{12.6}$$

Similarly,

$$\left\|B_{2j}(\iota)\right\| \leq \frac{2(1-\rho)}{(2-\rho)\psi(\rho)}\delta_2 \left\|B_{2j-1}(\iota)\right\| + \frac{2\rho}{(2-\rho)\psi(\rho)}\delta_2 \int_0^\iota \left\|B_{2j-1}(\lambda)\right\| d\lambda,$$

$$\left\|B_{3j}(\iota)\right\| \leq \frac{2(1-\rho)}{(2-\rho)\psi(\rho)}\delta_3 \left\|B_{3j-1}(\iota)\right\| + \frac{2\rho}{(2-\rho)\psi(\rho)}\delta_3 \int_0^\iota \left\|B_{3j-1}(\lambda)\right\| d\lambda,$$

$$\left\|B_{4j}(\iota)\right\| \leq \frac{2(1-\rho)}{(2-\rho)\psi(\rho)}\delta_4 \left\|B_{4j-1}(\iota)\right\| + \frac{2\rho}{(2-\rho)\psi(\rho)}\delta_4 \int_0^\iota \left\|B_{4j-1}(\lambda)\right\| d\lambda,$$

$$\left\|B_{5j}(\iota)\right\| \leq \frac{2(1-\rho)}{(2-\rho)\psi(\rho)}\delta_5 \left\|B_{5j-1}(\iota)\right\| + \frac{2\rho}{(2-\rho)\psi(\rho)}\delta_5 \int_0^\iota \left\|B_{5j-1}(\lambda)\right\| d\lambda. \tag{12.7}$$

This implies that the fractional-order COVID-19 model (12.2) has a solution.

Now to show that model (12.3) has unique solution, suppose that $S_1, E_1, I_1, R_1, Q_1$ has another solution of model (12.3) such that

$$\begin{aligned}\|S(\iota) - S_1(\iota)\| &= \left\| \frac{2(1-\rho)}{(2-\rho)\psi(\rho)}[A_1(\iota, S) - A_1(\iota, S_1)] \right. \\ &\quad \left. + \frac{2\rho}{(2-\rho)\psi(\rho)} \int_0^\iota [A_1(\lambda, S) - A_1(\lambda, S_1)] d\lambda \right\|, \\ &\leq \frac{2(1-\rho)}{(2-\rho)\psi(\rho)} \|[A_1(\iota, S) - A_1(\iota, S_1)]\|\end{aligned}$$

$$+\frac{2\rho}{(2-\rho)\psi(\rho)}\left\|\int_0^{\iota}[A_1(\lambda,S)-A_1(\lambda,S_1)]\mathrm{d}\lambda\right\|.$$

From Lipchitz condition of $A_1$ yields,

$$\|S(\iota)-S_1(\iota)\|\leq\frac{2(1-\rho)}{(2-\rho)\psi(\rho)}\delta_1\|S(\iota)-S_1(\iota)\|+\frac{2\rho}{(2-\rho)\psi(\rho)}\delta_1\iota\|S(\iota)-S_1(\iota)\|,$$

equivalently

$$\left(1-\frac{2(1-\rho)}{(2-\rho)\psi(\rho)}\delta_1-\frac{2\rho}{(2-\rho)\psi(\rho)}\delta_1\iota\right)\|S(\iota)-S_1(\iota)\|\leq 0. \tag{12.8}$$

**Theorem 4.1** *The solution of COVID-19 fractional-order CF model* (12.3) *has a unique solution if the following assumption holds*:

$$\left(1-\frac{2(1-\rho)}{(2-\rho)\psi(\rho)}\delta_1-\frac{2\rho}{(2-\rho)\psi(\rho)}\delta_1\iota\right)\geq 0. \tag{12.9}$$

***Proof*** From assumption (12.9) and Eq. (12.8), it is easy to arrive at

$$\left(1-\frac{2(1-\rho)}{(2-\rho)\psi(\rho)}\delta_1-\frac{2\rho}{(2-\rho)\psi(\rho)}\delta_1\iota\right)\|S(\iota)-S_1(\iota)\|=0.$$

Therefore, $\|S(\iota)-S_1(\iota)\|=0$ and this implies that $S(\iota)=S_1(\iota)$. Similarly, it is easy to prove

$$E(\iota)=E_1(\iota),\ I(\iota)=I_1(\iota),\ R(\iota)=R_1(\iota),\ Q(\iota)=Q_1(\iota).$$

This completes the proof.

## Equilibrium Points and Its Stability

The equilibrium points of disease-free and endemic states are calculated for the transmission of COVID-19. The transmission COVID-19 model is studied by varying reproduction number. The reproduction number $R_0$ is determined by applying the next generation matrix. Consequently, the disease-free and endemic equilibrium states are locally and globally asymptotically stable.

Since, the recovered class $R$ does not have a role in $S$, $E$, $I$ and $Q$ compartment, model (12.2) can be simplified and summarised to the following equivalent 4D model

$${}^{\mathrm{CF}}D_{\iota}^{\rho}S(\iota)=\varpi-\eta S(\iota)I(\iota)-\varepsilon S(\iota)Q(\iota)-\kappa S(\iota),$$

$$
\begin{aligned}
{}^{CF}D_{\iota}^{\rho}E(\iota) &= \eta S(\iota)I(\iota) + \varepsilon S(\iota)Q(\iota) - (1-\xi)\vartheta E(\iota) - \kappa E(\iota),\\
{}^{CF}D_{\iota}^{\rho}I(\iota) &= (1-\xi)\vartheta E(\iota) - \nu I(\iota) - \kappa I(\iota),\\
{}^{CF}D_{\iota}^{\rho}Q(\iota) &= \chi I(\iota) - \tau Q(\iota),
\end{aligned} \tag{12.10}
$$

with $S(0) = S_0$, $E(0) = E_0$, $I(0) = I_0$, $Q(0) = Q_0$ are the initial conditions and the feasible region of the model is

$$
\Xi = \left\{(S, E, I, Q) | S, E, I, Q \geq 0, S + E + I + Q \leq N \leq \frac{\varpi}{\kappa}\right\}.
$$

**Theorem 5.1** *Model (12.10) has disease-free and endemic equilibrium points, i.e.* $\Upsilon_0 = (S_0, E_0, I_0, Q_0) = \left(\frac{\varpi}{\kappa}, 0, 0, 0\right)$ *and* $\Upsilon_* = (S_*, E_*, I_*, Q_*)$, *where*

$$
S_* = \frac{\tau(\kappa+\nu)(\vartheta\xi - \vartheta - \kappa)}{\vartheta(\xi-1)(\eta\tau + \varepsilon\chi)},\ E_* = \frac{\tau\kappa(\kappa+\nu)}{\vartheta(\xi-1)(\eta\tau+\varepsilon\chi)} - \frac{\varpi}{(\vartheta\xi - \vartheta - \kappa)},
$$
$$
I_* = \frac{\vartheta\varpi(\xi-1)}{(\kappa+\nu)(\vartheta\xi-\vartheta-\kappa)} - \frac{\tau\kappa}{(\eta\tau+\varepsilon\chi)},\ Q_* = \frac{\chi}{\tau}I_*.
$$

***Proof*** By solving the model (12.10) while $I = 0$, disease-free equilibrium point $\Upsilon_0 = (S_0, E_0, I_0, Q_0) = \left(\frac{\varpi}{\kappa}, 0, 0, 0\right)$ is obtained. While $I \neq 0$, endemic equilibrium point $\Upsilon_* = (S_*, E_*, I_*, Q_*)$ is achieved. This ends the proof.

The reproduction number for the COVID-19 model (12.10) is computed with the help of next generation matrix [23]. From Theorem 5.1, model (12.10) has the disease-free equilibrium $\Upsilon_0 = (S_0, E_0, I_0, Q_0) = \left(\frac{\varpi}{\kappa}, 0, 0, 0\right)$ and considering the compartments $E, I, Q$ gives

$$
\mathrm{H} = \begin{bmatrix} 0 & \frac{\varpi\eta}{\kappa} & \frac{\varpi\varepsilon}{\kappa} \\ 0 & 0 & 0 \\ 0 & 0 & 0 \end{bmatrix}, \mathrm{T} = \begin{bmatrix} (1-\xi)\vartheta + \kappa & 0 & 0 \\ -(1-\xi)\vartheta & \nu+\kappa & 0 \\ 0 & -\chi & \tau \end{bmatrix},
$$

and the next generation matrix

$$
\Omega = \mathrm{HT}^{-1} = \begin{bmatrix} \frac{\varpi\vartheta(\xi-1)(\eta\tau+\varepsilon\chi)}{\tau\kappa(\kappa+\nu)(\vartheta\xi-\vartheta-\kappa)} & \frac{\varpi\eta\tau}{\kappa} & 0 \\ 0 & 0 & 0 \\ 0 & 0 & 0 \end{bmatrix}.
$$

Therefore, the spectral radius of $\Omega$ is $\mathrm{K}(\Omega) = \frac{\varpi\vartheta(\xi-1)(\eta\tau+\varepsilon\chi)}{\tau\kappa(\kappa+\nu)(\vartheta\xi-\vartheta-\kappa)}$.

According to [36], the basic number of reproduction of COVID-19 model (12.10) is

$$
R_0 = \frac{\varpi\vartheta(\xi-1)(\eta\tau+\varepsilon\chi)}{\tau\kappa(\kappa+\nu)(\vartheta\xi-\vartheta-\kappa)}.
$$

Local and global asymptotic stability of equilibrium points for disease-free and endemic states of the system (12.10) is studied in the following theorems.

**Theorem 5.2** *In a feasible region $\Xi$, the disease-free equilibrium point $\Upsilon_0 = (S_0, E_0, I_0, Q_0)$ is locally asymptotically stable when $R_0 < 1$.*

***Proof*** The Jacobian matrix and characteristic equation of the Jacobian matrix at $\Upsilon_0 = (S_0, E_0, I_0, Q_0)$ of the model (12.10) are given by

$$|J(\Upsilon_0) - \Theta I| = \begin{vmatrix} -\kappa - \Theta & 0 & -\frac{\varpi\eta}{\kappa} & -\frac{\varpi\varepsilon}{\kappa} \\ 0 & (\xi - 1)\vartheta - \kappa - \Theta & 0 & 0 \\ 0 & (1 - \xi)\vartheta & -(\nu + \kappa) - \Theta & 0 \\ 0 & 0 & \chi & -\tau - \Theta \end{vmatrix} = 0,$$

and

$$(\kappa + \Theta)(\kappa - (\xi - 1)\vartheta + \Theta)(\nu + \kappa + \Theta)(\tau + \Theta) = 0. \tag{12.11}$$

Using Hurwitz criterion [37], each root of (12.11) has no positive real roots. Hence, based on the stability theory [38], $\Upsilon_0$ is locally asymptotically stability when $R_0 < 1$. This concludes the proof.

**Theorem 5.3** *In a feasible region $\Xi$, the disease-free equilibrium point $\Upsilon_0 = (S_0, E_0, I_0, Q_0)$ is globally asymptotically stable when $R_0 < 1$.*

***Proof*** From $S(\iota)$ in the model (12.1) yields $S'(\iota) \leq \varpi - \kappa S(\iota)$, solving the equation leads to $S(\iota) \leq \frac{\varpi}{\kappa} + \left[S(0) - \frac{\varpi}{\kappa}\right]e^{-\kappa\iota}$.

Define a Lyapunov function:

$$L(\iota) = (1 - \xi)\vartheta E + [(1 - \xi)\vartheta + \kappa]I.$$

Thus,

$$\begin{aligned} L'(\iota) &= (1 - \xi)\vartheta E' + [(1 - \xi)\vartheta + \kappa]I', \\ &= (1 - \xi)\vartheta[\eta S(\iota)I(\iota) + \varepsilon S(\iota)Q(\iota) - (1 - \xi)\vartheta E(\iota) - \kappa E(\iota)] \\ &\quad + [(1 - \xi)\vartheta + \kappa][(1 - \xi)\vartheta E(\iota) - \nu I(\iota) - \kappa I(\iota)], \\ &= \left\{\left(\eta\vartheta(1 - \xi) + \varepsilon\vartheta(1 - \xi)\frac{\chi}{\tau}\right)S - \left[\nu\vartheta(1 - \xi) + \kappa\vartheta(1 - \xi) + \kappa\nu + \kappa^2\right]\right\}I, \\ L'(\iota) &\leq \left[\frac{\varpi\vartheta(1 - \xi)(\eta\tau + \chi\varepsilon)}{\kappa\tau} - (\kappa + \nu)(\kappa + \vartheta - \vartheta\xi)\right]I. \end{aligned}$$

Hence, $R_0 = \frac{\varpi\vartheta(\xi-1)(\eta\tau+\varepsilon\chi)}{\tau\kappa(\kappa+\nu)(\vartheta\xi-\vartheta-\kappa)} < 1$ concludes $L'(\iota) < 0$. Based on the LaSalle invariance principle [39], the disease-free equilibrium point $\Upsilon_0 = (S_0, E_0, I_0, Q_0)$ is globally asymptotically stable when $R_0 < 1$. The proof is completed.

**Theorem 5.4** *In a feasible region $\Xi$, the endemic equilibrium point $\Upsilon_* = (S_*, E_*, I_*, Q_*)$ is locally asymptotically stable when $R_0 > 1$.*

***Proof*** The Jacobian matrix and characteristic equation of the Jacobian matrix at $\Upsilon_* = (S_*, E_*, I_*, Q_*)$ of the model (12.10) are expressed by

$$|J(\Upsilon_0) - \Theta I| = \begin{vmatrix} -\eta I_* - \varepsilon Q_* - \kappa - \Theta & 0 & -\eta S_* & -\varepsilon S_* \\ \eta I_* + \varepsilon Q_* & (\xi - 1)\vartheta - \kappa - \Theta & 0 & 0 \\ 0 & (1-\xi)\vartheta & -(\nu + \kappa) - \Theta & 0 \\ 0 & 0 & \chi & -\tau - \Theta \end{vmatrix} = 0,$$

and

$$(\eta I_* + \varepsilon Q_* + \kappa + \Theta)(\kappa + (1-\xi)\vartheta + \Theta)(\nu + \kappa + \Theta)(\tau + \Theta) + \varepsilon S_*(\eta I_* + \varepsilon Q_*)(1-\xi)\vartheta(\chi + \tau + \Theta) = 0. \tag{12.12}$$

Using Hurwitz criterion [37], each root of (12.12) has no positive real roots. Hence, based on the stability theory [38], $\Upsilon_*$ is locally asymptotically stability when $R_0 > 1$. This concludes the proof.

## Stability Analysis via Iterative Scheme

A solution to the fractional COVID-19 system (12.2) is obtained by applying LT and using the fixed point theorem to analyse the stability via iterative technique. In order to achieve this, taking LT on both side of system (12.2), we obtain

$$\begin{aligned} L\big[{}^{\mathrm{CF}}D_\iota^\rho S(\iota)\big] &= L[\varpi - \eta S(\iota)I(\iota) - \varepsilon S(\iota)Q(\iota) - \kappa S(\iota)], \\ L\big[{}^{\mathrm{CF}}D_\iota^\rho E(\iota)\big] &= L[\eta S(\iota)I(\iota) + \varepsilon S(\iota)Q(\iota) - (1-\xi)\vartheta E(\iota) - \kappa E(\iota)], \\ L\big[{}^{\mathrm{CF}}D_\iota^\rho I(\iota)\big] &= L[(1-\xi)\vartheta E(\iota) - \nu I(\iota) - \kappa I(\iota)], \\ L\big[{}^{\mathrm{CF}}D_\iota^\rho R(\iota)\big] &= L[\nu I(\iota) - \kappa R(\iota)], \\ L\big[{}^{\mathrm{CF}}D_\iota^\rho Q(\iota)\big] &= L[\chi I(\iota) - \tau Q(\iota)]. \end{aligned} \tag{12.13}$$

Applying Definition 2.4. For LT of fractional-order CF operator in (12.13) yields

$$\begin{aligned} \frac{\zeta L[S(\iota)] - S(0)}{\zeta + \rho(1-\zeta)} &= L[\varpi - \eta S(\iota)I(\iota) - \varepsilon S(\iota)Q(\iota) - \kappa S(\iota)], \\ \frac{\zeta L[E(\iota)] - E(0)}{\zeta + \rho(1-\zeta)} &= L[\eta S(\iota)I(\iota) + \varepsilon S(\iota)Q(\iota) - (1-\xi)\vartheta E(\iota) - \kappa E(\iota)], \\ \frac{\zeta L[I(\iota)] - I(0)}{\zeta + \rho(1-\zeta)} &= L[(1-\xi)\vartheta E(\iota) - \nu I(\iota) - \kappa I(\iota)], \\ \frac{\zeta L[R(\iota)] - R(0)}{\zeta + \rho(1-\zeta)} &= L[\nu I(\iota) - \kappa R(\iota)], \end{aligned}$$

$$\frac{\zeta L[Q(\iota)] - Q(0)}{\zeta + \rho(1-\zeta)} = L[\chi I(\iota) - \tau Q(\iota)].$$

or

$$\begin{aligned}
L[S(\iota)] &= \frac{S(0)}{\zeta} + \left(\frac{\zeta + \rho(1-\zeta)}{\zeta}\right) L[\varpi - \eta S(\iota) I(\iota) - \varepsilon S(\iota) Q(\iota) - \kappa S(\iota)],\\
L[E(\iota)] &= \frac{E(0)}{\zeta} + \left(\frac{\zeta + \rho(1-\zeta)}{\zeta}\right) L[\eta S(\iota) I(\iota) + \varepsilon S(\iota) Q(\iota) - (1-\xi)\vartheta E(\iota) - \kappa E(\iota)],\\
L[I(\iota)] &= \frac{I(0)}{\zeta} + \left(\frac{\zeta + \rho(1-\zeta)}{\zeta}\right) L[(1-\xi)\vartheta E(\iota) - \nu I(\iota) - \kappa I(\iota)],\\
L[R(\iota)] &= \frac{R(0)}{\zeta} + \left(\frac{\zeta + \rho(1-\zeta)}{\zeta}\right) L[\nu I(\iota) - \kappa R(\iota)],\\
L[Q(\iota)] &= \frac{Q(0)}{\zeta} + \left(\frac{\zeta + \rho(1-\zeta)}{\zeta}\right) L[\chi I(\iota) - \tau Q(\iota)].
\end{aligned} \tag{12.14}$$

Using inverse LT of equation of (12.14) leads to

$$\begin{aligned}
S(\iota) &= S(0) + L^{-1}\left[\left(\frac{\zeta + \rho(1-\zeta)}{\zeta}\right) L[\varpi - \eta S(\iota) I(\iota) - \varepsilon S(\iota) Q(\iota) - \kappa S(\iota)]\right],\\
E(\iota) &= E(0) + L^{-1}\left[\left(\frac{\zeta + \rho(1-\zeta)}{\zeta}\right) L[\eta S(\iota) I(\iota) + \varepsilon S(\iota) Q(\iota) - (1-\xi)\vartheta E(\iota) - \kappa E(\iota)]\right],\\
I(\iota) &= I(0) + L^{-1}\left[\left(\frac{\zeta + \rho(1-\zeta)}{\zeta}\right) L[(1-\xi)\vartheta E(\iota) - \nu I(\iota) - \kappa I(\iota)]\right],\\
R(\iota) &= R(0) + L^{-1}\left[\left(\frac{\zeta + \rho(1-\zeta)}{\zeta}\right) L[\nu I(\iota) - \kappa R(\iota)]\right],\\
Q(\iota) &= Q(0) + L^{-1}\left[\left(\frac{\zeta + \rho(1-\zeta)}{\zeta}\right) L[\chi I(\iota) - \tau Q(\iota)]\right].
\end{aligned}$$

The series solutions are assumed as

$$S = \sum_{j=0}^{\infty} S_j,\ E = \sum_{j=0}^{\infty} E_j,\ I = \sum_{j=0}^{\infty} I_j,\ R = \sum_{j=0}^{\infty} R_j,\ Q = \sum_{j=0}^{\infty} Q_j.$$

The terms $SI$ and $SQ$ from system (12.2) are nonlinear which can be rewritten as

$$SI = \sum_{j=0}^{\infty} X_j, \quad SQ = \sum_{j=0}^{\infty} Y_j,$$

while $X_j$ and $Y_j$ are decomposed in the following form

$$X_j = \sum_{\ell=0}^{j} S_\ell \sum_{\ell=0}^{j} I_\ell - \sum_{\ell=0}^{j-1} S_\ell \sum_{\ell=0}^{j-1} I_\ell, \quad Y_j = \sum_{\ell=0}^{j} S_\ell \sum_{\ell=0}^{j} Q_\ell - \sum_{\ell=0}^{j-1} S_\ell \sum_{\ell=0}^{j-1} Q_\ell.$$

the recursive formula attained by initial conditions and is expressed in the form of

$$
\begin{aligned}
S_{j+1}(\iota) &= S_j(0) + L^{-1}\left[\left(\frac{\zeta+\rho(1-\zeta)}{\zeta}\right)L\left[\varpi - \eta S_j(\iota)I_j(\iota) - \varepsilon S_j(\iota)Q_j(\iota) - \kappa S_j(\iota)\right]\right],\\
E_{j+1}(\iota) &= E_j(0) + L^{-1}\left[\left(\frac{\zeta+\rho(1-\zeta)}{\zeta}\right)L\left[\eta S_j(\iota)I_j(\iota) + \varepsilon S_j(\iota)Q_j(\iota) - (1-\xi)\vartheta E_j(\iota) - \kappa E_j(\iota)\right]\right],\\
I_{j+1}(\iota) &= I_j(0) + L^{-1}\left[\left(\frac{\zeta+\rho(1-\zeta)}{\zeta}\right)L\left[(1-\xi)\vartheta E_j(\iota) - \nu I_j(\iota) - \kappa I_j(\iota)\right]\right],\\
R_{j+1}(\iota) &= R_j(0) + L^{-1}\left[\left(\frac{\zeta+\rho(1-\zeta)}{\zeta}\right)L\left[\nu I_j(\iota) - \kappa R_j(\iota)\right]\right],\\
Q_{j+1}(\iota) &= Q_j(0) + L^{-1}\left[\left(\frac{\zeta+\rho(1-\zeta)}{\zeta}\right)L\left[\chi I_j(\iota) - \tau Q_j(\iota)\right]\right].
\end{aligned} \tag{12.15}
$$

Consider the Banach space $(\Im, \|\bullet\|)$ and let $\wp$ be a self-map on $\Im$. Also, let $Z_{j+1} = \Phi\left(\wp, Z_j\right)$ be denote recursive formula. Suppose that $\Pi(\wp)$ is the fixed point set of $\wp$ with $\Pi(\wp) \neq \emptyset$ and $\lim_{j\to\infty} Z_j = Z \in \Pi(\wp)$. Moreover, $\{\iota_j\} \subset \Pi$ and $P_j = \left\|\iota_{j+1} - \Phi\left(\wp, Z_j\right)\right\|$. If $\lim_{j\to\infty} P_j = 0$, which implies $\lim_{j\to\infty} \iota_j = Z$, the procedure of recursive $Z_{j+1} = \Phi\left(\wp, Z_j\right)$ is $\wp$-stable. In this way, sequence $\{\iota_j\}$ has an upper bounded. If Picard's approximation iteration is satisfied all of the above conditions, then $Z_{j+1} = \wp Z_j$ is $\wp$-stable.

In order to verify the stability results of the specified C-FFD of COVID-19 model (12.2) is employed in the following results.

**Theorem 6.1** [40] *Assume that* $(\Im, \|\bullet\|)$ *is a Banach space with a self-map* $\wp$ *on* $\Im$ *and sustaining the following inequality*

$$
\|\wp_u - \wp_v\| \le \Lambda\|u - \wp_u\| + \hbar\|u - v\|,
$$

*for each,* $u, v \in \Im$ *with* $0 \le \Lambda, 0 \le \hbar < 1$. *Then* $\wp$ *is* $\wp$*-stable.*

*According to* (12.15), *the fractional-order COVID-19 transmission CF model* (12.2) *is associated with iterative method.*

**Theorem 6.2** *Assume that* $\wp$ *is a self-map defined as*

$$
\begin{aligned}
\wp\left(S_j(\iota)\right) &= S_{j+1}(\iota) = S_j(0)\\
&\quad + L^{-1}\left[\left(\frac{\zeta+\rho(1-\zeta)}{\zeta}\right)L\left[\varpi - \eta S_j(\iota)I_j(\iota) - \varepsilon S_j(\iota)Q_j(\iota) - \kappa S_j(\iota)\right]\right],\\
\wp\left(E_j(\iota)\right) &= E_{j+1}(\iota) = E_j(0)\\
&\quad + L^{-1}\left[\left(\frac{\zeta+\rho(1-\zeta)}{\zeta}\right)L\left[\eta S_j(\iota)I_j(\iota) + \varepsilon S_j(\iota)Q_j(\iota) - (1-\xi)\vartheta E_j(\iota) - \kappa E_j(\iota)\right]\right],\\
\wp\left(I_j(\iota)\right) &= I_{j+1}(\iota) = I_j(0) + L^{-1}\left[\left(\frac{\zeta+\rho(1-\zeta)}{\zeta}\right)L\left[(1-\xi)\vartheta E_j(\iota) - \nu I_j(\iota) - \kappa I_j(\iota)\right]\right],\\
\wp\left(R_j(\iota)\right) &= R_{j+1}(\iota) = R_j(0) + L^{-1}\left[\left(\frac{\zeta+\rho(1-\zeta)}{\zeta}\right)L\left[\nu I_j(\iota) - \kappa R_j(\iota)\right]\right],\\
\wp\left(Q_j(\iota)\right) &= Q_{j+1}(\iota) = Q_j(0) + L^{-1}\left[\left(\frac{\zeta+\rho(1-\zeta)}{\zeta}\right)L\left[\chi I_j(\iota) - \tau Q_j(\iota)\right]\right].
\end{aligned}
$$

*Then the iterative fractional-order COVID-19 CF model* (12.16) *is* $\wp$*-stable in* $L^{'}(u, v)$ *if*

$$\Big(1 - \eta H_1^{\diamond}\mathrm{T}_1(\iota) - \eta H_3^{\diamond}\mathrm{T}_2(\iota) - \varepsilon H_1^{\diamond}\mathrm{T}_3(\iota) - \varepsilon H_5^{\diamond}\mathrm{T}_4(\iota) - \kappa\mathrm{T}_5(\iota)\Big) < 1$$
$$\Big(1 + \eta H_1^{\diamond}\mathrm{T}_6(\iota) + \eta H_3^{\diamond}\mathrm{T}_7(\iota) + \varepsilon H_1^{\diamond}\mathrm{T}_8(\iota) + \varepsilon H_5^{\diamond}\mathrm{T}_9(\iota)$$
$$-(1-\varepsilon)\vartheta\mathrm{T}_{10}(\iota) - \kappa\mathrm{T}_{11}(\iota)) < 1,$$
$$(1 + (1-\varepsilon)\vartheta\mathrm{T}_{12}(\iota) - \nu\mathrm{T}_{13}(\iota) - \kappa\mathrm{T}_{14}(\iota)) < 1,$$
$$(1 + \nu\mathrm{T}_{15}(\iota) - \kappa\mathrm{T}_{16}(\iota)) < 1,$$
$$(1 + \chi\mathrm{T}_{17}(\iota) - \tau\mathrm{T}_{18}(\iota)) < 1.$$

***Proof*** We shall show that the nonlinear self-map $\wp$ has a fixed point, for each $\ell, j \in \mathbb{N}$

$$\begin{aligned}
&\wp(S_\ell(\iota)) - \wp\big(S_j(\iota)\big) = S_\ell(\iota) - S_j(\iota)\\
&+ L^{-1}\Bigg[\left(\frac{\zeta + \rho(1-\zeta)}{\zeta}\right)L[\varpi - \eta S_\ell(\iota)I_\ell(\iota) - \varepsilon S_\ell(\iota)Q_\ell(\iota) - \kappa S_\ell(\iota)]\Bigg]\\
&- L^{-1}\Bigg[\left(\frac{\zeta + \rho(1-\zeta)}{\zeta}\right)L\big[\varpi - \eta S_j(\iota)I_j(\iota) - \varepsilon S_j(\iota)Q_j(\iota) - \kappa S_j(\iota)\big]\Bigg],\\
&\wp(E_\ell(\iota)) - \wp\big(E_j(\iota)\big) = E_\ell(\iota) - E_j(\iota)\\
&+ L^{-1}\Bigg[\left(\frac{\zeta + \rho(1-\zeta)}{\zeta}\right)L[\eta S_\ell(\iota)I_\ell(\iota) + \varepsilon S_\ell(\iota)Q_\ell(\iota) - (1-\xi)\vartheta E_\ell(\iota) - \kappa E_\ell(\iota)]\Bigg]\\
&- L^{-1}\Bigg[\left(\frac{\zeta + \rho(1-\zeta)}{\zeta}\right)L\big[\eta S_j(\iota)I_j(\iota) + \varepsilon S_j(\iota)Q_j(\iota) - (1-\xi)\vartheta E_j(\iota) - \kappa E_j(\iota)\big]\Bigg],\\
&\wp(I_\ell(\iota)) - \wp\big(I_j(\iota)\big) = I_\ell(\iota) - I_j(\iota)\\
&+ L^{-1}\Bigg[\left(\frac{\zeta + \rho(1-\zeta)}{\zeta}\right)L\big[(1-\xi)\vartheta E_\iota(\iota) - \nu I_\iota(\iota) - \kappa I_\iota(\iota)\big]\Bigg]\\
&- L^{-1}\Bigg[\left(\frac{\zeta + \rho(1-\zeta)}{\zeta}\right)L\big[(1-\xi)\vartheta E_j(\iota) - \nu I_j(\iota) - \kappa I_j(\iota)\big]\Bigg],\\
&\wp(R_\ell(\iota)) - \wp\big(R_j(\iota)\big) = R_\ell(\iota)\\
&- R_j(\iota) + L^{-1}\Bigg[\left(\frac{\zeta + \rho(1-\zeta)}{\zeta}\right)L\big[\nu I_\iota(\iota) - \kappa R_\iota(\iota)\big]\Bigg]\\
&- L^{-1}\Bigg[\left(\frac{\zeta + \rho(1-\zeta)}{\zeta}\right)L\big[\nu I_j(\iota) - \kappa R_j(\iota)\big]\Bigg],\\
&\wp(Q_\ell(\iota)) - \wp\big(Q_j(\iota)\big) = Q_\ell(\iota) - Q_j(\iota)\\
&+ L^{-1}\Bigg[\left(\frac{\zeta + \rho(1-\zeta)}{\zeta}\right)L\big[\chi I_\iota(\iota) - \tau Q_\iota(\iota)\big]\Bigg]\\
&- L^{-1}\Bigg[\left(\frac{\zeta + \rho(1-\zeta)}{\zeta}\right)L\big[\chi I_j(\iota) - \tau Q_j(\iota)\big]\Bigg].
\end{aligned} \tag{12.16}$$

Taking norm on both sides of first equation of (12.16) leads to

$$\begin{aligned}\left\|\wp(S_\ell(\iota)) - \wp\big(S_j(\iota)\big)\right\| &= \Big\| S_\ell(\iota) - S_j(\iota) \\ &+ L^{-1}\left[\left(\frac{\zeta+\rho(1-\zeta)}{\zeta}\right)L[\varpi - \eta S_\ell(\iota)I_\ell(\iota) - \varepsilon S_\ell(\iota)Q_\ell(\iota) - \kappa S_\ell(\iota)]\right] \\ &- L^{-1}\left[\left(\frac{\zeta+\rho(1-\zeta)}{\zeta}\right)L\big[\varpi - \eta S_j(\iota)I_j(\iota) - \varepsilon S_j(\iota)Q_j(\iota) - \kappa S_j(\iota)\big]\right]\Big\| \\ &\le \left\|S_\ell(\iota) - S_j(\iota)\right\| + L^{-1}\Big[\left(\frac{\zeta+\rho(1-\zeta)}{\zeta}\right)L\big[\left\|-\eta S_j\big(I_\ell(\iota) - I_j(\iota)\big)\right\| \\ &+ \left\|-\eta I_\ell\big(S_\ell(\iota) - S_j(\iota)\big)\right\| + \left\|-\varepsilon S_j\big(Q_\ell(\iota) - Q_j(\iota)\big)\right\| \\ &+ \left\|-\varepsilon Q_\ell\big(S_\ell(\iota) - S_j(\iota)\big)\right\| + \left\|-\kappa\big(S_\ell(\iota) - S_j(\iota)\big)\right\|\big]\Big].\end{aligned} \tag{12.17}$$

Every solution has the same role as follows

$$\begin{aligned}\left\|S_\ell(\iota) - S_j(\iota)\right\| &\cong \left\|E_\ell(\iota) - E_j(\iota)\right\| \cong \left\|I_\ell(\iota) - I_j(\iota)\right\| \\ &\cong \left\|R_\ell(\iota) - R_j(\iota)\right\| \cong \left\|Q_\ell(\iota) - Q_j(\iota)\right\|.\end{aligned} \tag{12.18}$$

Using (12.18) in (12.17) yields

$$\begin{aligned}\left\|\wp(S_\ell(\iota)) - \wp\big(S_j(\iota)\big)\right\| &\le \left\|S_\ell(\iota) - S_j(\iota)\right\| \\ &+ L^{-1}\Big[\left(\frac{\zeta+\rho(1-\zeta)}{\zeta}\right)L\big[\left\|-\eta S_j\big(S_\ell(\iota) - S_j(\iota)\big)\right\| \\ &+ \left\|-\eta I_\ell\big(S_\ell(\iota) - S_j(\iota)\big)\right\| + \left\|-\varepsilon S_j\big(S_\ell(\iota) - S_j(\iota)\big)\right\| \\ &+ \left\|-\varepsilon Q_\ell\big(S_\ell(\iota) - S_j(\iota)\big)\right\| + \left\|-\kappa\big(S_\ell(\iota) - S_j(\iota)\big)\right\|\big]\Big].\end{aligned} \tag{12.19}$$

Hence $S_\ell$, $E_\ell$, $I_\ell$, $R_\ell$, $Q_\ell$ are bounded. Therefore for all $\iota$, there exist positive constants $\mathrm{H}_a^\diamond$, $a = 1, 2, 3, 4, 5$ such that

$$\begin{aligned}&\|S_\ell(\iota)\| < H_1^\diamond,\ \|E_\ell(\iota)\| < H_2^\diamond,\ \|I_\ell(\iota)\| < H_3^\diamond, \\ &\|R_\ell(\iota)\| < H_4^\diamond,\ \|Q_\ell(\iota)\| < H_5^\diamond\end{aligned} \tag{12.20}$$

From (12.19) and (12.20)

$$\begin{aligned}\left\|\wp(S_\ell(\iota)) - \wp\big(S_j(\iota)\big)\right\| &\le \Big(1 - \eta H_1^\diamond \mathrm{T}_1(\iota) - \eta H_3^\diamond \mathrm{T}_2(\iota) \\ &- \varepsilon H_1^\diamond \mathrm{T}_3(\iota) - \varepsilon H_5^\diamond \mathrm{T}_4(\iota) - \kappa \mathrm{T}_5(\iota)\Big)\left\|S_\ell(\iota) - S_j(\iota)\right\|,\end{aligned} \tag{12.21}$$

where $T_a(\iota)$, $a = 1, 2, 3, 4, 5$ are the functions of $L^{-1}\left[\left(\frac{\zeta+\rho(1-\zeta)}{\zeta}\right)L[\bullet]\right]$. Similarly

$$\begin{aligned}\left\|\wp(E_\ell(\iota)) - \wp\big(E_j(\iota)\big)\right\| &\le \Big(1 + \eta H_1^\diamond \mathrm{T}_6(\iota) + \eta H_3^\diamond \mathrm{T}_7(\iota) + \varepsilon H_1^\diamond \mathrm{T}_8(\iota) + \varepsilon H_5^\diamond \mathrm{T}_9(\iota) \\ &- (1-\varepsilon)\vartheta \mathrm{T}_{10}(\iota) - \kappa \mathrm{T}_{11}(\iota)\big)\left\|E_\ell(\iota) - E_j(\iota)\right\|,\ \left\|\wp(I_\ell(\iota)) - \wp\big(I_j(\iota)\big)\right\|\end{aligned}$$

$$\begin{aligned}
&\leq (1 + (1-\varepsilon)\vartheta \mathrm{T}_{12}(\iota) - \nu \mathrm{T}_{13}(\iota) - \kappa \mathrm{T}_{14}(\iota)) \| I_\ell(\iota) - I_j(\iota) \|, \\
&\| \wp(R_\ell(\iota)) - \wp\big(R_j(\iota)\big) \| \leq (1 + \nu \mathrm{T}_{15}(\iota) - \kappa \mathrm{T}_{16}(\iota)) \| R_\ell(\iota) - R_j(\iota) \|, \\
&\| \wp(Q_\ell(\iota)) - \wp\big(Q_j(\iota)\big) \| \leq (1 + \chi \mathrm{T}_{17}(\iota) - \tau \mathrm{T}_{18}(\iota)) \| Q_\ell(\iota) - Q_j(\iota) \|,
\end{aligned} \tag{12.22}$$

with

$$\begin{aligned}
&\Big(1 - \eta H_1^{\diamond} \mathrm{T}_1(\iota) - \eta H_3^{\diamond} \mathrm{T}_2(\iota) - \varepsilon H_1^{\diamond} \mathrm{T}_3(\iota) - \varepsilon H_5^{\diamond} \mathrm{T}_4(\iota) - \kappa \mathrm{T}_5(\iota)\Big) < 1 \\
&\Big(1 + \eta H_1^{\diamond} \mathrm{T}_6(\iota) + \eta H_3^{\diamond} \mathrm{T}_7(\iota) + \varepsilon H_1^{\diamond} \mathrm{T}_8(\iota) + \varepsilon H_5^{\diamond} \mathrm{T}_9(\iota) \\
&\quad - (1-\varepsilon)\vartheta \mathrm{T}_{10}(\iota) - \kappa \mathrm{T}_{11}(\iota)) < 1, \\
&(1 + (1-\varepsilon)\vartheta \mathrm{T}_{12}(\iota) - \nu \mathrm{T}_{13}(\iota) - \kappa \mathrm{T}_{14}(\iota)) < 1, \\
&(1 + \nu \mathrm{T}_{15}(\iota) - \kappa \mathrm{T}_{16}(\iota)) < 1, \\
&(1 + \chi \mathrm{T}_{17}(\iota) - \tau \mathrm{T}_{18}(\iota)) < 1.
\end{aligned}$$

Hence, the self-map $\wp$ has a fixed point. Assuming (12.21) and (12.22) hold, consider$\bar{\lambda} = (0, 0, 0, 0, 0)$ and

$$C = \begin{cases}
\Big(1 - \eta H_1^{\diamond} \mathrm{T}_1(\iota) - \eta H_3^{\diamond} \mathrm{T}_2(\iota) - \varepsilon H_1^{\diamond} \mathrm{T}_3(\iota) - \varepsilon H_5^{\diamond} \mathrm{T}_4(\iota) - \kappa \mathrm{T}_5(\iota)\Big), \\
\Big(1 + \eta H_1^{\diamond} \mathrm{T}_6(\iota) + \eta H_3^{\diamond} \mathrm{T}_7(\iota) + \varepsilon H_1^{\diamond} \mathrm{T}_8(\iota) + \varepsilon H_5^{\diamond} \mathrm{T}_9(\iota) - (1-\varepsilon)\vartheta \mathrm{T}_{10}(\iota) - \kappa \mathrm{T}_{11}(\iota)), \\
(1 + (1-\varepsilon)\vartheta \mathrm{T}_{12}(\iota) - \nu \mathrm{T}_{13}(\iota) - \kappa \mathrm{T}_{14}(\iota)), \\
(1 + \nu \mathrm{T}_{15}(\iota) - \kappa \mathrm{T}_{16}(\iota)), \\
(1 + \chi \mathrm{T}_{17}(\iota) - \tau \mathrm{T}_{18}(\iota)).
\end{cases}$$

Thus, every condition in Theorem 6.1 is satisfied. Therefore, $\wp$ is Picard $\wp$-stable. This completes the proof.

## Numerical Simulations

This section is devoted to the numerical computations for the transmission of COVID-19 model with fractional order over a period of time $\iota$. Consider a mathematical formulation of fractional-order SEIRQ COVID-19 model as follows:

$$\begin{aligned}
D_\iota^\rho S(\iota) &= \varpi - \eta S(\iota) I(\iota) - \varepsilon S(\iota) Q(\iota) - \kappa S(\iota), \\
D_\iota^\rho E(\iota) &= \eta S(\iota) I(\iota) + \varepsilon S(\iota) Q(\iota) - (1-\xi)\vartheta E(\iota) - \kappa E(\iota), \\
D_\iota^\rho I(\iota) &= (1-\xi)\vartheta E(\iota) - \nu I(\iota) - \kappa I(\iota), \\
D_\iota^\rho R(\iota) &= \nu I(\iota) - \kappa R(\iota), \\
D_\iota^\rho Q(\iota) &= \chi I(\iota) - \tau Q(\iota).
\end{aligned} \tag{12.23}$$

Numerical scheme of the above system (12.23) has the following form [41]:

$$S(\iota_n) = \left[\varpi - \eta S(\iota_{n-1})I(\iota_{n-1}) - \varepsilon S(\iota_{n-1})Q(\iota_{n-1}) - \kappa S(\iota_{n-1})\right]h^{\rho} - \sum_{r=t}^{n} c_r^{(\rho)} S(\iota_{n-r}),$$

$$E(\iota_n) = \left[\eta S(\iota_{n-1})I(\iota_{n-1}) + \varepsilon S(\iota_{n-1})Q(\iota_{n-1}) - (1-\xi)\vartheta E(\iota_{n-1}) - \kappa E(\iota_{n-1})\right]h^{\rho} - \sum_{r=t}^{n} c_r^{(\rho)} E(\iota_{n-r}),$$

$$I(\iota_n) = \left[(1-\xi)\vartheta E(\iota_{n-1}) - \nu I(\iota_{n-1}) - \kappa I(\iota_{n-1})\right]h^{\rho} - \sum_{r=t}^{n} c_r^{(\rho)} I(\iota_{n-r}),$$

$$R(\iota_n) = \left[\nu I(\iota_{n-1}) - \kappa R(\iota_{n-1})\right]h^{\rho} - \sum_{r=t}^{n} c_r^{(\rho)} R(\iota_{n-r}),$$

$$Q(\iota_n) = \left[\chi I(\iota_{n-1}) - \tau Q(\iota_{n-1})\right]h^{\rho} - \sum_{r=t}^{n} c_r^{(\rho)} Q(\iota_{n-r}),$$

with $n = 1, 2, 3, \ldots,$ N, for $N = [T_{\text{sim}}/h]$ and $(S(0), E(0), I(0), R(0), Q(0))$ are the initial point. The numerical solutions for $S(\iota)$, $E(\iota)$, $I(\iota)$, $R(\iota)$, $Q(\iota)$ for non-integers $\rho = 0.92, 0.94, 0.96, 0.98$ and $\rho = 1$ are calculated for the fractional-order COVID-19 model (12.23). Assume that the total population is $N = 100$. [42] The birth rate of India in 2019 is 20 per 1000 people, then $\varpi = n \times N = 2$, and the death rate is 6.2 per 1000 people, then $\kappa = \frac{6.2}{1000} \times 100 = 0.62$. Distinct positive parameter values $\eta = 0.05$, $\varepsilon = 0.45$, $\xi = 0.05$, $\vartheta = 0.19$, $\nu = 0.07$, $\chi = 0.5$ and $\tau = 0.15$ are considered to obtain the numerical results. Also, the initial points for model (12.23) are selected as $S(0) = 45$ $E(0) = 30$, $I(0) = 25$, $R(0) = 0$ and $Q(0) = 5$ from the total populations $N = 100$. The estimated basic reproduction number of the model (12.2), $R_0 = 1.6323$ is calculated from the above set of parameter values.

The behaviours of $SEIRQ$ classes are presented in Figs. 12.1 and 12.5 with time $\iota$ for distinct fractional order $\rho$. Figure 12.1 shows the susceptible population decreases and moves to some values with non-integer order $\rho$. Figure 12.2 displays the exposed population increases then decreases as order $\rho$ goes up with time $\iota$. The infected population decrease for various values of $\rho$, but as the value of $\rho$ increases, the number of infection decreases. It can be understood from Fig. 12.4 that population recovers quickly with a change of $\rho$. Finally, Fig. 12.5 demonstrates the recovered people, which increases and then decreases with various values of $\rho$ (Tables 12.1, 12.2, 12.3, 12.4 and 12.5).

## Fractional SEIRQ COVID-19 Model with Vaccination

Consider a mathematical formulation of a fractional SEIRQ COVID-19 system along with vaccinated population $[V(\iota)]$ is given as follows:

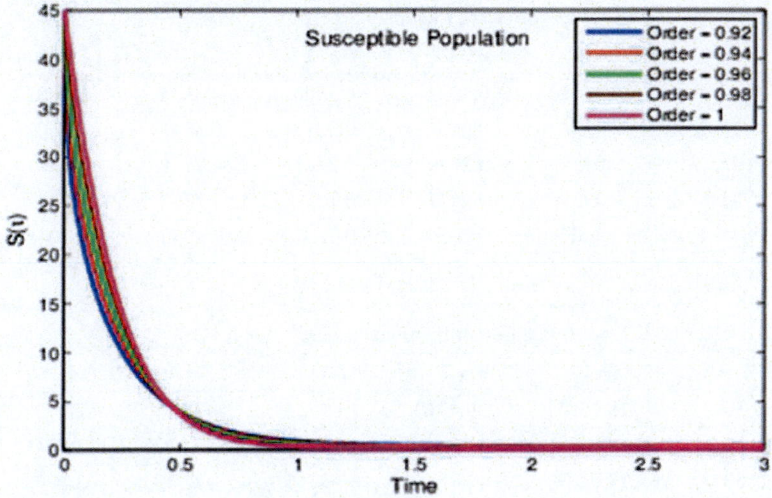


**Fig. 12.1** Illustration of susceptible class $S(\iota)$ for different fractional orders $\rho = 0.92, 0.94, 0.96, 0.98$ and $\rho = 1$. *Source* Own

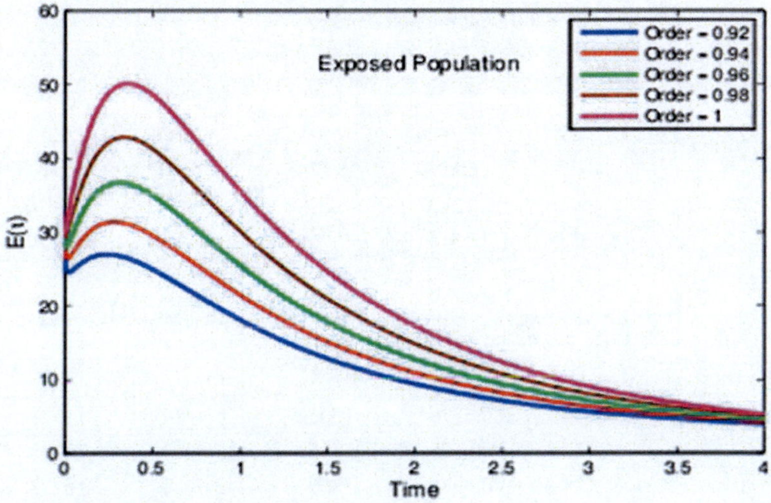


**Fig. 12.2** Illustration of exposed class $E(\iota)$ for different fractional orders $\rho = 0.92, 0.94, 0.96, 0.98$ and $\rho = 1$. *Source* Own

$$
\begin{aligned}
D_{\iota}^{\rho} S(\iota) &= \varpi - \eta S(\iota) I(\iota) - \varepsilon S(\iota) Q(\iota) - \alpha S(\iota) + \beta V(\iota) - \kappa S(\iota), \\
D_{\iota}^{\rho} E(\iota) &= \eta S(\iota) I(\iota) + \varepsilon S(\iota) Q(\iota) - (1 - \xi) \vartheta E(\iota) - \kappa E(\iota), \\
D_{\iota}^{\rho} I(\iota) &= (1 - \xi) \vartheta E(\iota) - \nu I(\iota) - \kappa I(\iota), \\
D_{\iota}^{\rho} R(\iota) &= \nu I(\iota) - \kappa R(\iota),
\end{aligned}
$$

**Table 12.1** Values of $S(\iota)$ for five different fractional orders

| $\rho$ | $\iota = 0$ | $\iota = 1$ | $\iota = 2$ | $\iota = 3$ | $\iota = 4$ | $\iota = 5$ |
|---|---|---|---|---|---|---|
| 0.92 | 45 | 0.9415 | 0.4234 | 0.3806 | 0.3826 | 0.4002 |
| 0.94 | 45 | 0.7924 | 0.3554 | 0.3217 | 0.3241 | 0.3402 |
| 0.96 | 45 | 0.6422 | 0.2970 | 0.2711 | 0.2735 | 0.2881 |
| 0.98 | 45 | 0.4965 | 0.2473 | 0.2277 | 0.2302 | 0.2432 |
| 1 | 45 | 0.3609 | 0.2054 | 0.1907 | 0.1932 | 0.2048 |

*Source* Own

**Table 12.2** Values of $E(\iota)$ for five different fractional orders

| $\rho$ | $\iota = 0$ | $\iota = 1$ | $\iota = 2$ | $\iota = 3$ | $\iota = 4$ | $\iota = 5$ |
|---|---|---|---|---|---|---|
| 0.92 | 30 | 18.1586 | 9.4710 | 5.7316 | 4.0165 | 3.1732 |
| 0.94 | 30 | 21.5433 | 10.9976 | 6.4243 | 4.3392 | 3.3341 |
| 0.96 | 30 | 25.5127 | 12.7838 | 7.2050 | 4.6743 | 3.4793 |
| 0.98 | 30 | 30.1643 | 14.8812 | 8.0888 | 5.0205 | 3.6029 |
| 1 | 30 | 35.6134 | 17.3515 | 9.0926 | 5.3752 | 3.6967 |

*Source* Own

**Table 12.3** Values of $I(\iota)$ for five different fractional orders

| $\rho$ | $\iota = 0$ | $\iota = 1$ | $\iota = 2$ | $\iota = 3$ | $\iota = 4$ | $\iota = 5$ |
|---|---|---|---|---|---|---|
| 0.92 | 25 | 9.4126 | 6.1210 | 4.0073 | 2.6993 | 1.9017 |
| 0.94 | 25 | 11.1237 | 7.2926 | 4.7255 | 3.1147 | 2.1332 |
| 0.96 | 25 | 13.1313 | 8.6860 | 5.5745 | 3.5931 | 2.3867 |
| 0.98 | 25 | 15.4841 | 10.3432 | 6.5789 | 4.1444 | 2.6630 |
| 1 | 25 | 18.2383 | 12.3141 | 7.7686 | 4.7803 | 2.9622 |

*Source* Own

**Table 12.4** Values of $R(\iota)$ for five different fractional orders

| $\rho$ | $\iota = 0$ | $\iota = 1$ | $\iota = 2$ | $\iota = 3$ | $\iota = 4$ | $\iota = 5$ |
|---|---|---|---|---|---|---|
| 0.92 | 0 | 0.6144 | 0.6796 | 0.5952 | 0.4786 | 0.3723 |
| 0.94 | 0 | 0.7116 | 0.8105 | 0.7153 | 0.5724 | 0.4391 |
| 0.96 | 0 | 0.8228 | 0.9656 | 0.8594 | 0.6846 | 0.5178 |
| 0.98 | 0 | 0.9498 | 1.1493 | 1.0322 | 0.8190 | 0.6107 |
| 1 | 0 | 1.0946 | 1.3664 | 1.2396 | 0.9802 | 0.7203 |

*Source* Own

**Table 12.5** Values of $Q(\iota)$ for five different fractional orders

| $\rho$ | $\iota = 0$ | $\iota = 1$ | $\iota = 2$ | $\iota = 3$ | $\iota = 4$ | $\iota = 5$ |
|---|---|---|---|---|---|---|
| 0.92 | 5 | 8.0403 | 9.8772 | 10.3342 | 10.0823 | 9.5024 |
| 0.94 | 5 | 9.2785 | 11.6838 | 12.3680 | 12.1213 | 11.4254 |
| 0.96 | 5 | 10.6944 | 13.8056 | 14.7911 | 14.5664 | 13.7342 |
| 0.98 | 5 | 12.3116 | 16.2949 | 17.6762 | 17.4979 | 16.5065 |
| 1 | 5 | 14.1568 | 19.2124 | 21.1091 | 21.0115 | 19.8353 |

*Source* Own

$$
\begin{aligned}
D_{\iota}^{\rho} Q(\iota) &= \chi I(\iota) - \tau Q(\iota), \\
D_{\iota}^{\rho} V(\iota) &= \alpha S(\iota) - \beta V(\iota) - \kappa V(\iota).
\end{aligned} \tag{12.24}
$$

Here, all the parameters $\varpi$, $\eta$, $\varepsilon$, $\kappa$, $\xi$, $\vartheta$, $\nu$, $\chi$, $\tau$ assume only non-negative values and have the same meaning as in model (12.1). Also $\alpha$, $\beta$ are positive and rate of vaccination, waning immunity rate of vaccinated population, respectively. In the following simulations, the same parameters as in Figs. 12.1, 12.2, 12.3, 12.4 and 12.5 along with $\alpha = 0.3$, $\beta = 0.001$ are chosen. Then, Figs. 12.6, 12.7 and 12.8 depict an illustration of the model with and without vaccination for $S(\iota)$,$I(\iota)$ and $R(\iota)$ through $\rho = 0.95$. It is observed that recovery is quicker in the presence of vaccination as shown in Fig. 12.8.

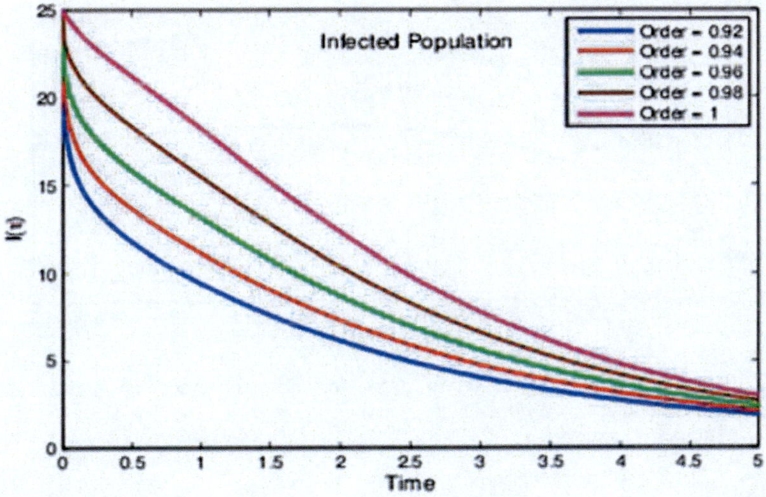


**Fig. 12.3** Illustration of infected class $I(\iota)$ for different fractional orders $\rho = 0.92, 0.94, 0.96, 0.98$ and $\rho = 1$. *Source* Own

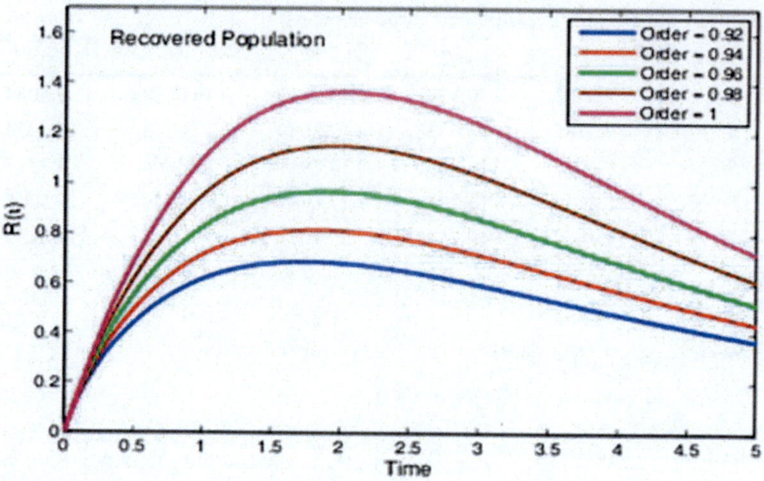


**Fig. 12.4** Illustration of recovered class $R(\iota)$ for different fractional orders $\rho = 0.92, 0.94, 0.96, 0.98$ and $\rho = 1$. *Source* Own

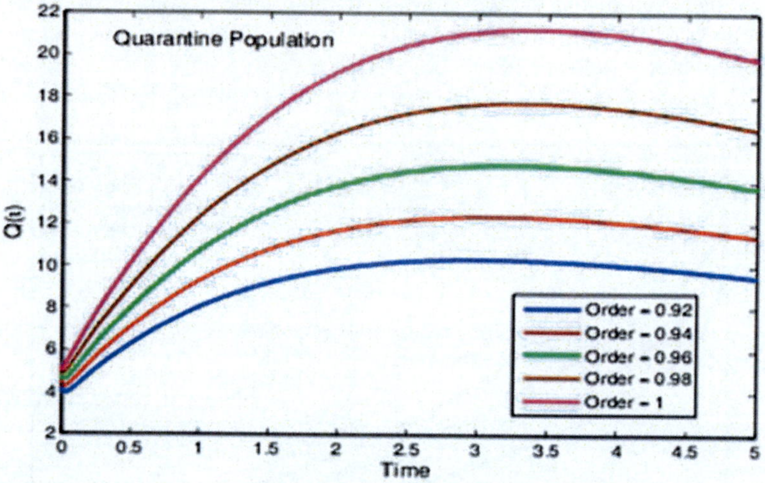


**Fig. 12.5** Illustration of quarantine class $Q(\iota)$ for different fractional orders $\rho = 0.92, 0.94, 0.96, 0.98$ and $\rho = 1$. *Source* Own

## Conclusion

In this work, a new dynamic SEIRQ compartment system of novel corona virus (COVID-19) is provided and analysed for dynamic behaviour based on the fractional-order Caputo–Fabrizio derivative. The Picard successive approximation technique is utilised in establishing the existence criteria of solution for proposed fractional-order

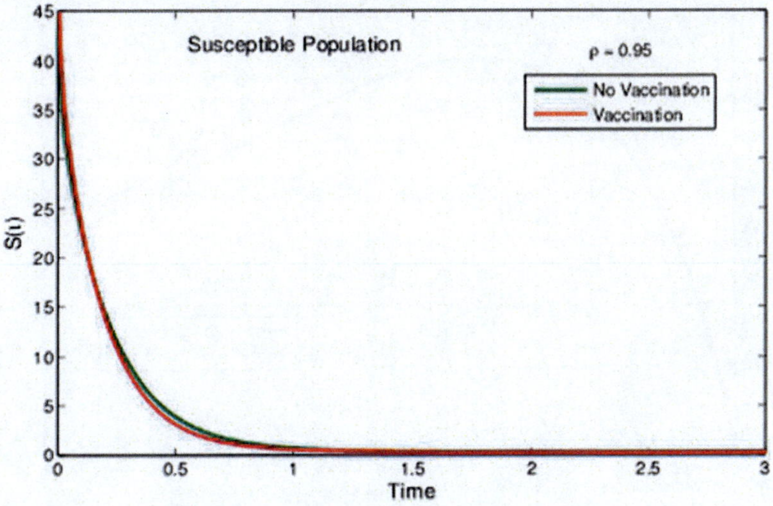


Fig. 12.6 Illustration of $S(\iota)$ with and without vaccination for $\rho = 0.95$. *Source* Own

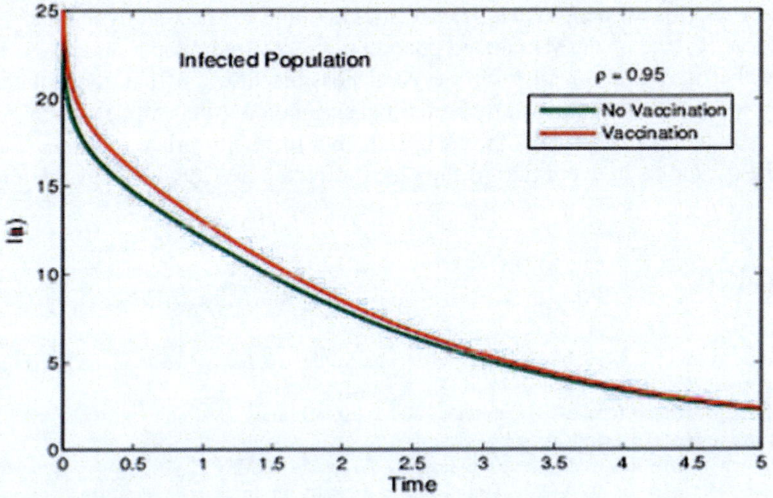


Fig. 12.7 Illustration of $I(\iota)$ with and without vaccination for $\rho = 0.95$. *Source* Own

CF model. Then the equilibrium points for disease-free and endemic are computed and verified the local and global stability properties with the aid of the basic number of reproductions $R_0$. Moreover, the stability results of the system (12.2) are established by using Banach's fixed point theory and an iterative scheme based on the LT involving the operator C-FFD. Finally, a numerical simulation of the fractional COVID-19 system is presented to analysis the dynamical behaviour of the solutions of the model during a time interval. Also, a fractional-order SEIRQ COVID-19

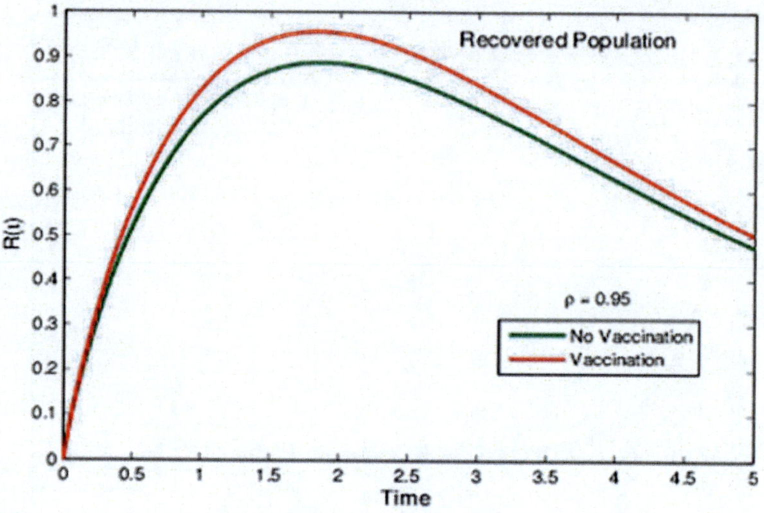


**Fig. 12.8** Illustration of $R(\iota)$ with and without vaccination for $\rho = 0.95$. *Source* Own

model with vaccinated people has been formulated and studied its dynamics. It is observed that instead of the vaccinated people, the recovered people increases very rapidly in the time interval. In future, several possible fitting of parameter values can be implemented and analysed by numerical simulations. This model is analysed from the estimated COVID-19 outbreak in India but it is appropriate to forecast and explore the dynamics in every part of the global affected by COVID-19 pandemic.

## References

1. Situation report-113: Available at https://www.who.int/docs/defaultsource/coronaviruse/situation-reports/.
2. "Coronavirus disease (covid-19) outbreak," Jan 2020. [Online]. Available: https://www.who.int/emergencies/diseases/novel-coronavirus-2019.
3. https://www.mohfw.gov.in/dashboard/index.php, Accessed on May 12, 2020.
4. Al-qaness M. A. A. et al. (2020). Optimization method for forecasting confirmed cases of COVID-19 in China. *Journal of Clinical Medicine*, **9**, 674.
5. Vyasarayani C. P. et al. (2020). New approximations, and policy implications, from a delayed dynamic model of a fast pandemic, arXiv preprint 2020: arXiv:2004.03878v1.
6. Biswas, K. et al. (2020). Covid-19 spread: Reproduction of data and prediction using a SIR model on Euclidean network, arXiv preprint 2020: arXiv:2003.07063v1.
7. Nita, H. (2020). Shah, Nisha Sheoran, and Ekta Jayswal, Z-Control on COVID-19-Exposed Patients in Quarantine. *International Journal of Differential Equations, 2020*, 1–11.
8. Ranjan, R. (2020). Predictions for COVID-19 outbreak in India using epidemiological models. medRxiv 2020: https://doi.org/10.1101/2020.04.02.20051466.
9. Mandal, S. et al., Prudent public health intervention strategies to control the coronavirus disease 2019 transmission in India: A mathematical model-based approach. *Indian Journal of Medical Research* (2020): https://doi.org/10.4103/ijmr.ijmr_504_20.

10. Shah, N. H., Suthar, A. H., & Satia, M. H. (2020). Modelling the impact of nationwide BCG vaccine recommendations on COVID-19 transmission, severity, and mortality. medRxiv preprint 2020. https://doi.org/10.1101/2020.05.10.20097121.
11. Shah, N. H. Sheoran N., & Jayswal, E. (2020) Modelling COVID-19 transmission in the United States through Interstate and Foreign travels and evaluating impact of Governmental Public Health Interventions. medRxiv preprint 2020. https://doi.org/10.1101/2020.05.23.20110999.
12. Prakash, M. K. et al. (2020). A minimal and adaptive prediction strategy for critical resource planning in a pandemic. medRxiv 2020: https://doi.org/10.1101/2020.04.08.20057414.
13. Singh, R. et al. (2020). Age-structured impact of social distancing on the COVID-19 epidemic in India. arXiv preprint 2020: arXiv:2003.12055v1.
14. Jakhar, M., Ahluwalia, P. K., & Kumar, A. (2020). COVID-19 epidemic forecast in different States of India using SIR model. medRxiv 2020: https://doi.org/10.1101/2020.05.14.20101725.
15. Nita H. Shah, Ankush H. Suthar, Ekta N. Jayswal, N. Shukla and J. Shukla, Modelling the impact of Plasma Therapy and Immunotherapy for Recovery of COVID-19 Infected Individuals, medRxiv preprint 2020 doi:https://doi.org/10.1101/2020.05.23.20110973.
16. Ghosh, P. et al. (2020). COVID-19 in India: State-wise analysis and prediction. medRxiv 2020: https://doi.org/10.1101/2020.04.24.20077792.
17. Siettos, C. I., & Russo, L. (2013). Mathematical modeling of infectious disease dynamics. *Virulence, 4*, 295–306. https://doi.org/10.4161/viru.24041.
18. Daley, D. J., & Gani, J. (1999). Epidemic modelling: An introduction. *Cambridge Studies in Mathematical Biology*, Cambridge University Press.
19. Khalil, R., Al Horani M., Yousef, A., & Sababheh, M. (2014). A new definition of fractional Derivative. *Journal of Computational and Applied Mathematics, 264*, 65–70.
20. Samko, G., Kilbas, A., & Marichev, O. (1993). *Fractional integrals and derivatives: Theory and applications,* Gordon and Breach.
21. Caputo, M., & Fabrizio, M. (2015). A new definition of fractional derivative without singular kernel. *Progress in Fractional Differentiation and Applications, 1*(2), 73–85.
22. Losada, J., & Nieto, J. J. (2015). Properties of the new fractional derivative without singular kernel. *Progress in Fractional Differentiation and Applications, 1*(2), 87–92.
23. Shaikh, A. M., Shaikh, I. N., & Nisar, K. S. (2020). A mathematical model of COVID-19 using fractional derivative: Outbreak in India with dynamics of transmission and control. *Advances in Difference Equations, 2020*(373), 1–19.
24. Kilbas, A. A., Srivastava, H. M., & Trujillo, J. J. (2006). *Theory and applications of fractional differential equations.* Amsterdam: Elsevier.
25. Podlubny, I. (1999). *Fractional differential equations.* San Diego: Academic Press.
26. Mouaouine, A., Boukhouima, A., Hattaf, K., & Yousfi, N. (2018). A fractional order SIR epidemic model with nonlinear incidence rate. *Advances in Difference Equations, 2018,* 160.
27. Sene, N. (2020). SIR epidemic model with Mittag-Leffler fractional derivative. *Chaos, Solitons & Fractals, 137,* 109833.
28. Ghanbari, B., & Djilali, S. (2020). Mathematical and numerical analysis of a three-species predator-prey model with herd behavior and time fractional-order derivative. *Mathematical Methods in the Applied Sciences, 43,* 1736–1752.
29. Kumar, S., Kumar, R., Cattani, C., & Samet, B. (2020). Chaotic behaviour of fractional predator-prey dynamical system. *Chaos, Solitons & Fractals, 135,* 109811.
30. Aguilar, J. F. G. et al. (2016). Analytical and numerical solutions of electrical circuits described by fractional derivatives. *Applied Mathematical Modelling, 40*, 9079–9094.
31. Yuzbasi, S. (2015). A collocation method for numerical solutions of fractional-order logistic population model. *International Journal of Biomathematics, 9*(2), 31–45.
32. Diethelm, K. (2013). A fractional calculus based model for the simulation of an outbreak of dengue fever. *Nonlinear Dynamics, 71,* 613–619.
33. Ghanbari, B., Kumar, S., & Kumar, R. (2020). A study of behaviour for immune and tumor cells in immunogenetic tumour model with non-singular fractional derivative. *Chaos, Solitons & Fractals, 133,* 109619.

34. Alshabanat, A., Jleli, M., Kumar, S., & Samet, B. (2020). Generalization of Caputo-Fabrizio fractional derivative and applications to electrical circuits. *Front. Phys.*
35. Khan, Y., Wu, Q., Faraz, N., Yildirim, A., & Madani, M. (2012). A new fractional analytical approach via a modified Riemann-Liouville derivative. *Applied Mathematics Letters, 25*, 1340–1346.
36. Allen, L. J. et al. (2008). *Mathematical epidemiology,* Berlin, Germany: Springer.
37. Barbashin, E. A. (1970). *Introduction to the theory of stability, Groningen.* The Netherlands: Walters-Noordhoff.
38. Robinson, R. C. (2004). *An introduction to dynamical systems: Continuous and discrete.* Englewood Cliffs, NJ, USA: Prentice-Hall.
39. LaSalle, J. P. (1976). *The stability of dynamical systems* (Regional Conference Series in Applied Mathematics). Philadelphia, PA, USA: SIAM.
40. Wang, J., Zhou, Y., & Medved, M. (2012). Picard and weakly Picard operator's technique for nonlinear differential equations in Banach spaces. *Journal of Mathematical Analysis and Applications, 389*(1), 261–274.
41. Petras, I. (2011). *Fractional-order nonlinear systems: Modeling, analysis and simulation.* New York, USA: Springer.
42. Sample Registration System (SRS) Bulletin, 53(1), (2020).

# Chapter 13
# Compartmental Modelling Approach for Accessing the Role of Non-Pharmaceutical Measures in the Spread of COVID-19

**Yashika Bahri, Sumit Kaur Bhatia, Sudipa Chauhan, and Mandeep Mittal**

**Abstract** Epidemic diseases are well known to be fatal and cause great loss worldwide—economically, socially and mentally. Even after around nine months, since the Coronavirus Disease 2019 (COVID-19) began to spread, people are getting infected all over the world. This is one of the areas where human medical advancements fail because by the time the disease is identified and its treatment is figured out, most of the population is already exposed to it. In such cases, it becomes easier to take steps if the dynamics of the disease and its sensitivity to various factors is known. This chapter deals with developing a mathematical model for the spread of Coronavirus disease, by employing a number of parameters that affect its spread. A compartmental modelling approach using ordinary differential equation has been used to formulate the set of equations that describe the model. We have used the next generation matrix method to find the basic reproduction number of the system and proved that the system is locally asymptotically stable at the disease-free equilibrium for $R_0 < 1$. Stability and existence of endemic equilibrium have been discussed, followed by sensitivity of infective classes to parameters like proportion of vaccinated individuals and precautionary measures like social distancing. It is expected that after the vaccine is developed and is available to use, as the proportion of vaccinated individuals will increase, the infection will decrease in the population which can gradually lead to herd immunity. Since, the vaccine is still under development, non-intervention measures play a major role in coping with the disease. The disease generally transmits when the water droplets from an infected individuals' mouth or nose are inhaled by a healthy individual. The best measures that should be adopted are social distancing, washing one's hands frequently, and covering one's mouth with mask, quarantine and lockdowns. Thus, as more and more precautionary measures are taken, it would gradually reduce the infection which has also been proved numerically by the sensitivity analysis of '*w*' in our dynamical analysis.

**Keywords** Compartmental modelling · Reproduction number · Non-pharmaceutical measures · Equilibrium points · Sensitivity analysis

---

Y. Bahri · S. K. Bhatia (✉) · S. Chauhan · M. Mittal
Amity Institute of Applied Sciences, Amity University, Uttar Pradesh, India
e-mail: skbhatia@amity.edu



N. H. Shah and M. Mittal (eds.), *Mathematical Analysis for Transmission of COVID-19*, Mathematical Engineering, https://doi.org/10.1007/978-981-33-6264-2_13

**Mathematics Subject Classification** 34A34 · 34B09 · 34D05 · 34D20

## Introduction

Epidemic diseases have always been a major concern worldwide. In addition to many people losing their lives, outbreaks also impose financial, social, and mental strains on individuals and nations as a whole. History has many examples of such brutal outbreaks like Russian plague, Flu pandemic, Spanish Flu, Asian Flu, H1N1 Swine Flu pandemic, West African Ebola epidemic, Zika virus epidemic and many more. Quite a chunk of these have been able to be suppressed by vaccinations, ofcourse only after causing the mass destructions, while some diseases still have no vaccinations. A current such threat is COVID-19 epidemic, which started spreading from the Wuhan city of China in December 2019 and is now brutally taking lives worldwide. Since, people were unaware of this new virus initially, they kept on travelling internationally and the virus soon spread out from China to all across the world. One of the worst affected countries were Italy and Spain. These were the first to suffer at the hands of the virus and had massive death tolls. Most other countries had a buffer period and did not experience an outbreak immediately, so steps like passenger screening were adopted at airports, assuming that the virus could only affect an individual if the person had a travel history to affected countries like China or Italy. Soon cases started appearing wherein the patients did not had a travel history but came in contact with someone who had and it did not take long for the community spread to begin leading to outbreak all over the world. With no vaccination at hand and no prior information about the virus or it's treatment, it had been difficult to handle the disease initially and even months after the spread the number of cases are still increasing. Various researches are being carried out by several countries to develop a vaccination for the disease and some have already come up with one, like Russia.

But a major fact about epidemic diseases is that they mostly cause an outbreak and are unable to control initially because of lack of prior knowledge. Even though the human race has aced in providing themselves health stability over the years, by getting deeper into medical research, it takes time and effort to deal with epidemics. That is where epidemiology comes into play. In order to deal with epidemics, epidemiologists bring together the real life prospects of the disease into a mathematical model and using real life data one can make estimates regarding length and extent of the transmission and also various control measures that can be adopted to control the disease in long and short run. In the presence of a vaccine, one can even make estimates regarding how much of the population has to be vaccinated before achieving herd immunity. For instance, Ochoche and Gweryina in their study on measles using mathematical modelling concluded that atleast 94 % of the population must be vaccinated in order to achieve herd immunity [1]. Similarly, Kassa et al. in their research on COVID-19 have discussed various mitigation strategies to cope with the disease [7]. Different possible scenarios are being considered by researchers in order to be as accurate as possible [5, 8–10, 14–17]. Ngonghala et al. came up with a complex mathematical model for COVID-19 to study the impact of non-pharmaceutical interventions in [4]. Chang et al. have discussed the impact of media coverage on the spread of the disease in their research [6]. Researchers like Paul et al. have done prediction analysis for various South Asian countries focussing on the impact of various precautionary measures in [3].

In the sections that follow, we have attempted to study the spread of an epidemic with the help of mathematical modelling, bringing together the various stages involved in the spread of an epidemic and analysing various factors affecting this spread. We have first dealt with a general model for an epidemic, incorporating a variety of parameters that affects it's spread. Scenarios like vaccination have been added that can be used based on whether vaccination is available for a particular disease or not. Later on we have used the model to analyse the spread of COVID-19 and the intensity of the affect certain parameters have on it. Section Model Formulation deals with model formulation. In Sect. Stability Analysis, we have discussed about the stability and existence of the two equilibrium points - the disease-free and the endemic equilibrium based on the basic reproduction number. Section Sensitivity Analysis consists of sensitivity analysis of COVID-19 model parameters. Lastly, we have concluded the results in Sect. Conclusion discussing a few mitigation strategies and summarising the importance of epidemiology.

## Model Formulation

### *Description*

The model is based on the $S$-$E$-$I$-$R$-$S$ deterministic compartmental modelling approach. The total population($N$) is divided amongst seven compartments listed in Table 13.1.

**Table 13.1** The seven compartments in the model (Source: own)

| Compartment | Description |
|---|---|
| $S$ | Susceptible individuals (those at a risk of being infected) |
| $E$ | Exposed individuals (those exposed to the infection) |
| $I$ | Symptomatic individuals (infected individuals who show symptoms) |
| $I_a$ | Asymptomatic individuals (infected individuals who do not show any symptoms) |
| $I_q$ | Quarantined individuals (infected individuals who only show mild symptoms) |
| $H$ | Hospitalized individuals |
| $R$ | Recovered individuals |

In the beginning the model assumes that the entire population is susceptible. The susceptible individuals are exposed to the disease after coming in contact with the infectious population. The force of infection is defined as $\beta S I_a + \frac{\beta S I}{1+\alpha I}$. Keeping in mind the fact that people tend to avoid direct contact with symptomatic individuals and hence the infection due to symptomatic individuals is less than that of asymptomatic individuals, the inhibitory parameter $\alpha$ has been incorporated in the expression. Once the disease starts spreading people tend to take precautionary measures like washing their hands frequently, using face masks and social distancing which is being taken care by the parameter $\omega$.

The exposed individuals move to the infectious compartments $(I, I_a, I_q)$ at a rate of $\sigma$ depending on the intensity of infection, in the proportions $\rho_1$, $\rho_2$ and $\rho_3$, respectively. Individuals showing very high symptoms move to the 'Symptomatic' compartment, those showing no symptoms at all move to 'Asymptomatic' compartment and individuals showing mild symptoms move to 'Quarantined' compartment.

The infectious individuals move to 'Hospitalized' and 'Recovered' compartments depending upon their health status. Further at any stage individuals might die a natural death or a disease induced death depending upon their compartmental position. The model also incorporates new births at a rate b.

To provide flexibility to the modelling process parameters p and $\Delta$ have been incorporated in the model that denote the proportion of vaccinated individuals and the rate at which the recovered individuals become susceptible again, respectively. Both of these scenarios are not always sure to occur and hence can be set to 0 whenever the model does not allow for vaccination or when recovery ensures immunity. Figure 13.1 shows the flow diagram of the model, depicting the various stages involved in it.

Based on the above description the model can be represented by the following system of differential equations:

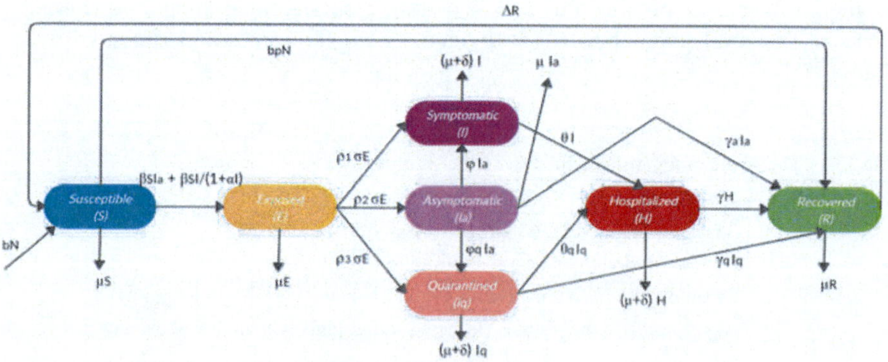


**Fig. 13.1** Flow diagram for model (*Source* own)

$$
\begin{aligned}
\frac{dS}{dt} &= bN(1-p) - \beta SI_a - \frac{\beta SI}{1+\alpha I} - \mu S + \Delta R \\
\frac{dE}{dt} &= \beta SI_a(1-\omega) + \frac{\beta SI}{1+\alpha I}(1-\omega) - (\sigma+\mu)E \\
\frac{dI}{dt} &= \rho_1 \sigma E + \phi I_a - (\theta+\mu+\delta)I \\
\frac{dI_a}{dt} &= \rho_2 \sigma E - (\phi+\phi_q+\gamma_a+\mu)I_a \\
\frac{dI_q}{dt} &= \rho_3 \sigma E + \phi_q I_a - (\theta_q+\gamma_q+\mu+\delta)I_q \\
\frac{dH}{dt} &= \theta I + \theta_q I_q - (\gamma+\mu+\delta)H \\
\frac{dR}{dt} &= bpN + \gamma_a I_a + \gamma_q I_q + \gamma H - (\Delta+\mu)R
\end{aligned} \tag{13.1}
$$

Table 13.2 lists all the parameters used in the model and their description.

**Table 13.2** Description of model parameters (Source: own)

| Parameter | Description |
|---|---|
| $b$ | Birth rate |
| $\mu$ | Natural death rate |
| $\delta$ | Disease-induced death rate |
| $p$ | Proportion of vaccinated individuals |
| $\rho_1$ | Proportion of exposed individuals moving to Symptomatic class |
| $\rho_2$ | Proportion of exposed individuals moving to Asymptomatic class |
| $\rho_3$ | Proportion of exposed individuals moving to Quarantined class |
| $\alpha$ | Parameter measuring inhibitory effect |
| $\beta$ | Transmission rate |
| $\sigma$ | Transition rate to infected classes from Exposed class |
| $\theta$ | Transition rate to Hospitalized class from Symptomatic class |
| $\theta_q$ | Transition rate to Hospitalized class from Quarantined class |
| $\gamma$ | Transition rate to Recovered class from Symptomatic class |
| $\gamma_a$ | Transition rate to Recovered class from Asymptomatic class |
| $\gamma_q$ | Transition rate to Recovered class from Quarantined class |
| $\phi$ | Transition rate to Symptomatic class from Asymptomatic class |
| $\phi_q$ | Transition rate to Quarantined class from Asymptomatic class |
| $\Delta$ | Transition rate to Susceptible class from Recovered class |
| $\omega$ | Parameter capturing the effect of precautionary measures like washing hands frequently, using masks and social distancing |

### *Well Orderedness*

The feasible region for the system (13.1) is:

$$\tau = \{(S, E, I, I_a, I_q, H, R) : S + E + I + I_A + I_q + H + R \leqslant \frac{bN}{\mu}, \\ S > 0, E \geqslant 0, I \geqslant 0, I_a \geqslant 0, I_q \geqslant 0, H \geqslant 0, R \geqslant 0\} \quad (13.2)$$

## Stability Analysis

### *Disease-Free Equilibrium ($E_0$)*

Disease-free equilibrium (DFE) is defined as the point at which the disease is completely eradicated from the system. Hence, all the infectious classes become constant at zero and the non-infectious classes attain a constant non-zero level.
In order to compute DFE of system in (13.1), we first set the sum of equations equal to zero:

$$0 = \frac{dS}{dt} + \frac{dE}{dt} + \frac{dI}{dt} + \frac{dI_a}{dt} + \frac{dI_q}{dt} + \frac{dH}{dt} + \frac{dR}{dt}$$

$$0 = bN - (S + E + I + I_a + I_q + H + R)\mu - (I + I_q + H)\delta - \beta S I_a \omega - \frac{\beta S I \omega}{1 + \alpha I}$$

Next we substitute, $E = I = I_a = I_q = H = R = 0$:

$$0 = bN - S\mu$$

$$S = \frac{bN}{\mu}$$

Thus we have the DFE,

$$E_0 = (S^*, E^*, I^*, I_a^*, I_q^*, H^*, R^*) = \left(\frac{bN}{\mu}, 0, 0, 0, 0, 0, 0\right) \quad (13.3)$$

### *Basic Reproduction Number ($R_0$)*

Basic reproduction number is defined as the average number of new infectious individuals produced when a single infectious individual is exposed to the susceptible popultaion. $R_0$ plays a major role in stability analysis of epidemic diseases. It is one of the first parameters to be calculated while modelling any epidemic because

it determines important factors like whether the system would be stable or not, the disease will persist or will be eradicated, and what perecentage of population should be vaccinated in order to achieve herd immunity.
Hence, if $R_0$ is known we can make the following conclusions:

1. $R_0$ <1 implies that on an average each new infectious will give rise to less than one new infectious upon contact and hence the DFE is stable which implies that the disease will be eradicated.
2. $R_0$ >1 implies that on an average each new infectious will give rise to more than one infectious upon contact and hence the DFE is unstable which implies that the disease will not be eradicated and will lead to an outbreak.

In order to compute $R_0$ we will bring into use the next generation matrix method [1]. $R_0$ is equal to the spectral radius of the next generation matrix.

**Step 1**: We first express system (13.1) as:

$$\mathcal{H} = (E, I, I_a, I_q, H, R, S)$$

which can be rewritten as:

$$\begin{aligned} \mathcal{H}' &= \mathscr{F}(x) - \mathscr{V}(x) \\ &= \mathscr{F}(x) - [\mathscr{V}^{-}(x) - \mathscr{V}^{+}(x)] \end{aligned}$$

$$\text{where, } \mathscr{F}(x) = \begin{bmatrix} \beta S(1-\omega)(I_a + \frac{I}{1+\alpha I}) \\ 0 \\ 0 \\ 0 \\ 0 \\ 0 \\ 0 \end{bmatrix} \text{ and } \mathscr{V}(x) = \begin{bmatrix} (\sigma+\mu)E \\ (\theta+\mu+\delta)I - \rho_1\sigma E - \phi I_a \\ (\phi+\phi_q+\gamma_a+\mu)I_a - \rho_2\sigma E \\ (\theta_q+\gamma_q+\mu+\delta)I_q - \rho_3\sigma E - \phi_q I_a \\ (\gamma+\mu+\delta)H - \theta I - \theta_q I_q \\ (\mu+\Delta)R - bpN - \gamma_a I_a - \gamma_q I_q - \gamma H \\ bN(1-p) + \beta S(I_a + \frac{I}{1+\alpha I}) + \mu S - \Delta R \end{bmatrix}$$

$\mathscr{F}(x)$ is a column vector whose each entry is the collection of terms which result in new infectious in each compartment and $\mathscr{V}(x)$ is a column vector consisting of remaining terms.

**Step 2**: Next we find the Jacobian matrix of $\mathscr{F}(x)$:

$$D\mathscr{F} = \begin{bmatrix} 0 & \frac{\beta S(1-\omega)}{(1+\alpha I)^2} & \beta S(1-\omega) & 0 & 0 & 0 & \beta(1-\omega)(I_a + \frac{I}{1+\alpha I}) \\ 0 & 0 & 0 & 0 & 0 & 0 & 0 \\ 0 & 0 & 0 & 0 & 0 & 0 & 0 \\ 0 & 0 & 0 & 0 & 0 & 0 & 0 \\ 0 & 0 & 0 & 0 & 0 & 0 & 0 \\ 0 & 0 & 0 & 0 & 0 & 0 & 0 \\ 0 & 0 & 0 & 0 & 0 & 0 & 0 \end{bmatrix}$$

The Jacobian matrix of $\mathscr{F}(x)$ at DFE is:

$$D\mathscr{F}(E_0) = \begin{bmatrix} 0 & \frac{b\beta N(1-\omega)}{\mu} & \frac{b\beta N(1-\omega)}{\mu} & 0\,0\,0\,0 \\ 0 & 0 & 0 & 0\,0\,0\,0 \\ 0 & 0 & 0 & 0\,0\,0\,0 \\ 0 & 0 & 0 & 0\,0\,0\,0 \\ 0 & 0 & 0 & 0\,0\,0\,0 \\ 0 & 0 & 0 & 0\,0\,0\,0 \\ 0 & 0 & 0 & 0\,0\,0\,0 \end{bmatrix} = \begin{bmatrix} F & 0 \\ 0 & 0 \end{bmatrix}$$

where,

$$F = \begin{bmatrix} 0 & \frac{b\beta N(1-\omega)}{\mu} & \frac{b\beta N(1-\omega)}{\mu} & 0 \\ 0 & 0 & 0 & 0 \\ 0 & 0 & 0 & 0 \\ 0 & 0 & 0 & 0 \end{bmatrix}$$

**Step 3**: Next we find the Jacobian matrix of $\mathscr{V}(x)$:

$$D\mathscr{V} = \begin{bmatrix} (\sigma+\mu) & 0 & 0 & 0 & 0 & 0 & 0 \\ -\rho_1\sigma & (\theta+\mu+\delta) & -\phi & 0 & 0 & 0 & 0 \\ -\rho_2\sigma & 0 & (\phi+\phi_q+\gamma_a+\mu) & 0 & 0 & 0 & 0 \\ -\rho_3\sigma & 0 & -\phi_q & (\theta_q+\gamma_q+\mu+\delta) & 0 & 0 & 0 \\ 0 & -\theta & 0 & -\theta_q & (\gamma+\mu+\delta) & 0 & 0 \\ 0 & 0 & -\gamma_a & -\gamma_q & -\gamma & (\Delta+\mu) & 0 \\ 0 & \frac{\beta S}{(1+\alpha I)^2} & \beta S & 0 & 0 & -\Delta & \mu+\beta(I_a+\frac{I}{1+\alpha I}) \end{bmatrix}$$

The Jacobian matrix of $\mathscr{V}(x)$ at DFE is:

$$D\mathscr{V}(E_0) = \begin{bmatrix} (\sigma+\mu) & 0 & 0 & 0 & 0 & 0 & 0 \\ -\rho_1\sigma & (\theta+\mu+\delta) & -\phi & 0 & 0 & 0 & 0 \\ -\rho_2\sigma & 0 & (\phi+\phi_q+\gamma_a+\mu) & 0 & 0 & 0 & 0 \\ -\rho_3\sigma & 0 & -\phi_q & (\theta_q+\gamma_q+\mu+\delta) & 0 & 0 & 0 \\ 0 & -\theta & 0 & -\theta_q & (\gamma+\mu+\delta) & 0 & 0 \\ 0 & 0 & -\gamma_a & -\gamma_q & -\gamma & (\Delta+\mu) & 0 \\ 0 & \frac{\beta bN}{\mu} & \frac{\beta bN}{\mu} & 0 & 0 & -\Delta & \mu \end{bmatrix}$$

$$= \begin{bmatrix} V & 0 \\ J_1 & J_2 \end{bmatrix}$$

where,

$$V = \begin{bmatrix} (\sigma+\mu) & 0 & 0 & 0 \\ -\rho_1\sigma & (\theta+\mu+\delta) & -\phi & 0 \\ -\rho_2\sigma & 0 & (\phi+\phi_q+\gamma_a+\mu) & 0 \\ -\rho_3\sigma & 0 & -\phi_q & (\theta_q+\gamma_q+\mu+\delta) \end{bmatrix}$$

**Step 4**: The last thing we need to generate the next generation matrix is $V^{-1}$, which after calculations is as below:

$$V^{-1} = \begin{bmatrix} \frac{1}{(\sigma+\mu)} & 0 & 0 & 0 \\ (\frac{\sigma}{(\sigma+\mu)(\theta+\mu+\delta)})(\rho_1 + \frac{\rho_2\phi}{(\phi+\phi_q+\gamma_a+\mu)}) & \frac{1}{(\theta+\mu+\delta)} & \frac{\phi}{(\theta+\mu+\delta)(\phi+\phi_q+\gamma_a+\mu)} & 0 \\ \frac{\rho_2\sigma}{(\sigma+\mu)(\phi+\phi_q+\gamma_a+\mu)} & 0 & \frac{1}{(\phi+\phi_q+\gamma_a+\mu)} & 0 \\ (\frac{\sigma}{(\sigma+\mu)(\theta_q+\gamma_q+\mu+\delta)})(\frac{\rho_2\phi}{(\phi+\phi_q+\gamma_a+\mu)} + \rho_3) & 0 & \frac{\phi_q}{(\theta_q+\gamma_q+\mu+\delta)(\phi+\phi_q+\gamma_a+\mu)} & \frac{1}{(\theta_q+\gamma_q+\mu+\delta)} \end{bmatrix}$$

**Step 5**: Now, we shall compute next generation matrix:

$$FV^{-1} = [f_{ij}]_{7X7}$$

where,

$$f_{11} = \frac{\beta bN\sigma(1-\omega)\sigma}{\mu(\sigma+\mu)}\left(\frac{\rho_1}{(\theta+\mu+\delta)} + \frac{\rho_2\phi}{(\theta+\mu+\delta)(\phi+\phi_q+\gamma_a+\mu)} + \frac{\rho_2}{(\phi+\phi_q+\gamma_a+\mu)}\right);$$

$$f_{12} = \frac{\beta bN(1-\omega)}{\mu(\theta+\mu+\delta)};$$

$$f_{13} = \frac{\beta bN(1-\omega)}{\mu(\phi+\phi_q+\gamma_a+\mu)};$$

and all other entries are zero.

Spectrum of $FV^{-1}$ is the set of eigen values of the matrix: $\{\frac{\beta bN\sigma(1-\omega)}{\mu(\sigma+\mu)}(\frac{\rho_1}{(\theta+\mu+\delta)} + \frac{\rho_2\phi}{(\theta+\mu+\delta)(\phi+\phi_q+\gamma_a+\mu)} + \frac{\rho_2}{(\phi+\phi_q+\gamma_a+\mu)}); 0; 0; 0\}$

Spectral radius of $FV^{-1}$ is the maximum eigen value of the matrix: $\frac{\beta bN\sigma(1-\omega)}{\mu(\sigma+\mu)}\left(\frac{\rho_1}{(\theta+\mu+\delta)} + \frac{\rho_2\phi}{(\theta+\mu+\delta)(\phi+\phi_q+\gamma_a+\mu)} + \frac{\rho_2}{(\phi+\phi_q+\gamma_a+\mu)}\right)$

Hence, $R_0 = \frac{\beta bN\sigma(1-\omega)}{\mu(\sigma+\mu)}\left(\frac{\rho_1}{(\theta+\mu+\delta)} + \frac{\rho_2\phi}{(\theta+\mu+\delta)(\phi+\phi_q+\gamma_a+\mu)} + \frac{\rho_2}{(\phi+\phi_q+\gamma_a+\mu)}\right)$

or

$$R_0 = \frac{\beta bN\sigma(1-\omega)(\rho_1(\phi+\phi_q+\gamma_a+\mu) + \rho_2(\phi+\theta+\mu+\delta))}{\mu(\sigma+\mu)(\theta+\mu+\delta)(\phi+\phi_q+\gamma_a+\mu)} \tag{13.4}$$

## *Local Stability of Disease-Free Equilibrium*

The stability of DFE ensures that the disease can be removed from the system over a finite period of time. We will now derive conditions for local stability of system (13.1). The Jacobian matrix for the system in (13.1) is given by:

$$J = \begin{bmatrix} -\mu-\beta_* & 0 & -\frac{\beta S}{(1+\alpha I)^2} & -\beta S & 0 & 0 & \Delta \\ \beta_*(1-\omega) & -(\sigma+\mu) & \frac{\beta S(1-\omega)}{(1+\alpha I)^2} & \beta S(1-\omega) & 0 & 0 & 0 \\ 0 & \rho_1\sigma & -(\theta+\mu+\delta) & \phi & 0 & 0 & 0 \\ 0 & \rho_2\sigma & 0 & -(\phi+\phi_q+\gamma_a+\mu) & 0 & 0 & 0 \\ 0 & \rho_3\sigma & 0 & \phi_q & -(\theta_q+\gamma_q+\mu+\delta) & 0 & 0 \\ 0 & 0 & \theta & 0 & \theta_q & -(\gamma+\mu+\delta) & 0 \\ 0 & 0 & 0 & \gamma_a & \gamma_q & \gamma & -(\Delta+\mu) \end{bmatrix} \tag{3.3}$$

where, $\beta_* = \beta(I_a + \frac{I}{1+\alpha I})$

The Jacobian matrix for the system in (13.1) at DFE is given by:

$$J(E_0) = \begin{bmatrix} -\mu & 0 & -\beta S^* & -\beta S^* & 0 & 0 & \Delta \\ 0 & -(\sigma+\mu) & \beta S^*(1-\omega) & \beta S^*(1-\omega) & 0 & 0 & 0 \\ 0 & \rho_1\sigma & -(\theta+\mu+\delta) & \phi & 0 & 0 & 0 \\ 0 & \rho_2\sigma & 0 & -(\phi+\phi_q+\gamma_a+\mu) & 0 & 0 & 0 \\ 0 & \rho_3\sigma & 0 & \phi_q & -(\theta_q+\gamma_q+\mu+\delta) & 0 & 0 \\ 0 & 0 & \theta & 0 & \theta_q & -(\gamma+\mu+\delta) & 0 \\ 0 & 0 & 0 & \gamma_a & \gamma_q & \gamma & -(\Delta+\mu) \end{bmatrix}$$

where, $S^* = \dfrac{bN}{\mu}$

Now, the system in (13.1) is locally assymptotically stable if the eigen values of the above jacobian matrix are all real and negative, which is true if the following conditions are met:

1. $(\theta+\mu+\delta) + (\sigma+\mu) + (\phi+\phi_q+\gamma_a+\mu) > 0$
2. $-\beta S^*\sigma(1-\omega)(\rho_1+\rho_2) + (\theta+\mu+\delta)[(\sigma+\mu) + (\phi+\phi_q+\gamma_a+\mu)] + (\sigma+\mu)(\phi+\phi_q+\gamma_a+\mu) > 0$
3. $-\beta S^*\sigma(1-\omega)[\rho_1(\phi+\phi_q+\gamma_a+\mu) + \rho_2(\phi+\theta+\mu+\delta)] + (\sigma+\mu)(\theta+\mu+\delta)(\phi+\phi_q+\gamma_a+\mu) > 0$

It can be observed that condition 2 holds true for $R_0 < 1$ and it can be deduced from the third condition that:

$$\frac{\beta S^*\sigma(1-\omega)[\rho_1(\phi+\phi_q+\gamma_a+\mu) + \rho_2(\phi+\theta+\mu+\delta)]}{(\sigma+\mu)(\theta+\mu+\delta)(\phi+\phi_q+\gamma_a+\mu)} < 1$$

$$\frac{\beta bN\sigma(1-\omega)[\rho_1(\phi+\phi_q+\gamma_a+\mu) + \rho_2(\phi+\theta+\mu+\delta)]}{\mu(\sigma+\mu)(\theta+\mu+\delta)(\phi+\phi_q+\gamma_a+\mu)} < 1$$

which implies that:

$R_0 < 1$

Hence, we have the following theorem.

**Theorem 1** *The system (13.1) is locally asymptotically stable at the disease-free equilibrium point if* $R_0 < 1$.

### *Endemic Equilibirum ($E_1$)*

Endemic equilibrium is defined as the point at which the disease is not completely eradicated but it approaches to a constant level in the population. Let the endemic equilibrium for system (13.1) be denoted by:

$$E_1 = (S^{**}, E^{**}, I^{**}, I_a^{**}, I_q^{**}, H^{**}, R^{**})$$

To solve for endemic equilibrium, we assume force of infection to be defined as:

$$\lambda^{**} = \beta I_a^{**} + \frac{\beta I^{**}}{1 + \alpha I^{**}}$$

Each of the equation in system (13.1) is then equated to zero to get the following solution:

$$\begin{aligned} E_1 = (&\frac{M}{\mu + J\lambda^{**}}, A\lambda^{**}S^{**}, \frac{\sigma A\lambda^{**}S^{**}G}{BC}, \frac{\rho_2\sigma A\lambda^{**}S^{**}}{B}, \\ &\frac{\sigma A\lambda^{**}S^{**}L}{BD}, \frac{\sigma A\lambda^{**}S^{**}(\theta DG + \theta_q CL)}{BCDE}, \frac{bpN}{F} + \frac{(1-J)\lambda^{**}S^{**}}{\Delta}) \end{aligned} \tag{13.5}$$

Here, $\lambda^{**}$ ca the below equation:

$$a_0(\lambda^{**})^2 + b_0\lambda^{**} + c_0 = 0 \tag{13.6}$$

where,

$$\begin{aligned} a_0 &= \frac{J^2}{\mu} + \frac{\alpha\sigma AGM}{\mu BC} \\ b_0 &= 2J + \frac{\alpha\sigma AGM}{BC} - \frac{\alpha\beta\sigma^2\rho_2 A^2 GM^2}{\mu B^2 C} - J(1 - p + \frac{\Delta p}{F})R_0 \\ c_0 &= -\mu[(1 - p + \frac{\Delta p}{F})R_0 - 1] \end{aligned}$$

Further $A, B, C, D, E, F, G, L, M, J$ used in the above expressions can be calculated as:

$$
\begin{aligned}
A &= \frac{1-\omega}{\sigma+\mu} \\
B &= \phi+\phi_q+\gamma_a+\mu \\
C &= \theta+\mu+\delta \\
D &= \theta_q+\gamma_q+\mu+\delta \\
E &= \mu+\gamma+\delta \\
F &= \mu+\Delta \\
G &= B\rho_1+\phi\rho_2 \\
L &= B\rho_3+\phi_q\rho_2 \\
M &= bN(1-p+\frac{\Delta p}{F}) \\
J &= 1-(\frac{\Delta\sigma A}{FB})(\gamma_a\rho_2+\frac{\gamma[\theta DG+\theta_q CL]}{CDE}+\frac{\gamma_q L}{D})
\end{aligned}
$$

It can be noted, that corresponding to each value of $\lambda^{**}$ in equation (13.6) we get a value for endemic equilibrium. Therefore, the number of endemic equilibria of the system (13.1) is equal to the number of positive roots of the equation (13.6). Clearly in equation (13.6), $a_0 > 0$ while $c_0 > 0$ when $R_0 < 1$ and $c_0 < 0$ when $R_0 > 1$. Using Descartes' rule of signs we have the following theorem on the existence of the endemic equilibrium.

**Theorem 2** *The system (13.1) has:*

1. *exactly one unique endemic equilibrium, if* $c_0 < 0$ *(i.e.* $R_0 > 1$*).*
2. *exactly one unique endemic equilibrium, if* $b_0 < 0$*, and* $c_0 = 0$ *(i.e.* $R_0 = 1$*) or* ${b_0}^2 - 4a_0c_0 = 0$.
3. *exactly two endemic equilibria, if* $c_0 > 0$ *(i.e.* $R_0 < 1$*),* $b_0 < 0$ *and* ${b_0}^2 - 4a_0c_0 > 0$.

***Remark*** It can be seen from the above theorem, making $R_0 < 1$ is not sufficient for controlling the disease. Therefore, some extra measures should be taken so that the diease can be controlled.

## *Local Stability of Endemic Equilibrium*

Now we discuss the local stability of endemic equilibrium. Similar to the process followed in subsection (3.3) for the stability of disease-free equilibrium, we start by finding the jacobian matrix of system (13.1) at endemic equilibrium point, which can be obtained by substituing the value of $E_1$ from equation (13.5) into equation (3.3):

$$J(E_1) = \begin{bmatrix} -\mu-\beta_{**} & 0 & \frac{-\beta S^{**}}{(1+\alpha I^{**})^2} & -\beta S^{**} & 0 & 0 & \Delta \\ \beta_{**}(1-\omega) & -(\sigma+\mu) & \frac{\beta(1-\omega)S^{**}}{(1+\alpha I^{**})^2} & \beta(1-\omega)S^{**} & 0 & 0 & 0 \\ 0 & \rho_1\sigma & -(\theta+\mu+\delta) & \phi & 0 & 0 & 0 \\ 0 & \rho_2\sigma & 0 & -(\phi+\phi_q+\gamma_a+\mu) & 0 & 0 & 0 \\ 0 & \rho_3\sigma & 0 & \phi_q & -(\theta_q+\gamma_q+\mu+\delta) & 0 & 0 \\ 0 & 0 & \theta & 0 & \theta_q & -(\gamma+\mu+\delta) & 0 \\ 0 & 0 & 0 & \gamma_a & \gamma_q & \gamma & -(\Delta+\mu) \end{bmatrix}$$

where, $\beta_{**} = \beta(I_a^{**} + \frac{I^{**}}{1+\alpha I^{**}})$
To find the eigen values for the above jacobian matrix, we put:

$$|J(E_1) - \lambda I| = 0$$

which implies,

$$\begin{vmatrix} -(\mu+\lambda+\beta_{**}) & 0 & \frac{-\beta S^{**}}{(1+\alpha I^{**})^2} & -\beta S^{**} & 0 & 0 & \Delta \\ \beta_{**}(1-\omega) & -(\sigma+\mu+\lambda) & \frac{\beta(1-\omega)S^{**}}{(1+\alpha I^{**})^2} & \beta(1-\omega)S^{**} & 0 & 0 & 0 \\ 0 & \rho_1\sigma & -(\theta+\mu+\delta+\lambda) & \phi & 0 & 0 & 0 \\ 0 & \rho_2\sigma & 0 & -(\phi+\phi_q+\gamma_a+\mu+\lambda) & 0 & 0 & 0 \\ 0 & \rho_3\sigma & 0 & \phi_q & -(\theta_q+\gamma_q+\mu+\delta+\lambda) & 0 & 0 \\ 0 & 0 & \theta & 0 & \theta_q & -(\gamma+\mu+\delta+\lambda) & 0 \\ 0 & 0 & 0 & \gamma_a & \gamma_q & \gamma & -(\Delta+\mu+\lambda) \end{vmatrix} = 0 \quad (13.7)$$

where, $\beta_{**} = \beta(I_a^{**} + \frac{I^{**}}{1+\alpha I^{**}})$

**Theorem 3** *The system (13.1) is locally asymtotically stable at the endemic equilibrium point if all the seven eigen values, obtained by solving the determinant in equation (13.7), have negative real parts.*

## Sensitivity Analysis

Sensitivity analysis helps us to determine the affect on model results due to change in parameter values and the extent to which this change affects the model results. It plays a crucial role because if the sensitivity to model parameters is known, we can use this in real life to control the spread of the disease. For instance, if we know a certain parameter when decreased leads to the reduction in infectious classes, certain measures can be employed which reduces it's effect in the real life. In this section, we have used our model to analyse the COVID-19 epidemic. We have first calculated the Sensitivity Indices of $R_0$ with respect to various parameters and then have done the sensitivity analysis using a few parameters.
The first step is to assign the values to model parameters. For simplicity, we have assumed a few parameter values and picked the rest from the existing studies that fit best with our model. The list of parameter estimates and their sources are summarized in Table 13.3 below.
Now, we study the impact on $R_0$ due to changing paramaters. To do this we obtain the Sensitivity Index of $R_0$ with respect to different model parameters, which measures the change in $R_0$ in response to the change in a parameter.
The sensitivity index of $R_0$ with respect to any parameter $Z$ is calculated as follows (refer to [11]):

**Table 13.3** List of parameter values

| Parameter | Value | Source | Sensitivity index |
|---|---|---|---|
| $b$ | 0.002 | Assumed | 1.0000 |
| $N$ | 60,000 | Assumed | 1.0000 |
| $\mu$ | 0.0098 | Assumed | −1.0726 |
| $\delta$ | 0.0175 $day^{-1}$ | [2] | –0.0242 |
| $p$ | 0 | Assumed | NA |
| $\rho_1$ | 0.5210 | [2] | 0.0428 |
| $\rho_2$ | 0.2740 | [2] | 0.5011 |
| $\rho_3$ | 0.2050 | [2] | NA |
| $\alpha$ | 0.25 | Assumed | NA |
| $\beta$ | 0.02 | Assumed | 1.0000 |
| $\sigma$ | 0.219 $day^{-1}$ | Assumed | 0.0428 |
| $\theta$ | 0.2174 $day^{-1}$ | [2] | −0.5778 |
| $\theta_q$ | 0.1429 $day^{-1}$ | [2] | NA |
| $\gamma$ | 0.701 $day^{-1}$ | Assumed | NA |
| $\gamma_a$ | 0.13978 $day^{-1}$ | [2] | −0.2307 |
| $\gamma_q$ | 0.11624 $day^{-1}$ | [2] | NA |
| $\phi$ | 0.139 $day^{-1}$ | Assumed | −0.1061 |
| $\phi_q$ | 0.019 $day^{-1}$ | Assumed | −0.0314 |
| $\Delta$ | 0.001 | Assumed | NA |
| $\omega$ | 0.2 | Assumed | -0.25 |

*Source* own

$$\varphi_{R_0}{}^{Z} = \frac{\partial R}{\partial Z}\frac{Z}{R_0}$$

Table 13.3 lists the sensitivity index of $R_0$ with respect to various parameters appearing in the formula for $R_0$, while Fig. 13.2 gives a pictorial representation of the same. The positivity or negativity of the index determines the relationship (direct or inverse) of $R_0$ with the parameter whereas its magnitude determines the strength of dependence of $R_0$ on the parameter. In Fig. 13.2, for every parameter with index extending towards the right, $R_0$ increases as the parameter increases while the one with index extending towards the left, $R_0$ decreases as it increases. Sensitivity Index of $R_0$ with respect to $\beta$ is +1.0 (i.e. $\varphi_{R_0}{}^{\beta} = +1.000$), which means that a 1% increase in $\beta$ results in a 1% increase in $R_0$. Similarly, $\varphi_{R_0}{}^{\theta} = -0.5778$, which implies that a 1% increase in $\theta$ will result in a 0.5778% decrease in $R_0$.

A pictorial representation of the COVID-19 compartmental model, fitted with parameter values is demonstrated in Fig. 13.3. Since, no vaccination has been brought into use yet we have set the proportion of vaccinated individuals ($p$) to zero. Also, since there is no concrete evidence that recovery ensures permanent immunity, therefore, we have allowed for the transition from recovered to susceptible state with a very

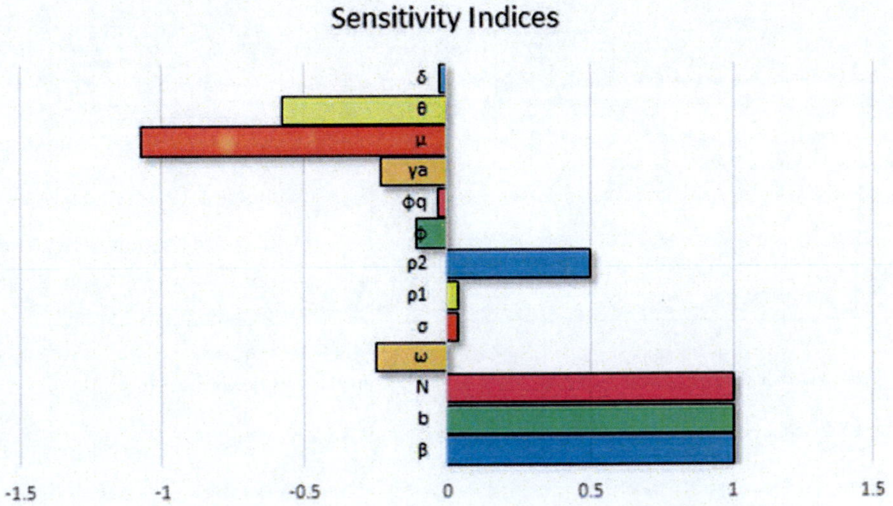


Fig. 13.2 Sensitivity Indices of $R_0$ corresponding to various parameters (*Source* own)

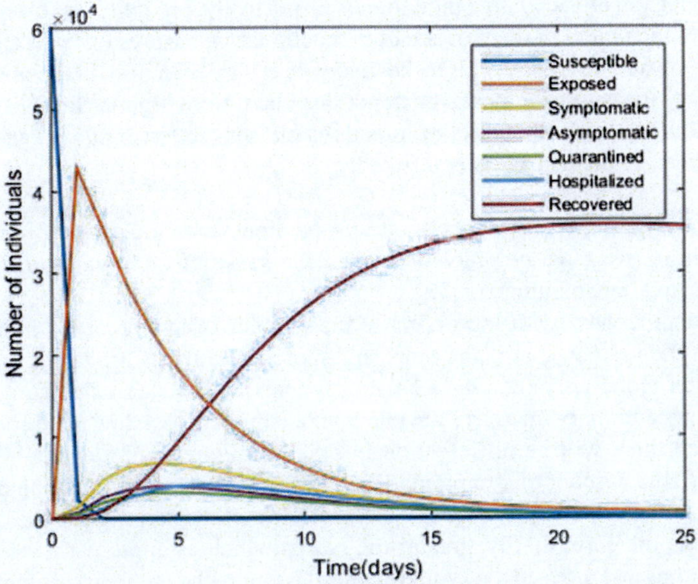


Fig. 13.3 The S-E-I-R-S compartmental model for COVID-19 epidemic (*Source* own)

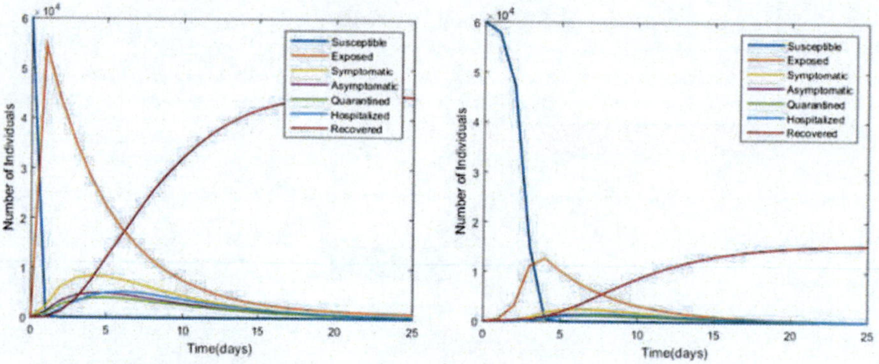


(a) Model without precautionary measures ($\omega = 0$) (b) Model with vaccination ($p = 0.72$)

**Fig. 13.4** Effect of precautionary measures and vaccination (*Source* own)

small possibility. We have analysed the sensitivity of the model to the parameters $\omega$ and $p$, which captures the effect of precautionary measures on the spread of the disease and the proportion of vaccinated individuals, respectively. Figure 13.4a, b depicts the behaviour of the model in the absence of precautionary measures ($\omega = 0$) and in the presence of a vaccine ($p = 0.72$), respectively. It can be seen how in the absence of precautionary measures the peaks of infectious classes are higher. Similarly, in the presence of the vaccine the peaks of infectious classes are lower and it can be seen how the spread of the disease is controlled.

Next, we have analysed the impact of various parameters on the spread of the infection. Therefore, we have graphically studied the sensitivity of symptomatic class with respect to a few parameters [18].
Fig. 13.5a depicts how with increasing $\omega$ the number of symptomatic individuals decrease in the system, that is, as more and more precautions are taken the infection keeps on reducing. When $\omega = 1$ the curve lies on the $x$-axis depicting that in case strict precautionary measures are taken, like complete lockdown and social distancing, then there will be no infection in system. Figure 13.5b depicts how with increasing $\beta$, the number of symptomatics increase, that is, greater the rate of transmission more the infection. Figure 13.5c depicts how with increasing $\phi$ the count of symptomatics increase, that is, greater the rate at which asymptomatic individuals show symptoms and move to the symptomatic class more the number of symptomatic individuals. Figure 13.5d depicts how with increasing $\theta$ the count of symptomatics decrease, that is greater the rate at which symptomatic individuals are hospitalized lesser the number of susceptible individuals.
This depicts the importance of steps like lockdown and social distancing, along with precautionary measures like using face masks and washing hands frequently, as these steps can help in controlling the spread of the disease by leading to a lesser number of infected individuals in the system. Similarly, better the medical facilities and the treatment, lesser will be the spread of the disease. As more and more infected people

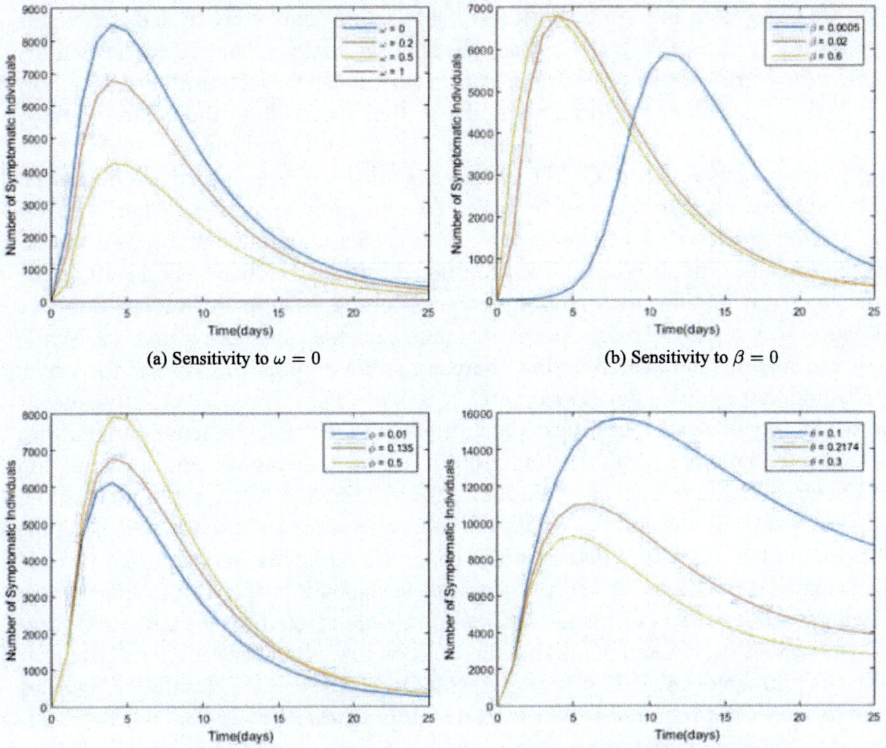


(a) Sensitivity to $\omega = 0$

(b) Sensitivity to $\beta = 0$

(c) Sensitivity to $\phi = 0$

(d) Sensitivity to $\theta = 0$

**Fig. 13.5** Sensitivity of Symptomatic class to changing parameters (*Source* own)

will be treated, leading to lower number of infectives in the system, this will result in lower risk of getting exposed and hence control the spread of the disease.

## Conclusion

We have defined a general $S$-$E$-$I$-$R$-$S$ compartmental model for epidemic diseases and derived important mathematical results like the reproduction number, the diseased-free equilibrium and the endemic equilibrium. While the reproduction number deals with how rapidly the disease will spread, the two equilibriums, in real sense, corresponds to the levels at which the disease would completely be eradicated or atleast will remain at constant levels within the population. We have also derived conditions for the stability of the two equilibrium points, along with the conditions for the existence of the endemic equilibrium and have established that $R_0$ <1 would imply local asymptotic stability of DFE.

Later, we have fitted the model with a suitable set of parameters to deduce predictions regarding the COVID-19 disease. Firstly, we have calculated the sensitivity indices of $R_0$ corresponding to various parameters. It can be seen from Fig. 13.2 how $R_0$ is highly dependent on parameters like $\beta$, thus, controlling the values of these parameters can help cope with the disease. Then, we have used the model to do sensitivity analysis for the COVID-19 epidemic disease. We studied the behaviour of symptomatic class in the presence of a few changing parameters. Then we have analyzed the impact of $\omega = 0$ and $p = 0.72$. It is clear that in the presence of a vaccine the spread of the disease can be minimized. We can see from Fig. 13.4b, in the presence of vaccination the infection decreases in the system. But vaccination as a mitigation strategy is not as simple as it sounds and is a time taking process. While many vaccinations are available now there are still a number of diseases for which vaccination has not been developed yet.

The availability of a vaccine alone do not bring down the list of challenges linked with it. One such challenge being the impossibility of vaccinating the entire population of a country. Ofcourse instead of targeting the vaccination of an entire population, epidemics deal with the concept of herd immunity, which can be estimated using the reproduction number (once determined). But even achieving herd immunity could take years in some cases due to the required vaccinated proportion falling almost around the whole of it. For instance, small pox which lasted for around 3000 years required an 80% vaccinated population in each country which was achieved over several years with an effort led by World Health Organization [13]. Similarly, measles lasted for years and required 96% vaccinated population [1].

The next challenge linked to vaccination is that even if a vaccine is available along with all the estimates regarding achieving the herd immunity, not all the countries can afford developing it or some might even face issues in purchasing it. In any case, coming up with a first and safe dose of a vaccine during any epidemic takes time and implementing it would take years.

Until a vaccine is made available, precautionary measures play an important role in eliminating the disease. As reported by WHO, the COVID-19 virus spreads due to the transfer of liquid droplets from an infected individual's mouth or nose while the person sneezes or coughs [12]. It can even stay on non-living bodies for varied amount of times. So, for instance, if a person sneezes and the infected droplets fall on a cereal packet in a super market, the next person touching the packet will be exposed to the disease if he/she touches his/her mouth, nose, or even eyes leading to the virus entering his/her body. Therefore, steps like social distancing, covering one's mouth and nose using a mask and washing one's hands frequently is playing a major role in combating with the disease. It can be noticed from Fig. 13.5a that while $\omega$ is set to 1 the infectious classes rest at zero, which means, in case of no contact at all the spread of disease would stop because the virus won't get new living bodies to cling onto and transfer any further. But, obviously the condition of no contact at all cannot be achieved, as even though lockdowns have been imposed all over the world to achieve this as much as possible, people still have to get out of their places for every day essentials and even for providing essential services. So, precautionary measures like lockdown cannot be treated as long term solution as it drastically affects the economy. But precautionary measures do help in slowing down the process and help in reducing the spread of the disease until long term solutions are explored. Figure 13.4a clearly shows how in the absence of any precautionary measures the infectious trajectories attain high peaks than usual, thus, depicting their importance.

Thus, it can be argued that even though vaccination is a permanent and a long term solution for an epidemic but until a vaccination comes in to play precautionary measures can help a great deal in controlling the disease. Since, all these mitigation strategies and the extent to which they should be allowed for are derived from a close insight into the epidemic model, it can be said the dynamic analysis of an epidemic is the base for it's future elimination. Also, it is to be noted that once sufficient data related to the impact of vaccination is available, from countries like Russia, where vaccination is now available, the model developed in this paper can be used to do future predictions regarding the long term effects of vaccination in the fight against COVID-19.

## References

1. Ochoche, J. M., & Gweryina, R. I. (2014). A mathematical model of Measels with vaccination and two phases of infectiousness. *IOSR Journal of Mathematics, 10*, 95–105. https://doi.org/10.9790/5728-101495105.
2. Senapati, A., Rana, S., Das, T., & Chattopadhyay, J. (2020). *Impact of intervention on the spread of COVID-19 in India: A model based study*. https://arxiv.org/abs/2004.04950.
3. Paul, A., Chatterjee, S., & Bairagi, N. (2020). *Prediction on COVID-19 epidemic for different countries:Focusing on South Asia under various precautionary measures*. https://doi.org/10.1101/2020.04.08.20055095.
4. Ngonghala, C. N., Iboi, E., Eikenberry, S., Scoth, M., MacIntyre, R. C., Bonds, M.H., & Gumel, A. B. (2020). *Mathematical assessment of the impact of non-pharmaceutical interventions on curtailing the 2019 novel Coronavirus, Mathematical Biosciences*. https://doi.org/10.1016/j.mbs.2020.108364.
5. Aguilar, J. B., Faust, J. S., Westafer, L. M., & Gutierrez, J. B. (2020). *Investigating the impact of asymptomatic carriers on COVID-19 transmission*. https://doi.org/10.1101/2020.03.18.20037994.
6. Chang, X., Liu, M., Jin, Z., & Wang, J. (2020). Studying on the impact of media coverage on the spread of COVID-19 in Hubei Province, China. *Mathematical Biosciences and Engineering, 17*(4), 3147–3159. http://dx.doi.org/10.3934/mbe.2020178.
7. Kassa, S. M., Njagarah, H. J. B., & Terefe, Y. A. (2020). Analysis of the mitigation strategies for COVID-19: from mathematical modelling perspective. *Chaos, Solitons and Fractals, 138*, 1–16, Article ID 109968. https://doi.org/10.1016/j.chaos.2020.109968.
8. Shah, N. H., Suthar, A. H., & Jayswal, E. N. (2020). Control strategies to curtail transmission of COVID-19. *International Journal of Mathematics and Mathematical Sciences, 2020*, 1–12, Article ID 2649514. https://doi.org/10.1155/2020/2649514.
9. Zaman, G., Kang, Y. H., & Jung, I. H. (2009). Optimal treatment of an SIR epidemic model with delay. *BioSystems, 98*(1), 43–50. https://doi.org/10.1016/j.biosystems.2009.05.006.
10. Zaman, G., Kang, Y. H., Cho, G., & Jung, I. H. (2017). Optimal strategy of vaccination and treatment in an SIR epidemic model. *Mathematics and Computers in Simulation, 136*, 63–77. https://doi.org/10.1016/j.matcom.2016.11.010.

11. Samsuzzoha, M., Singh, M., & Lucy, D. (2013). Uncertainty and sensitivity analysis of the basic reproduction number of a vaccinated epidemic model of influenza. *Applied Mathematical Modelling, 37*(3), 903–915. https://doi.org/10.1016/j.apm.2012.03.029
12. World Health Organization (WHO) website, COVID-19 information: https://www.who.int/emergencies/diseases/novel-coronavirus-2019/advice-for-public.
13. World Health Organization (WHO) website, Small-pox information: https://www.who.int/csr/disease/smallpox/vaccines/en/.
14. Hang, J., Yang, Z., Pawelek, K. A., & Liu, S. (2019). Optimal control strategies for a two-group epidemic model with vaccination-resource constraints. *Applied Mathematics and Computation, 371*, Article ID 124956. https://doi.org/10.1016/j.amc.2019.124956.
15. Chatterjee, A. N., Saha, S., Roy, P. K., Basir, F. A., Khailov, E., & Grigorieva, E. (2020). *Insight of COVID-19/ SARS-CoV-2 and its Probable Treatment—A Mathematical Approach*. https://doi.org/10.21203/rs.3.rs-34519/v1.
16. Kaustuv Chatterjee, Kaushik Chatterjee, A. Kumar, S. Shankar, Healthcare impact of COVID-19 epidemic in India: A stochastic mathematical model, Medical Journal Armed Forces India, vol. 76(2), pp. 147-155, 2020. https://doi.org/10.1016/j.mjafi.2020.03.022
17. Cintra, P. H. P., Citeli, M. F., & Fontinele, F. N. (2020). *Mathematical Models for Describing and Predicting The COVID-19 Pandemic Crisis*. https://arxiv.org/abs/2006.02507.
18. A.M. Mishra, S.D. Purohit, K.M. Owolabi, Y.D. Sharma, A nonlinear epidemiological model considering asymptotic and quarantine classes for SARS CoV-2 virus, Chaos, Solitons and Fractals, vol. 138, Article ID 109953, pp. 1-10, 2020. https://doi.org/10.1016/j.chaos.2020.109953

# Chapter 14
# Impact of 'COVID-19' on Education and Service Sectors

**Mansi Aggarwal and Vijay Kumar**

**Abstract** Coronavirus disease is an infectious disease which is caused by a virus called coronavirus. The people who are infected with this disease will experience respiratory illness. This disease has been declared pandemic by 'World Health Organization (WHO)'. There are many sectors that have been affected due to the lockdown practiced in the entire country, among agricultural sector, manufacturing sector, service sector, education sector, business sector, etc. In our research, we examined the impact of COVID-19 in India vis-a-vis different sectors. For the purposes of this research, we shortlisted two particular sectors, i.e. education and service sectors. These two sectors form the backbone of our country. While the impact on education sector has led to many young minds and vulnerable school kids being affected in an adverse way and has left them to cope with new practices such as online classes during the lockdown period, and on the other hand, in the service sector, employees are working from home which in some case has had an impact on the effectiveness and efficiency of their work. In this paper, we assessed the impact on these two sectors on Indian economy by analyzing the responses given by our respondents through the questionnaires (Google form), and we then combined the data points to study how students and employees are being affected during the present lockdown period, imposed due to COVID-19. This chapter will help the readers to get to know more about the thinking of the students and employees in lockdown, and how much they are affected by this pandemic. Towards the trailing part of our research, we have discussed about the possible steps that can be adopted in future, by the employers and educational institutions, in order to limit the damage to the sector and to make recovery in the future.

**Keywords** COVID-19 · Education sector · Service sector · SPSS · MS excel · Data analysis

M. Aggarwal · V. Kumar (✉)
Department of Mathematics, Amity Institute of Applied Sciences, Amity University Uttar Pradesh, Noida, India
e-mail: vijay_parashar@yahoo.com

M. Aggarwal
e-mail: aggarwalmansi80@gmail.com

 
N. H. Shah and M. Mittal (eds.), *Mathematical Analysis for Transmission of COVID-19*, Mathematical Engineering, https://doi.org/10.1007/978-981-33-6264-2_14

## Introduction

Coronavirus disease 2019 (COVID-19) is a communicable/transferable disease which is caused by acute respiratory syndrome 'coronavirus-2 (SARA-CoV-2)'. The disease was initially recognized in December 2019 'Wuhan', the capital of China's Hubei province, and since then, it has been increasing in numbers which lead to the current pandemic. The time from contact to the beginning of symptoms is about five days but it may also vary from 2-14 days. This virus initially spreads among human beings through close contact mostly through small droplets generated by coughing, sneezing or talking. This virus can last on the surfaces for about 72 h. Historically, one of the most wrecking pandemic was the 'Black Death', which killed vaguely 75–200 million people in the fourteenth century. Later, HIV/AIDS is the only pandemic which started in 1980s. The coronavirus has been declared a pandemic on 11th March, 2020 by 'World Health Organization'. Due to this pandemic, there are many precautions that had been taken all over the world:

- Japan closed all of its schools for 1.3 crores children till 31 May.
- China already did it for 20 crores students.
- Saudi Arabia blocked foreign pilgrims from entering their country.
- Tokyo Olympics have been cancelled.
- Australia forbidden foreign visitors arriving from mainland China.
- Italy quarantined its towns.
- San Francisco declared emergency.
- India declared lockdown till 31 May.

In India, the first case of COVID-19 was reported on 30th January, 2020 followed by two similar cases on 2nd and 3rd February. All the three cases had a travel history of Wuhan, China. A month later on 2nd March more two new cases were reported—one each from New Delhi and Hyderabad. A sharp increase in the number of cases since then followed. To contain the spread, the Ministry of Health and Family Welfare (MoHFW) immediately took an action and issued a travel advisory as such restrictions had previously remained efficacious during the outbreak of SARS, Ebola and bubonic plague. All international travellers entering the country were asked to self-quarantine for 14 days. All travel visas to other countries were cancelled until 5th April, 2020. All states were asked to invoke the epidemic disease act which allowed officials to quarantine suspected cases and to close down public places. An intensive campaign was rolled out, and guidelines were developed for personal hygiene, surveillance, contact tracing, quarantine, diagnosis, laboratory tests and management. People were advised not to visit farms, animals markets or places, where they are slaughtered to avoid mass gatherings. All the healthcare facilities were asked to stop regular out-patient and in-patient services and to continue with solely emergency services. Doctors were encouraged to use telemedicine services. Arogya Setu app has been launched to connect people with essential health services. This app desired to reach and inform the users of the risks, practices, etc., relating to the contamination of COVID-19. Amenities like hotels, colleges, railway train, coaches, etc., got converted into quarantine centres, whereas large public places such as stadiums were converted into isolation wards to handle an anticipated increased number of cases. Some of the states converted their existing hospitals to exclusively handle COVID-19 patients. On 22nd March, Prime Minister Narendra Modi initiated the lockdown process with a 14-hour 'Janta Curfew' followed by a lockdown in seventy five COVID-19 affected districts and then calling on for a national lockdown of 3 weeks. A containment plan involving the state and twenty ministers was set up. Also a round-the-clock control room was set up at the headquarters of the General Director of Health Services (DGHS) to address the virus-related queries.

## Literature Review

There are some research papers and articles on COVID-19 and its impact from which we took help in our project report. In an article, Khanna et al. [1] expressed that how much countries are being affected due to pandemic, how they are responding by maintaining social distancing, lockdowns and improving hygienic practices. Gumber and Bulsari [2] in their paper explained how coronavirus had hit the economy hard and how the innovations in technology like artificial intelligence resultant in increase in the employment rate. McKibbin and Fernando [3] explained in their paper how a pandemic outbreak could significantly impact the economy globally in a short period of time. Fetzer et al. [4] in their paper expressed the rapid increase in economic anxiety during and after the initial global spreading of the coronavirus through Google search activity. Viner et al. [5] explained in their paper that how much schools are effecting if restrictive social distancing policies will be implemented for long periods. Sahu [6] in their paper explained how the outbreak impacted on the education and mental health of students and academic staff. Singhal [7] reviewed on Coronavirus Disease-2019, how it has impacted different sectors in China, especially it's neighbouring countries. Fernandes [8] estimated GDP growth in 2020 under different scenarios. Bartik et al. [9] insighted the importance of different policies measures and how they can impact on small businesses. Nicola et al. [10] summarised the socio-economic effects of Coronavirus disease on different aspects of the economy of the world. Akash and Mala [11] highlighted the Impact of COVID-19 on Educational Institutions and several issues students are facing. Ray et al. [12] explained the predictions and effects of National Lockdown in the country and the role of interventions. Fotios and Sypros [13] in their paper described the timeline of live forecasting exercise with huge implications on providing objective forecasts for confirmed cases of the pandemic. The Government of India [14] website gave the regular track of state wise status of COVID-19 cases by using Aarogya Setu application. WHO [15] website gave the information about the number of cases and the situation in the worldwide due to pandemic.

Our research deals with the 'Impact of COVID-19 on education and service sectors'. Here, in this study, we will discuss how much their education institution is equipped with technology, problems they are facing during classes and their opinions about the present scenario on education system. And for service sector we'll discuss, the problems employees are facing in doing work from home, current information regarding their salaries, jobs, etc. To determine all the important challenger/issues faced by both the sectors, we designed an open online survey forms of both the sectors separately. These forms are available at-

Education Sector—"https://docs.google.com/forms/d/16u6aNPxaHhCHD_5alnLASm0f_5tOBIhJjpu7YnIC4_8/edit"

Service Sector—"https://docs.google.com/forms/d/e/1FAIpQLScQTlqYZFfom9A9jw-flKODpNwMjZHzJd2r9V7Z9mgpyP3rJQ/viewform?usp=sf_link"

This study provides an empirical evidence to support above stated questions (issues). It also provides more perception into the present thinking of students, employees in this pandemic. The modules were created over Google form which is an online survey form. In this module, we have all compulsory questions. This module contains only multiple-choice questions; students and employees can access only one answer. And every question has minimum three options. COVID-19 has impacted to all the sectors of the economy like education sector, service sector, agriculture sector, business sector, healthcare sector, entertainment sector, pharmaceuticals and tourism, etc. In this paper, we will talk about the impact on two specific sectors, i.e. education and service sectors. We have started collecting data from 4th April, 2020, and it took us 1 week to collect the whole data.

## Review Aim

The aim of this research was to gain insight of students and private employees of the impact on their lives due to COVID-19. Two questionnaires were prepared for education and service sector separately.

For education sector—The questionnaire was filled by students which was based on the experiences and problems in online classes organized by their schools, universities, etc.

For service sector—The questionnaire was filled by the employees of private sector based on the problems, experiences they are facing while working from home.

The aim was to demographic estimation of the problems, experiences faced by both education and service sector due to the lockdown in the entire country.

## Methods

### *Design*

To attend this module, it is necessary that the school and college students are attending online classes, and the private employees, employers are working from home during the lockdown in the entire country due to this pandemic. The review is conducted over Google form which is an online survey format.

## *Respondents*

We took 100 respondents from each module. The respondents in education sector were school and college students from Amity University (most) and Delhi University (least). And the respondents in service sector were employees working in private companies (mostly in IT sector). All the respondents were either taking online classes or working from home by using web applications like Zoom, Microsoft Teams, Skype, etc.

## *Tools*

We used different to calculate our variable. Those that describe the impact on students and employees personally were measured by various mathematical concepts:

'*Mean*'—In these modules, our scenario is of 'Discrete Probability Distribution of Random Variable' $X$ (=100). The mean is equivalent to the sum over every possible value weighted by the probability of that value. It is further calculated by taking the product of each possible value $x$ of $X$ and its probability $p(x)$. And then adding all the products together:

$$\mu = \sum x.p(x) \tag{14.1}$$

'*Standard Deviation*'—It is the measure of the amount of variation or dispersion of a set of values. When we get a low standard deviation, then it indicates that the values are tend to be near to the mean of given set but if we get a high standard deviation, then it indicates that the values are cover over a wider range of the particular set. In these modules, we have computed the standard deviation of all questions and further analyzed them.

$$\sigma = \sqrt{\frac{\sum(x_i - \mu)^2}{N}} \tag{14.2}$$

$\sigma$ = population standard deviation

$N$ = the size of population

$x_i$ = each value from the population

$\mu$ = the population mean.

*'Correlations'*—It is calculated between two, three related questions. 'Spearman's Correlations' revealed that all parameters are being correlated significantly at the 0.01 level. This indicates at which level/degree the parameters are related to each other. There are four degrees of correlation, i.e. high, low, moderate and perfect degree that depends on the coefficient value of correlations.

### *Questionnaire*

The modules of both the sectors consist of 17 questions. All the questions were compulsory, and questions had been framed according to the experiences, and problems everyone is facing during online classes or work from home.

## Data Analysis

To calculate the significant relationship, we will use different types of graphs like bar graph, histogram, line graph, pie chart, etc., to do the analyses between the related questions.

### *Service Sector*

We made a questionnaire of 17 questions with the help of Google forms and asked the employees of the private companies to share their experiences with us.

- In Fig. 14.1, the impact of COVID-19 on the workers personally and their companies is shown below. According to the data collected, more than 55% of them are

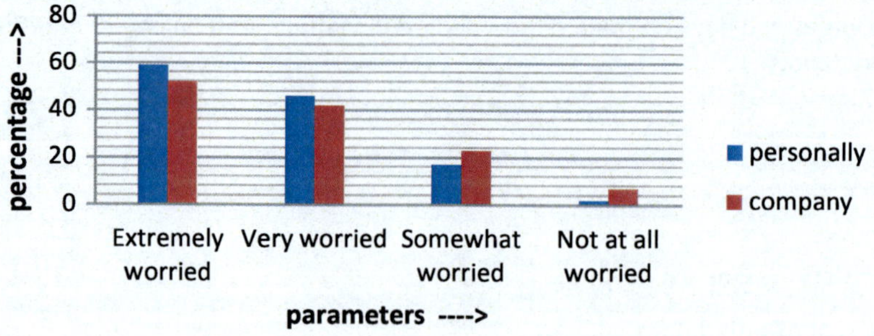


**Fig. 14.1** Bar graph showing the impact of COVID-19 on employees personally and on their company in percentage. *Source* Own

extremely worried of the impact of COVID-19 on them personally as it will affect the employment rate. And more than 50% are extremely worried of the impact on their companies as coronavirus has stopped the services given by the companies to different sectors.

- In Fig. 14.2, only 23.4% respondents believed that it is difficult to manage work from their homes and more than 50% of them found it neither easy nor difficult to work from home and this impacted the productivity of the work.
- In Fig. 14.3, the most common problem faced by people while working from home is lack of physical workspace as due to lockdown, all the family members are stuck at home. People are not getting proper workspace to work.
- In Fig. 14.4, 60% of employees think that work environment will be comfortably sustained up to couple of weeks as people are facing many problems in working from home.

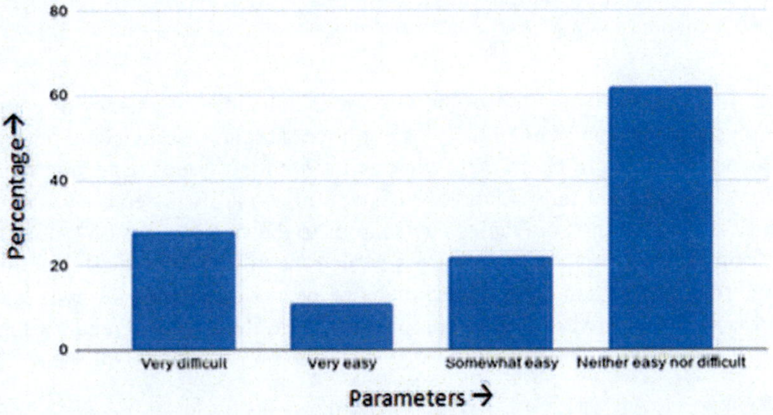


**Fig. 14.2** Histogram showing the percentage of impact of work productivity. *Source* Own

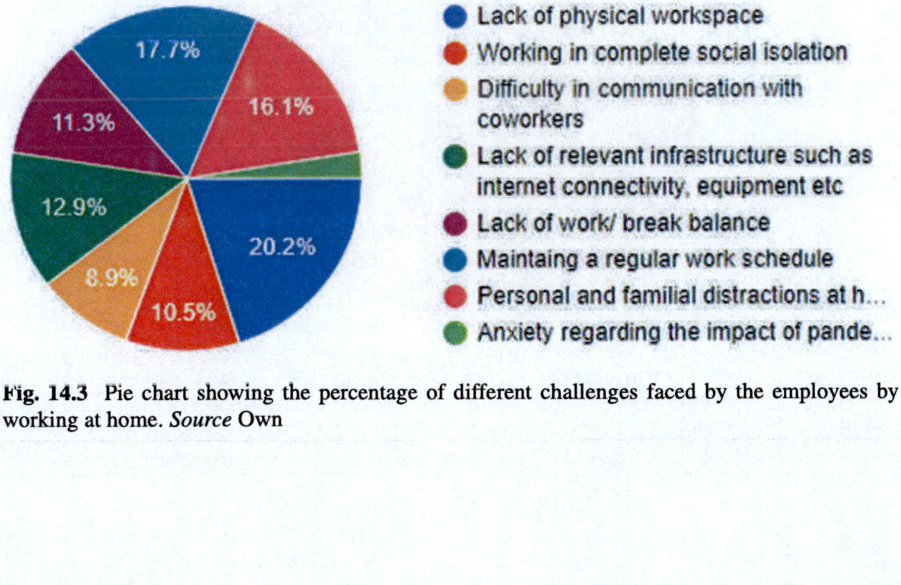


**Fig. 14.3** Pie chart showing the percentage of different challenges faced by the employees by working at home. *Source* Own

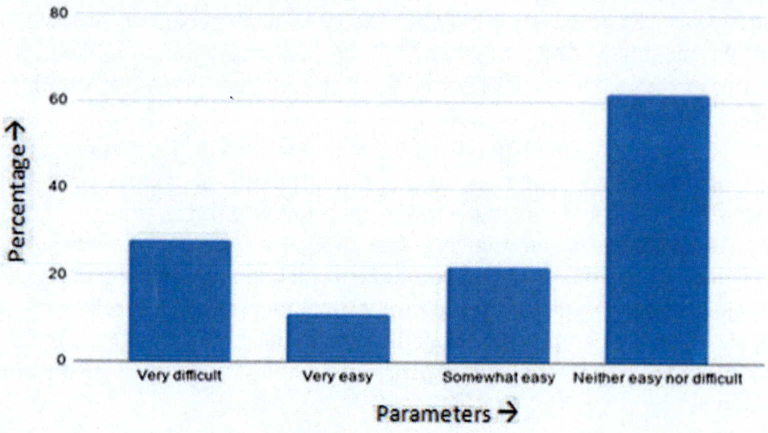


**Fig. 14.4** Histogram showing the percentage of sustainable work environment for employees. *Source* Own

- In Fig. 14.5, according to 46% of the respondents the ideal frequency of communication should be maintained daily as employer should be connected with their employees to keep a track on their work and solve their issues regarding work.
- In Fig. 14.6, there are more than 45% of people who are worried about their job security as usually, the companies pay salaries to their employees on the strength of working staffs, so when the numbers reduce the billing will be affected too.
- In Fig. 14.7, more than 50% of respondents agreed this pandemic will lead to decrease in the turnover of their companies. More than 45% agreed on having

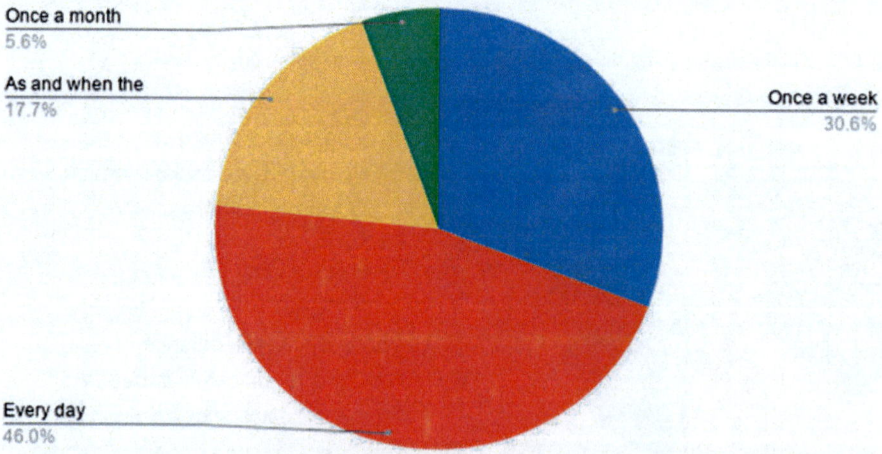


**Fig. 14.5** Pie chart showing the percentage of ideal frequency of communication between employee and leadership team. *Source* Own

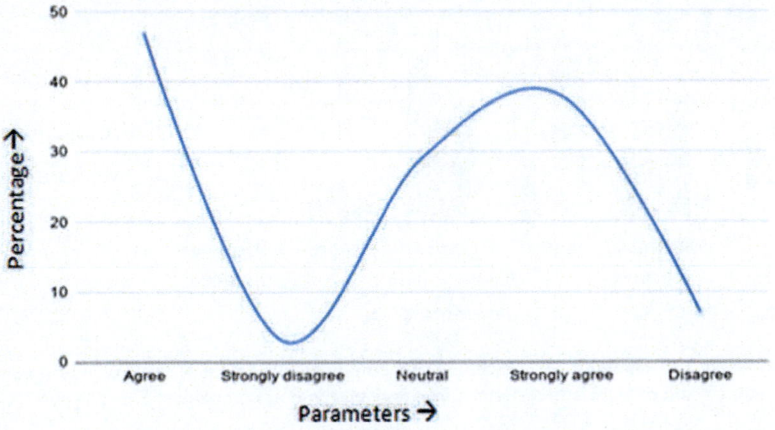


**Fig. 14.6** Line graph showing the percentage of employees worried about their job security. *Source* Own

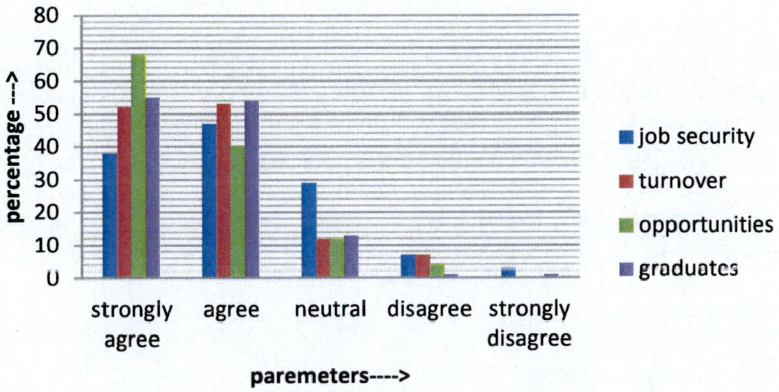


**Fig. 14.7** Bar graph showing the percentage of employees worried about the turnover of their companies, less job opportunities in future, effect on this year's graduates and anxiety regarding their jobs. *Source* Own

anxiety related to job security due to present environment. More than 65% respondents strongly agreed that there will be less job opportunities in future. And more than 50% strongly agreed that university graduates will be severely affected due to this pandemic as there will be withdrawal of job offers as COVID-19 will lead to global recession.

- In Fig. 14.8, according to 45% of the respondents have been provided various resources and benefits from there company which helps them in working efficiently. According to more than 50% of employees, the guidance regarding time management method, work from home, etc. has been provided support by their

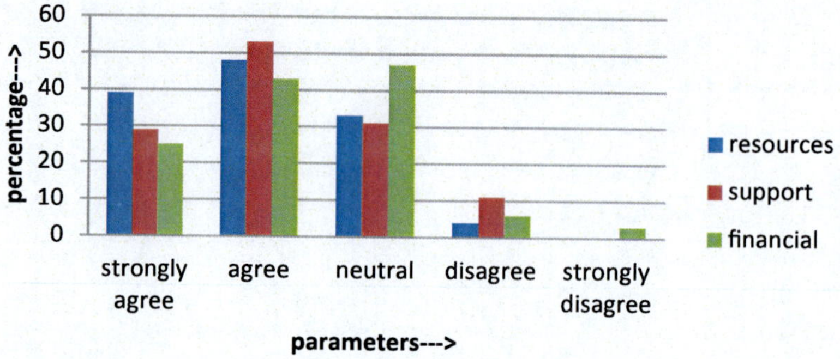


**Fig. 14.8** Bar graph showing the percentage of employee's agreement level on the resources, financial support and beyond professional support given by their companies. *Source* Own

company, and only 25% of respondent strongly agreed that their company has given them the information regarding the financial condition of there company.

- In Fig. 14.9, more than 60% of respondents agreed that there will be increase in the layoffs of the retail sector company. According to more than 50% of respondents, the car making industries, textiles, machineries, etc. are dependent on China, South Korea and Italy for the input and are vulnerable to disruption. Approximately, 50% of the respondents, strongly agreed that there will be increase in the demand of data centre services and video conferencing software's.
- In Fig. 14.10, only 33% of employees had been provided with the information about the pay cut by there leadership team which could lead the revision of there pay structure.

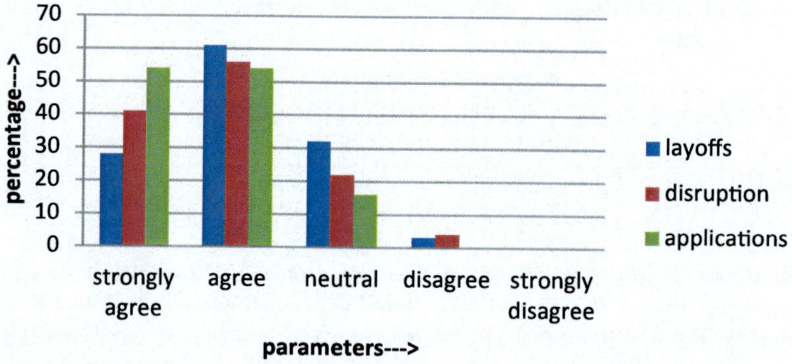


**Fig. 14.9** Bar graph showing the percentage of employee's agreement level with the increase in layoffs in their companies increase in disruption in textiles, machinery, etc. companies and the increase in the services of web applications and services. *Source* Own

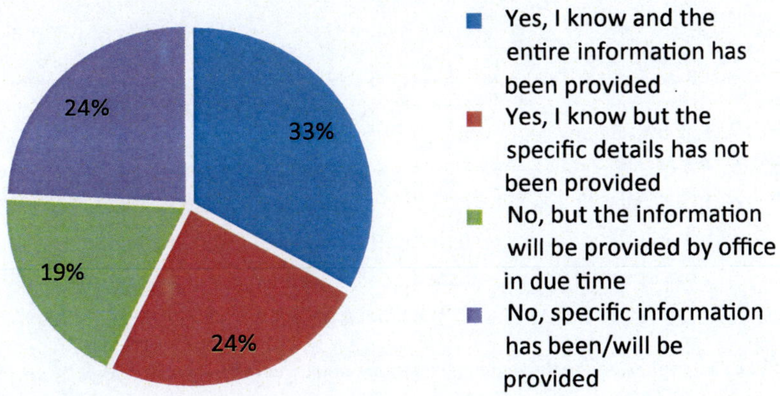


**Fig. 14.10** Pie chart showing the percentage of respondent's agreement level on the information given by company regarding their pay cuts. *Source* Own

## *Education Sector*

We made a questionnaire of 17 questions with the help of Google forms and asked the school and college students to share their experiences with us.

- In Fig. 14.11, only 49% of educational institutions are equipped with technology to provide online classes due to their low incomes.
- In Fig. 14.12, more than 64% students are well equipped in using technology related to online classes as nowadays almost everyone can easily access to Internet and are using technology for doing regular school, college assignments and projects.
- In Fig. 14.13, according to our respondents, only 50% of the students are comfortable using electronic communication sites such as teams, zoom, Skype, etc. As students who live in remote areas are facing problems in accessing these websites

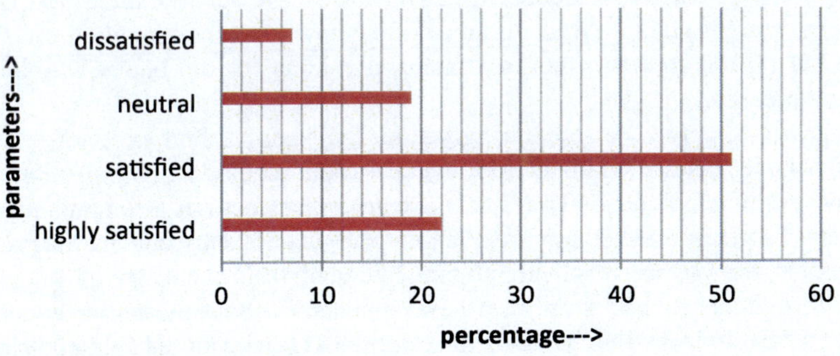


**Fig. 14.11** Bar graph showing the percentage of student's agreement level of being their educational institution well equipped. *Source* Own

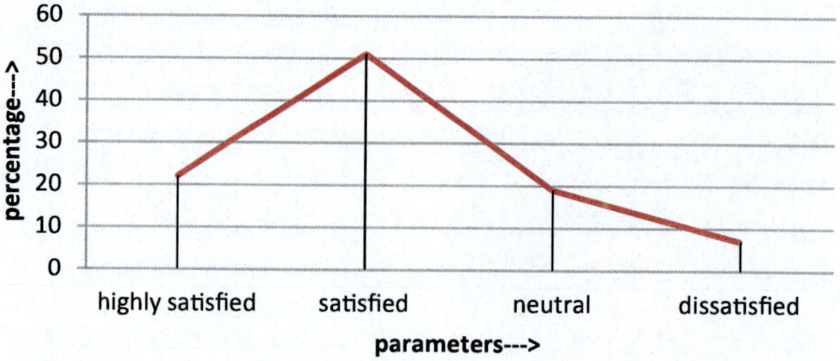


**Fig. 14.12** Line graph showing the percentage satisfaction level of using technology. *Source* Own

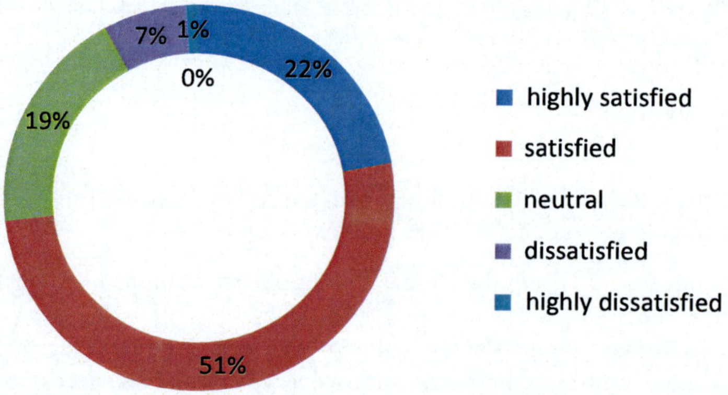


**Fig. 14.13** Doughnut chart showing the percentage of comfort level of students with electronic communication. *Source* Own

due to the need of strong Internet connectivity required to use these sites and applications.

- In Fig. 14.14, most of the students agreed on that the impact of online education on present nature is neutral. Students are learning new things, saving their travelling time but on the contrary, they are facing problems for the books, working in laboratories, etc.
- In Fig. 14.15, only 20% students agreed on the basis of their experience of the online classes conducted that traditional classroom method can be substituted by online teaching method, and physical presence of instructor is also required. 48% students have responded that virtual laboratories cannot substitute learning in the classroom for various practical subjects like statistics, chemistry, physics, etc. Also this pandemic had given us a new perspective about the online education which saves students time, gives them comfort and is convenient to attend classes sitting at home as students college students have to travel in buses, metros, cars, etc. to reach their college.

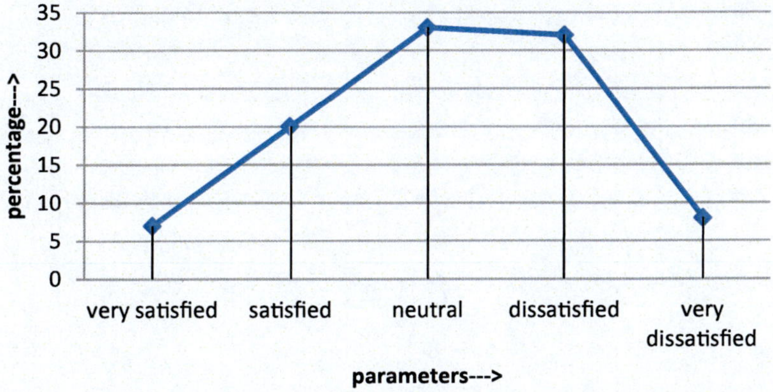


**Fig. 14.14** Line graph showing the percentage of impact of online classes on students. *Source* Own

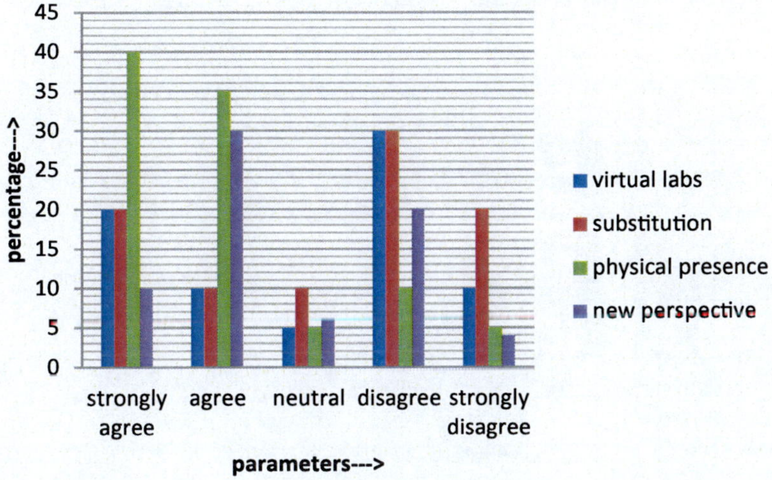


**Fig. 14.15** Bar graph showing the percentage of the agreement level of students of requirement of physical presence of instructor, new perspective on online teaching, substitution of virtual laboratories and traditional methods of teaching. *Source* Own

- In Fig. 14.16, only 37% educational institutions took steps to familiarize the students with the technology structure used to attend online classes. As due to low income and lack of resources, some institutions were not able to conduct online classes. Also, lack of physical resources like books, library access, etc. had adversely impacted the learning of students because college students were not able to find extra study material online and were facing problems.
- In Fig. 14.17, according to more than 50% of the students the use of technology such as laptop, mobile phone, tablet in online classes for learning leads to lapses in concentration and learning disruption of students due to poor Internet connectivity.

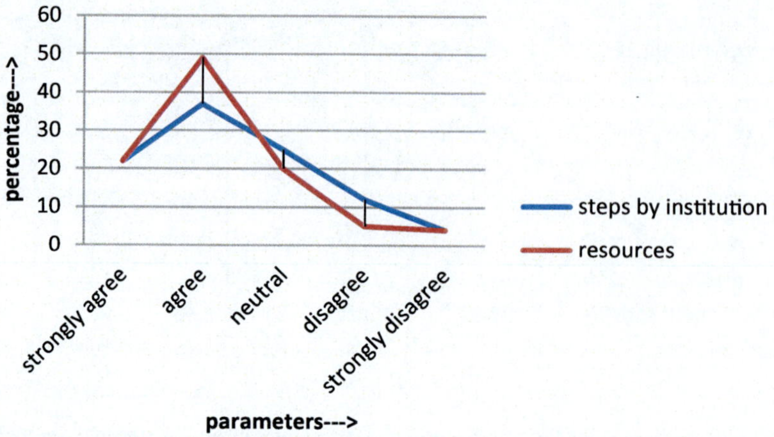


**Fig. 14.16** Line graph showing the percentage of student's agreement level on the steps taken by their institution for online classes and affect due to lack of physical resources. *Source* Own

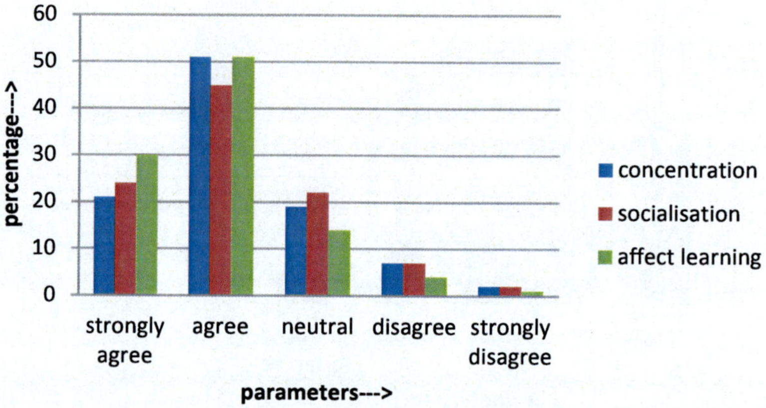


**Fig. 14.17** Bar graph showing the percentage of agreement level of students in less concentration level affects in socialization and affects the learning during online classes. *Source* Own

This also leads to affect in the social life of the students as they will not be able to get connected with people daily and experience day to day activities, dramas, theatre, sports, etc. And more than 50% students agreed that missing school will adversely affect the development of their social skills and awareness of outside world.

- In Fig. 14.18, according to more than 32% of students, the educational activities such as yoga, group discussion, chess, etc. cannot be adequately performed through online classes as the physical presence of their fellow mates, instructor make things more interesting and enjoyable.

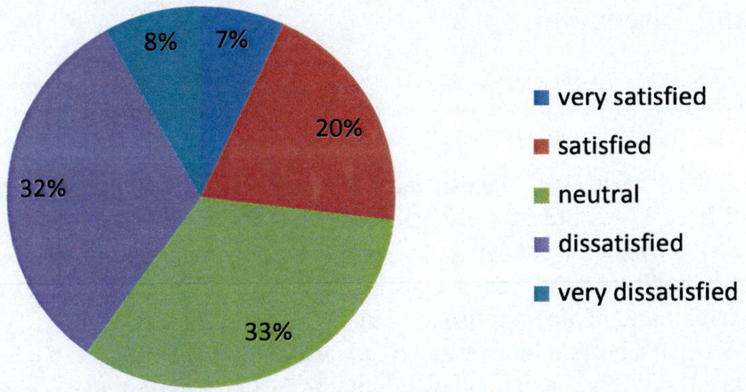


**Fig. 14.18** Pie chart showing the percentage of student's satisfaction level on educational activities being conducted in online classes. *Source* Own

- In Fig. 14.19, according to more than 50% of the students agreed that the graduates of this year may be severely affected by this pandemic as they have experienced major teaching interruption in their final semester, and they are most likely to graduate at the beginning of a major global recession. Employment rate will be affected negatively due to this pandemic. Also, many assessments have been cancelled or moved online which will create interruption in the long term which leads to increase in the inequality.

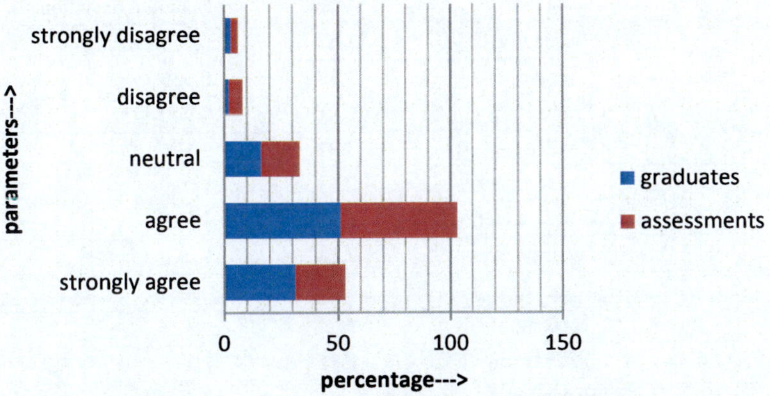


**Fig. 14.19** Bar graph showing the percentage of student's agreement level on the effect of online assessment and less job opportunities for this year's graduates. *Source* Own

## Results and Discussion

### *Service Sector*

In this sector, we have calculated mean and standard deviations of all the questions as shown in Table 14.1. According to the data collected, the ideal mean (i.e. highest) is of the question in which the challenges are being faced by the employees by working at home. ($M = 3.92$), and the ideal standard deviation (i.e. lowest) is in the question of increase in demand of web applications in the present scenario (0.665).

Now, Spearman's correlations revealed that all scales are correlated significantly at the 0.01 level. We calculated it by averaging the 2, 3 and 4 subscales according to our data collected, and these are the following:

- Impact of COVID-19 on employee personally and on the company shown in Table 14.2. In these two parameters, we have high degree of correlation as coefficient value is 0.537 which lies between ±0.50 and ±1. So, if an employee of the company is affected with the pandemic, then this will directly affect his/her work which leads to the impact on the company as well.
- As shown in Table 14.3, we have moderate degree of correlation as coefficient value of parameters job security and company turnover is 0.366 which lies between

**Table 14.1** Mean and standard deviation of all the questions of service sector

| S. No. | Questions | Mean | Standard deviation |
|---|---|---|---|
| 1. | Personally | 1.68 | 0.777 |
| 2. | Company | 1.88 | 0.935 |
| 3. | Productivity | 2.81 | 0.940 |
| 4. | Challenges | 3.92 | 2.196 |
| 5. | Environment | 2.35 | 1.019 |
| 6. | Communication | 1.91 | 1.083 |
| 7. | Security | 2.12 | 1.066 |
| 8. | Turnover | 1.74 | 0.787 |
| 9. | Opportunities | 1.64 | 0.828 |
| 10. | Information | 2.31 | 1.116 |
| 11. | Resources | 2.04 | 0.909 |
| 12. | Support | 2.20 | 0.865 |
| 13. | Financial | 2.29 | 0.988 |
| 14. | Layoffs | 1.91 | 0.753 |
| 15. | Graduates | 1.78 | 0.786 |
| 16. | Disruption | 1.82 | 0.809 |
| 17. | Applications | 1.69 | 0.665 |

*Source* Own

**Table 14.2** Spearman's correlation between impact of COVID-19 on employees personally and on their companies

| | Personally | Company |
|---|---|---|
| Personally | – | 0.537 |
| Company | 0.537 | – |

*Source* Own

**Table 14.3** Spearman's correlation between employees worried about their job security, turnover of the company, fewer opportunities of jobs in future and affect in career of recent graduates

| | Turnover | Job security | Opportunities | Graduates |
|---|---|---|---|---|
| Turnover | – | 0.366 | 0.336 | 0.216 |
| Job security | 0.336 | – | 0.272 | 0.174 |
| Opportunities | 0.295 | 0.272 | – | 0.344 |
| Graduates | 0.216 | 0.174 | 0.344 | – |

*Source* Own

±0.30 and ±0.49 as their is a correlation between the present scenarios in the company which resulted in the anxiety of the job which leads to decrease in the turnover of the company. The correlation between graduates and job opportunities is 0.344 as due to less job opportunities in the coming months; the careers of this year's university graduates will affect. The correlation between turnovers and opportunities is 0.295, between job security and opportunities is 0.272 and between graduates and turnover is 0.216 which is moderate degree of correlation. As due to decrease in the turnover of the company, less job opportunities will be their in future; due to less job opportunities, job security of the employees will be affected; and due to low turnover of the company, the career of the graduates will be affected. And the correlation between job security and graduates is of low degrease. 0.174 as coefficient value lies below ±0.29 as due to less jobs in future recent graduates will face the problems in jobs, and it lead to low job security.

- As shown in Table 14.4, we have high degree of correlation between resources and support, i.e. 0.505 and between support and financial assistance, i.e. 0.516 as correlation coefficient value lies between ±0.50 and ±1. The company has provided the resources and benefits to their employees for efficiently working.

**Table 14.4** Spearman's correlation between resources, support and financial support given by the companies

| | Resources | Support | Financial |
|---|---|---|---|
| Resources | – | 0.505 | 0.382 |
| Support | 0.505 | – | 0.516 |
| Financial | 0.382 | 0.516 | – |

*Source* Own

**Table 14.5** Spearman's correlation between layoffs, disruption in the companies and increase in demand of web applications

| | Layoffs | Disruption | Web applications |
|---|---|---|---|
| Lay offs | – | 0.213 | 0.222 |
| Disruption | 0.213 | – | 0.320 |
| Web applications | 0.222 | 0.320 | – |

*Source* Own

The degree of correlation is moderate between the financial support and resources given by companies, i.e. 0.382 as the correlation coefficient value lies between ±0.30 and ±0.49. Employees have their employer/ leadership team provided less information about financial conditions of the company.

- As shown in the Table 14.5, we have low degree of correlation as coefficient value between layoffs, disruption and web applications, layoffs as it lies below ±0.29. It is predicted that their will be layoffs in the companies of retail sector, car making industries, electronics as they are more dependent on the inputs from China. Italy, South Korea which may lead to disruption and increase in the demand of data centre services and video conferencing software's. The degree of correlation is moderate between the web applications and disruption of companies, i.e. 0.320 as the correlation coefficient lies between ±0.30 and ±0.49. So, due to the disruption in the other sectors, the web applications sectors will increase in demand.

## *Education Sector*

In this sector, as shown in Table 14.6, we have calculated mean and standard deviations of all the questions. According to the data collected, the ideal mean (i.e. highest) is of the question in which they have responded that virtual laboratories cannot substitute traditional classroom learning ($M = 3.57$), and the ideal standard deviation (i.e. lowest) is in the question that almost all the students are well equipped in using technology for online classes and increase in demand of web applications in the present scenario (0.785).

Spearman's correlations revealed that all scales are correlated significantly at the 0.01 level. We calculated it by averaging the 2, 3 and 4 subscales according to our data collected, and these are the following:

- As shown in Table 14.7, between virtual laboratories and substitution of traditional teaching methods, we have high degree of correlation, i.e. 0.526 as correlation coefficient value lies between ±0.50 and ±1. Online teaching students by using virtual laboratories lead to the substitution of the traditional teaching methods. We have perfect correlation as coefficient value +1 between physical presence of educator and substitution of virtual laboratories. As to teach subjects like chemistry, physics, statistics, there will be advantage to students if the instructor is

**Table 14.6** Mean and standard deviation of all the questions of education sector

| S. No. | Questions | Mean | Standard deviation |
|---|---|---|---|
| 1. | Institution equipped | 1.76 | 0.933 |
| 2. | Students equipped | 1.50 | 0.785 |
| 3. | Comfort level | 2.13 | 0.895 |
| 4. | Institutional steps | 2.37 | 1.060 |
| 5. | Substitution | 3.25 | 1.086 |
| 6. | Impact | 2.74 | 0.917 |
| 7. | Virtual laboratories | 3.57 | 1.047 |
| 8. | Physical presence | 1.98 | 1.073 |
| 9. | Physical resources | 2.17 | 0.933 |
| 10. | Educational activities | 3.09 | 1.055 |
| 11. | Concentration level | 2.22 | 0.970 |
| 12. | New perspective | 2.62 | 1.042 |
| 13. | Socialization | 2.16 | 0.950 |
| 14. | Graduates | 1.98 | 0.899 |
| 15. | Affect learning | 1.98 | 0.841 |
| 16. | Assessments | 1.16 | 0.940 |

*Source* Own

**Table 14.7** Spearman's correlation between substitution of traditional teaching methods, use of virtual laboratories, physical presence of educators and new perspective on online teaching

| | Substitution | Virtual laboratories | Physical presence | New perspective |
|---|---|---|---|---|
| Substitution | – | 0.526 | 0.252 | 0.318 |
| Virtual laboratories | 0.526 | – | 1.000 | 0.149 |
| Physical presence | 0.252 | 1.000 | – | 0.021 |
| New perspective | 0.318 | 0.149 | 0.021 | – |

*Source* Own

physically present there as compared to virtual classes. We have moderate degree of correlation, i.e. 0.318 as the correlation coefficient lies between ±0.30 and ±0.49 between new perspective on online teaching and substitution of traditional teaching methods. As these online classes gave a new perspective about the system of traditional teaching which leads to substitution of traditional teaching methods. We have low degree of correlation between physical presence, new perspective, i.e. 0.021 and substitution, physical presence, i.e. 0.252 as the correlation coefficient lies below ±0.29. This pandemic has given us a new perspective that online teaching can be useful for the students in terms of convenience, comfort, saves students time which can affect the physical presence of educator in the classroom which leads to substitution of traditional teaching method.

**Table 14.8** Spearman's correlation between steps taken by institutions and availability of physical resources

| | Steps | Resources |
|---|---|---|
| Steps | – | 0.150 |
| Resources | 0.150 | – |

- As shown in Table 14.8, we have low degree of correlation, i.e. 0.150 as coefficient value lies below ±029. As few institutions took the steps for conducting online classes which have very low correlation of the need of physical resources by students.
- As shown in Table 14.9, the correlation between concentration level and socialization is 0.236, and concentration level and affect in learning is 0.236. We have low degree of correlation as coefficient value lies below ±0.29 as students use technology like laptop, mobile phone, tablet in online classes for learning which leads to frequent lapses in the concentration but they have very less relation with socialization of the students, and it affects the learning of students at a low rate. There is moderate degree of correlation between socialization and affect in learning of students, i.e. 0.327 as the coefficient lies between ±0.30 and ±0.49. Online classes lead to affect children's social life and learning.
- As shown in Table 14.10, between the cancellation/online assessments and this year's university graduates career, we have high degree of correlation, i.e. 0.481 as coefficient value lies between ±0.50 and ±1. COVID-19 pandemic may severely affect the graduates as they will be graduating in the beginning of major global recession, and this interruption in online assessment can have long-term consequences.

**Table 14.9** Spearman's correlation between concentration level, socialization and learning effect of students

| | Concentration | Socialization | Affect learning |
|---|---|---|---|
| Concentration | – | 0.236 | 0.236 |
| Socialization | 0.236 | – | 0.327 |
| Affect learning | 0.236 | 0.327 | – |

*Source* Own

**Table 14.10** Spearman's correlation between online/cancellation of assessments and consequences on recent graduates

| | Assessments | Graduates |
|---|---|---|
| Assessments | – | 0.481 |
| Graduates | 0.481 | – |

*Source* Own

## Conclusion and Future Scope

The aim of this study was to investigate the impact of COVID-19 on the lives of students studying in schools and college as well as the private employees working in IT companies. The results showed that both the sectors were independent of each other. We found that many educational institutions are not well equipped with proper technology to provide online classes to the students mainly due to low income and lack of access to e-learning solutions. According to the data collected, most of the students from developed areas are well equipped in using the technology and have an easy access to Internet for attending their online classes, whereas students who live in remote areas are facing problems due to poor Internet connectivity and lack of skills to use the applications, sites in order to attend the online classes. The service respondents were extremely worried about the impact of this pandemic personally as they are getting more anxious by staying at their homes, and they are also worried about their companies as this pandemic has halted the services given by the companies of different sectors. A large number of companies will, however, struggle to survive. They have to pay rents, salaries, debts, etc., and also their revenues will steadily keep falling due to changes in lifestyles and cutting on the expenditures by the people. But it is also estimated that there will be an increase in demand of data centre services and video conferencing software's. Also, the university graduates of the current year will be affected too by this upcoming economic global recession. This pandemic has caused large job losses. The actual and potential job losing numbers are frightful in the initial staff itself. So, there will be decline in demand with widespread closures and job losses in India's large service sectors.

This study represents a vision towards the ways that how the educational institutions need to be well equipped to teach students online during such crisis. The COVID-19 pandemic has consequently affected the educational sector and has forced every element of the sector to evolve and adapt. The educational institutes have had to embrace and update their technological infrastructure to meet the need of the hour and facilitate remote learning for the students with greater flexibility. The role of the educator is also under the spotlight that is bound to formulate new methods of teaching, evaluation, interaction with students and peers. What has been established in stone is the fact that the role of technology in education is only going to increase from here on and will be an essential aspect of education for our future generations. In fact, this disruptive crisis can be viewed as a field test for the evolution of technology in the field of education and a perfect time to field test the same to work towards a more technologically bound classes which could bring more and more students towards classroom often from the comfort and security of their own homes. This evolution of learning will also be a breeding ground for some important skills that employers would attempt to identify and place significance over including, creativity, communication, collaboration, alongside empathy and emotional intelligence. Meanwhile, software companies providing collaboration tools. While COVID-19 will have immediate and short–term effect on employment, the effect of automation would be relatively long term. Therefore, this would affect private demand on a sustained basis compared to that of during the pandemic. This study gives us a glimpse that how much impact does this pandemic COVID-19 has and will be having on our lives.

## References


1. Khanna, R. C., Cicinelli, M. V., Gilbert, S. S., Honavar, S. G., & Murthy, G. S. (2020). COVID-19 pandemic: Lessons learned and future directions. *Indian Journal of Ophthalmology, 68*(5), 703–710.
2. Gumber, A., & Bulsari, S. (2020). COVID-19 Impact on Indian Economy and Health: The Emergence of Corona-Economics, 1–4.
3. McKibbin, W. J., & Fernando, R. (2020). The global macroeconomic impacts of COVID-19: Seven scenarios, 1–27.
4. Fetzer, T., Hensel, L., Hermle, J., & Roth, C. (2020). Perceptions of coronavirus mortality and contagiousness weaken economic sentiment. arXiv preprint arXiv:2003.03848, 1–23.
5. Viner, R. M., Russell, S. J., Croker, H., Packer, J., Ward, J., Stansfield, C. … Booy, R. (2020). School closure and management practices during coronavirus outbreaks including COVID-19: a rapid systematic review. *The Lancet Child & Adolescent Health,* 1–7. https://doi.org/10.1016/S2352-4642(20)30095-X
6. Sahu, P. (2020). Closure of universities due to coronavirus disease 2019 (COVID-19): Impact on education and mental health of students and academic staff. *Cureus, 12*(4), e7541 (April 04), 1–6. https://doi.org/10.7759/cureus.7541
7. Singhal, T. (2020). A review of coronavirus disease-2019 (COVID-19). *Indian Journal of Pediatrics, 87,* 281–286. https://doi.org/10.1007/s12098-020-03263-6.
8. Fernandes, N. (2020). Economic effects of coronavirus outbreak (COVID-19) on the world economy. Available: SSRN 3557504, 2–26.
9. Bartik, A. W., et al. (2020). The impact of COVID-19 on small business outcomes and expectations. *Proceedings of the National Academy of Sciences, 117*(30), 17656–17666, 2–11.
10. Nicola, M., Alsafi, Z., Sohrabi, C., Kerwan, A., Al-Jabir, A., Iosifidis, C. … Agha, R. (2020). The socio-economic implications of the coronavirus pandemic (COVID-19): A review. *International Journal of Surgery (London, England), 78,* 185.1–185.9. https://doi.org/10.1016/j.ijsu.2020.04.018.
11. Gupta, A., & Goplani, M. (2020). Impact of COVID-19 on Educational Institution in India. *Purakala Journal U (CARE Listed), 31*(21), 1–11.
12. Ray, D., Salvatore, M., Bhattacharyya, R., Wang, L., Du, J., Mohammed, S. … Kleinsasser, M. (2020). Predictions, role of interventions and effects of a historic national lockdown in India's response to the COVID-19 pandemic: data science call to arms. *Harvard Data Science Review,* (Suppl 1), 1–33. https://doi.org/10.1162/99608f92.60e08ed5
13. Petropoulos, F., & Makridakis, S. (2020). Forecasting the novel coronavirus COVID-19. *PLoS ONE, 15*(3), e0231236. https://doi.org/10.1371/journal.pone.0231236.
14. Government of India, https://www.mygov.in/covid-19.
15. WHO, https://www.who.int/emergencies/diseases/novel-coronavirus-2019/question-and-answers-hub.

# Chapter 15
# Global Stability Analysis Through Graph Theory for Smartphone Usage During COVID-19 Pandemic

**Nita H. Shah, Purvi M. Pandya, and Ekta N. Jayswal**

**Abstract** During the pandemic due to coronavirus disease-19 (COVID-19), technology is regarded as a boon as well as a curse to human life which has a great impact on surroundings, people and the society. One of the innovative, however, perilous (if misused) inventions of humans is the smartphone which is becoming more and more alarmingly common yet an urgent question to be addressed. A wide application of smartphone technology is observed during this pandemic. It has both positive as well as negative impact on the prominent areas which include education, business, health, social life and furthermore. Moreover, the impact of such an addiction is observed not only among youngsters but has influenced all age groups. This scenario is modelled in this research through non-linear ordinary differential equations, where individuals susceptible to smartphone use will be either positively or negatively infected/addicted, may suffer from health issues procuring medication. Threshold is calculated using the next generation matrix method. Stability analysis is done using graph theory, and for the validation of data, numerical simulation is carried out. This study gives results explaining positive and negative issues on health due to excessive use of smartphone.

**Keywords** COVID-19 · Smartphone use/addiction · Positively or negatively infected · Health issues · Threshold · Stability · Graph theory

**MSC** 37nxx

N. H. Shah · P. M. Pandya (✉) · E. N. Jayswal
Department of Mathematics, Gujarat University, Ahmedabad, Gujarat 380009, India
e-mail: pandya091@gmail.com

N. H. Shah
e-mail: nitahshah@gmail.com

E. N. Jayswal
e-mail: jayswal.ekta1993@gmail.com

 
N. H. Shah and M. Mittal (eds.), *Mathematical Analysis for Transmission of COVID-19*,
Mathematical Engineering, https://doi.org/10.1007/978-981-33-6264-2_15

## Introduction

Technology plays an immense role in every walk of human life, be it for transportation, education, health care or communication, etc. With every advancement in technology, it is becoming an alarmingly urgent question that necessitates the investigation of their negative impacts too. Innovations such as smartphones have become ubiquitous in everyone's life, which encompasses Internet facilities and a huge collection of apps. The use of such facilities differs accordingly in various age groups, and also the symptoms and the risks attributed to addiction of smartphone use may differ during various life stages. During the COVID-19 pandemic, smartphone usage has reached to its extremes, be it for digital tracking for COVID-19 cases near them, grocery shopping, entertainment, business, teaching/educating students, webinars, buying electronic items, etc. This has not only proved to be a boon for people during this pandemic but it also has some side effects to their health. This has proved to be a greater risk physically, mentally as well as socially. The risks are in terms of emotional/mental or psychological such as mood disorder, sleep disorders, depression, isolation from society, anxiety, etc.; physical like obesity, back and neck problems, etc.; also, social and workplace problems, etc. Proper medication or consultation is needed to prevent oneself from such health problems. However, they may again be addicted, and the problems may relapse.

Many more analyses and surveys are done to study the positive as well as negative impact of smartphone use. Positive use of smartphones among students can be seen, where it encourages them in the learning process, stimulates the mind, feeling, also gives positive values and bridges media literacy among them [24]. Also, smartphone applications such as WhatsApp and Facebook are proved to be a great help for online business. For this, survey of 100 participants was targeted which includes university students, housewives and random public at various places which proved that the applications give a positive impact, i.e. success to online business [7]. During COVID-19 pandemic, physical activities, smartphone use, sleep patterns were objectively assessed for young adults in Spain [19]. Also, several applications of smartphone technology are studied [8] along with social media addiction [9] and problematic online gaming [12] among individuals in such pandemic phase. Moreover, a consensus guidance was provided for preventing problematic Internet use during COVID-19 pandemic [13]. From early childhood, smartphones have affected a lot, as parents are busy working or they themselves are addicted to smartphones and don't have enough time to take care of their children. In one of the researches, a model of smartphone addiction of early childhood consisting of antecedents as parents and child variables and consequence referring to mental and physical development was considered and analyzed [17]. Moreover, the impact of this addiction on 200 college students between age 18 and 22 years [16], on 210 university students along with its implications for learning [14] was examined. Also, a data analysis with a survey identifying the impact of smartphones on academic of students and on adults who have finished their education through different algorithms was conducted [1]. Likewise, a study to understand both positive as well as negative impacts of smartphones on education, business, health sectors, human psychology and social life was conducted, and the results with recommended solutions to reduce negative impact on society were concluded [20]. Additionally, the health issues prevailing due to smartphone addiction were found and studied. Issues such as stress, related to academic performance and satisfaction with life [18]; anxiety and family relations [6]; social anxiety and loneliness [3] were observed and surveyed. Such addiction has also affected sleep quality and some behavioural problems in adolescents. So, a survey on a sample of 587 students was conducted to assess such problems attributed to depression, anxiety, stress prevailed due to such addiction [21]. The overall impact of mobile devices and smartphones on health and life was assessed and studied [15]. To save people from risks of such health issues, self-regulating mobile apps was proved to be very useful according to the survey taken from 284 college students [22]. Also, joining different programs to reduce smartphone overdependence and improve peer relationships and self-efficacy showed positive results [10]; also there are other methods to rehabilitate from smartphone addiction [11].

The mathematical model to study the impact of smartphone addiction is designed and formulated using non-linear differential equations along with computation of threshold in Sect. Mathematical Model. Global stability through graph theory is discussed in Sect. Stability followed by numerical simulation to validate the data in Sect. Numerical Simulation. Finally, concluding the findings in Sect. Conclusion.

## Mathematical Model

People enjoy great comfort with the advancement in technology such as smartphones. While during the COVID-19 pandemic, it is considered to be a boon as well as curse since it gives rise to different health related issues. So, to scrutinize the rate of health impacts due to addiction of smartphone use, be it positively or negatively is modelled in this section. As shown in Fig. 15.1, it comprises six compartments, namely susceptible individuals ($S_I$), i.e. individuals susceptible to smartphone use, exposed individuals ($E$). They are either positively infected or addicted ($I_P$) or negatively infected or addicted ($I_N$) who may suffer from different health issues ($H_I$) as mentioned a forehand. They need proper medication ($M$) or consultation if diagnosed by such health problems. But there are chances that they may again be addicted to smartphones after getting cured. In some cases, it is observed that positively infected individuals may turn into negatively infected and vice versa. In Table 15.1, the notations and parametric values used in this research to formulate the flow of individuals in the model are described.

Using all the parameters and the flow of the model shown in Fig. 15.1, the system of non-linear differential equations [4] is formulated as follows:

$$\frac{\mathrm{d}S_I}{\mathrm{d}t} = B - \alpha_1 S_I - \mu S_I$$

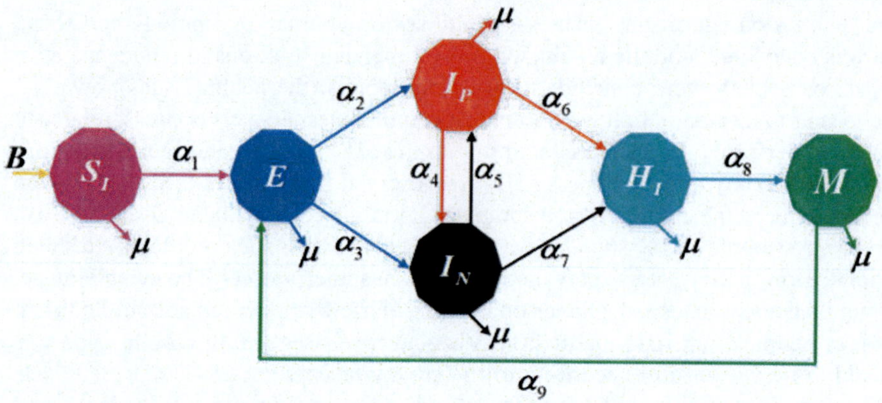


**Fig. 15.1** Mathematical model. *Source* Own

**Table 15.1** Notations and parametric values

| Notation | Description | Parametric values |
|---|---|---|
| $B$ | Natural birth rate | 0.5 |
| $\alpha_1$ | Rate at which individuals gets exposed to smartphone use | 0.4 |
| $\alpha_2$ | Rate of exposed individuals gets positively infected | 0.2 |
| $\alpha_3$ | Rate at which exposed individuals gets negatively infected | 0.7 |
| $\alpha_4$ | Rate at which positively infected individuals moves towards negatively infected | 0.2 |
| $\alpha_5$ | Rate of negatively infected individuals turns into positively infected | 0.1 |
| $\alpha_6$ | Rate at which positively infected individuals suffers from health issues | 0.1 |
| $\alpha_7$ | Rate of negatively infected individuals experience health issues | 0.3 |
| $\alpha_8$ | Rate at which individuals procures medication | 0.6 |
| $\alpha_9$ | Rate of individuals yet being exposed to smartphone use | 0.47 |
| $\mu$ | Natural escape rate | 0.15 |

*Source* Own

$$\frac{\mathrm{d}E}{\mathrm{d}t} = \alpha_1 S_I - \alpha_2 E I_P - \alpha_3 E I_N + \alpha_9 M - \mu E$$
$$\frac{\mathrm{d}I_P}{\mathrm{d}t} = \alpha_2 E I_P - \alpha_4 I_P + \alpha_5 I_N - \alpha_6 I_P - \mu I_P$$
$$\frac{\mathrm{d}I_N}{\mathrm{d}t} = \alpha_3 E I_N + \alpha_4 I_P - \alpha_5 I_N - \alpha_7 I_N - \mu I_N$$
$$\frac{\mathrm{d}H_I}{\mathrm{d}t} = \alpha_6 I_P + \alpha_7 I_N - \alpha_8 H_I - \mu H_I$$

$$\frac{\mathrm{d}M}{\mathrm{d}t} = \alpha_8 H_I - \alpha_9 M - \mu M \tag{15.1}$$

where $S_I + E + I_P + I_N + H_I + M \leq N$. Also, $S_I > 0$; $E, I_P, I_N, H_I, M \geq 0$

Therefore, consider the feasible region of the model as

$$\Lambda = \left\{ (S_I, E, I_P, I_N, H_I, M) \in R^6 : S_I + E + I_P + I_N + H_I + M \leq \frac{B}{\mu} \right\}.$$

Now, the issue-free equilibrium point of the model is $Y_0 = \left( \frac{B}{\alpha_1+\mu}, \frac{B\alpha_1}{(\alpha_1+\mu)\mu}, 0, 0, 0, 0 \right)$, and the endemic point is where $S_I^* = \frac{B}{\alpha_1+\mu}$, $E^* = r$,

$$I_P^* = \frac{\alpha_4(\alpha_8+\mu)\left(r\alpha_1\mu + r\mu^2 - B\alpha_1\right)(\alpha_9+\mu)}{\mu(\alpha_1+\mu)\begin{pmatrix} -\mu^3 + (\alpha_2 r - \alpha_4 - \alpha_5 - \alpha_6 - \alpha_7 - \alpha_8 - \alpha_9)\mu^2 \\ + \begin{pmatrix} (\alpha_2 r - \alpha_4 - \alpha_5 - \alpha_6 - \alpha_7 - \alpha_9)\alpha_8 + (\alpha_2 r - \alpha_5 - \alpha_6)\alpha_7 - \alpha_4\alpha_6 \\ + (\alpha_2 r - \alpha_4 - \alpha_5 - \alpha_6 - \alpha_7)\alpha_9 \end{pmatrix}\mu \\ + ((\alpha_2 r - \alpha_4 - \alpha_5 - \alpha_6)\alpha_9 + (\alpha_2 r - \alpha_5 - \alpha_6)\alpha_7 - \alpha_4\alpha_6)\alpha_8 \\ + ((\alpha_2 r - \alpha_5 - \alpha_6)\alpha_7 - \alpha_4\alpha_6)\alpha_9 \end{pmatrix}},$$

$$I_N^* = \frac{(-\alpha_3\mu + (\alpha_2 r - \alpha_5 - \alpha_6)\alpha_3 + \alpha_2\alpha_5)(\alpha_8+\mu)\left(B\alpha_1 - r\alpha_1\mu - r\mu^2\right)(\alpha_9+\mu)}{\mu(\alpha_1+\mu)\begin{pmatrix} (\alpha_2-\alpha_3)\mu^3 + (\alpha_2-\alpha_3)(\alpha_4+\alpha_5+\alpha_6+\alpha_7+\alpha_8+\alpha_9)\mu^2 \\ + \mu\begin{pmatrix} \begin{pmatrix} r(\alpha_7-\alpha_6)\alpha_2 - (\alpha_4+\alpha_5+\alpha_6+\alpha_7+\alpha_9)\alpha_8 \\ - (\alpha_4+\alpha_5+\alpha_6+\alpha_7)\alpha_9 - (\alpha_4+\alpha_7)\alpha_6 - \alpha_5\alpha_7 \end{pmatrix}\alpha_3 \\ + \begin{pmatrix} (\alpha_4+\alpha_5+\alpha_6+\alpha_7+\alpha_9)\alpha_8 + (\alpha_4+\alpha_5+\alpha_6+\alpha_7)\alpha_9 \\ + \alpha_5\alpha_7 + \alpha_6(\alpha_4+\alpha_7) \end{pmatrix}\alpha_2 \end{pmatrix} \\ + \begin{pmatrix} r(\alpha_8+\alpha_9)(\alpha_7-\alpha_6)\alpha_2 - ((\alpha_4+\alpha_7)\alpha_6 + \alpha_7\alpha_5)\alpha_9 \\ - ((\alpha_4+\alpha_5+\alpha_6)\alpha_9 + (\alpha_4+\alpha_7)\alpha_6 + \alpha_5\alpha_7)\alpha_8 \end{pmatrix}\alpha_3 \\ + \alpha_2\begin{pmatrix} ((\alpha_4+\alpha_5+\alpha_7)\alpha_9 + (\alpha_4+\alpha_7)\alpha_6 + \alpha_5\alpha_7)\alpha_8 \\ + \alpha_9(\alpha_5\alpha_7 + (\alpha_4+\alpha_7)\alpha_6) \end{pmatrix} \end{pmatrix}},$$

$$H_I^* = \frac{\left(-r\alpha_1\mu - r\mu^2 + B\alpha_1\right)\begin{pmatrix} (\alpha_2\alpha_6 - \alpha_3\alpha_7)\mu - (\alpha_5\alpha_7 + (\alpha_4+\alpha_7)\alpha_6)\alpha_3 \\ + (r(\alpha_7-\alpha_6)\alpha_3 + (\alpha_4+\alpha_7)\alpha_6 + \alpha_5\alpha_7)\alpha_2 \end{pmatrix}(\alpha_9+\mu)}{\mu(\alpha_1+\mu)\begin{pmatrix} (\alpha_2-\alpha_3)\mu^3 + (\alpha_2-\alpha_3)(\alpha_4+\alpha_5+\alpha_6+\alpha_7+\alpha_8+\alpha_9)\mu^2 \\ + \mu\begin{pmatrix} \begin{pmatrix} r(\alpha_7-\alpha_6)\alpha_3 + (\alpha_4+\alpha_7+\alpha_8+\alpha_9)\alpha_6 + (\alpha_5+\alpha_8+\alpha_9)\alpha_7 \\ + (\alpha_4+\alpha_5+\alpha_8)\alpha_9 + \alpha_8(\alpha_4+\alpha_5) \end{pmatrix}\alpha_2 \\ - \begin{pmatrix} (\alpha_4+\alpha_7+\alpha_8+\alpha_9)\alpha_6 + (\alpha_5+\alpha_8+\alpha_9)\alpha_7 \\ + (\alpha_4+\alpha_5+\alpha_8)\alpha_9 + \alpha_8(\alpha_4+\alpha_5) \end{pmatrix}\alpha_3 \end{pmatrix} \\ + \begin{pmatrix} r(\alpha_8+\alpha_9)(\alpha_7-\alpha_6)\alpha_3 + (\alpha_8+\alpha_9)(\alpha_4+\alpha_7)\alpha_6 \\ + ((\alpha_8+\alpha_5)\alpha_9 + \alpha_8\alpha_5)\alpha_7 + \alpha_8\alpha_9(\alpha_4+\alpha_5) \end{pmatrix}\alpha_2 \\ - \alpha_3\begin{pmatrix} ((\alpha_8+\alpha_9)\alpha_7 + (\alpha_8+\alpha_4)\alpha_9 + \alpha_8\alpha_4)\alpha_6 \\ + \alpha_5(\alpha_8+\alpha_9)\alpha_7 + \alpha_8\alpha_9(\alpha_4+\alpha_5) \end{pmatrix} \end{pmatrix}}$$

and

$$M^* = \frac{\alpha_8\begin{pmatrix}\alpha_7\mu^4 + \alpha_7(-\alpha_2 r + \alpha_1 + \alpha_4 + \alpha_5 + \alpha_6 + \alpha_7)\mu^3 \\ + \left((-\alpha_2 r + \alpha_1 + \alpha_5 + \alpha_6)\alpha_7^2 + \begin{pmatrix}(\alpha_1 + \alpha_4)\alpha_6 - \alpha_2 r\alpha_4 \\ + (-\alpha_2 r + \alpha_4 + \alpha_5)\alpha_1\end{pmatrix}\alpha_7 + r\alpha_3\alpha_4\alpha_6\right)\mu^2 \\ - \left((\alpha_2 r - \alpha_5 - \alpha_6)\alpha_7^2 + (\alpha_2 r\alpha_4 + B\alpha_3 - \alpha_4\alpha_6)\alpha_7 - r\alpha_3\alpha_4\alpha_6\right)\alpha_1\mu \\ + B\alpha_3((\alpha_2 r - \alpha_5 - \alpha_6)\alpha_7 - \alpha_4\alpha_6)\alpha_1\end{pmatrix}}{\mu\alpha_3(\alpha_1 + \mu)\begin{pmatrix}-\mu^3 + (\alpha_2 r - \alpha_4 - \alpha_5 - \alpha_6 - \alpha_7 - \alpha_8 - \alpha_9)\mu^2 \\ + \begin{pmatrix}(\alpha_2 r - \alpha_5 - \alpha_6 - \alpha_8 - \alpha_9)\alpha_7 - (\alpha_8 + \alpha_9 + \alpha_4)\alpha_6 \\ - (\alpha_8 + \alpha_9)\alpha_4 + (\alpha_2 r - \alpha_5 - \alpha_9)\alpha_8 + \alpha_9(\alpha_2 r - \alpha_5)\end{pmatrix}\mu \\ + (\alpha_2 r - \alpha_5 - \alpha_6)(\alpha_8 + \alpha_9)\alpha_7 + ((-\alpha_8 - \alpha_9)\alpha_4 - \alpha_8\alpha_9)\alpha_6 \\ + \alpha_8\alpha_9(\alpha_2 r - \alpha_4 - \alpha_5)\end{pmatrix}}$$

where

$$r = \text{Root of}\begin{pmatrix}\alpha_2\alpha_3 Z^2 + ((-\alpha_2 - \alpha_3)\mu + (-\alpha_4 - \alpha_7)\alpha_2 - \alpha_3(\alpha_5 + \alpha_6))Z \\ + \mu^2 + (\alpha_4 + \alpha_5 + \alpha_6 + \alpha_7)\mu + (\alpha_4 + \alpha_7)\alpha_6 + \alpha_5\alpha_7\end{pmatrix}$$

Now, to analyze the behaviour of individuals susceptible to smartphone use, the basic reproduction number also known as threshold ($R_0$) is evaluated using next generation matrix method [2].

Let

$$X' = (S_I, E, I_P, I_N, H_I, M)' \quad \text{and} \quad X' = \frac{dX}{dt} = F(X) - V(X)$$

where $F(X)$ denotes the rate of arrival of new individuals in the compartment, and

$V(X)$ denotes the rate of health issues due to smartphone addiction which is given by

$$F(X) = \begin{bmatrix}\alpha_2 E I_P \\ \alpha_3 E I_N \\ 0 \\ 0 \\ 0 \\ 0\end{bmatrix} \quad \text{and} \quad V(X) = \begin{bmatrix}\alpha_4 I_P - \alpha_5 I_N + \alpha_6 I_P + \mu I_P \\ -\alpha_4 I_P + \alpha_5 I_N + \alpha_7 I_N + \mu I_N \\ -\alpha_6 I_P - \alpha_7 I_N + \alpha_8 H_I + \mu H_I \\ -\alpha_8 H_I + \alpha_9 M + \mu M \\ -B + \alpha_1 S_I + \mu S_I \\ -\alpha_1 S_I + \alpha_2 E I_P + \alpha_3 E I_N - \alpha_9 M + \mu E\end{bmatrix}$$

Now, $DF(X_0) = \begin{bmatrix} f & 0 \\ 0 & 0\end{bmatrix}$ and $DV(X_0) = \begin{bmatrix} v & 0 \\ J_1 & J_2\end{bmatrix}$, where $f$ and $v$ are $6 \times 6$ matrices defined as $f = \left[\frac{\partial F_i(X_0)}{\partial X_j}\right]$ and $v = \left[\frac{\partial V_i(X_0)}{\partial X_j}\right]$.

Finding $f$ and $v$, we get

$$f = \begin{bmatrix} \alpha_2 E & 0 & 0\,0\,0 & \alpha_2 I_P \\ 0 & \alpha_3 E & 0\,0\,0 & \alpha_3 I_N \\ 0 & 0 & 0\,0\,0 & 0 \\ 0 & 0 & 0\,0\,0 & 0 \\ 0 & 0 & 0\,0\,0 & 0 \\ 0 & 0 & 0\,0\,0 & 0 \end{bmatrix}$$

and

$$v = \begin{bmatrix} \alpha_4+\alpha_6+\mu & -\alpha_5 & 0 & 0 & 0 & 0 \\ -\alpha_4 & \alpha_5+\alpha_7+\mu & 0 & 0 & 0 & 0 \\ -\alpha_6 & -\alpha_7 & \alpha_8+\mu & 0 & 0 & 0 \\ 0 & 0 & -\alpha_8 & \alpha_9+\mu & 0 & 0 \\ 0 & 0 & 0 & 0 & \alpha_1+\mu & 0 \\ \alpha_2 E & \alpha_3 E & 0 & -\alpha_9 & -\alpha_1 & \alpha_2 I_P+\alpha_3 I_N+\mu \end{bmatrix}$$

Here, $v$ is non-singular matrix.

For this model, the basic reproduction (threshold) $R_0$ at $Y_0$ is obtained as the spectral radius of matrix $fv^{-1}$.

$$R_0 = \frac{\alpha_1 B \left[ \begin{array}{l} \alpha_2(\alpha_5+\alpha_7+\mu)+\alpha_3(\alpha_4+\alpha_6+\mu) \\ + \sqrt{\alpha_3^2(\alpha_4+\alpha_6+\mu)^2 - 2\alpha_2\alpha_3 \left( \begin{array}{r} \alpha_4(\alpha_7-\alpha_5+\mu) \\ +(\alpha_6+\mu)(\alpha_5+\alpha_7+\mu) \end{array} \right) + \alpha_2^2(\alpha_5+\alpha_7+\mu)^2} \end{array} \right]}{2\mu(\alpha_1+\mu)\big(\alpha_4\alpha_7+\alpha_6(\alpha_5+\alpha_7)+(\alpha_4+\alpha_5+\alpha_6+\alpha_7)\mu+\mu^2\big)} \tag{15.2}$$

## Stability

Using graph theoretical results, global stability is carried out with thorough information [5, 23].

Out-degree $(d^+(i))$ refers to the number of edges incident from initial vertex $i$, and the number of edges away from the vertex $i$is denoted by in-degree $(d^-(i))$. A weighted graph directed graph $(G, A)$ where $A = (a_{ij})_{n\times n}$, $a_{ij} > 0$ are weight of each edge with numerical positive weight if it exists otherwise zero. Product of weight on all edges gives the weight $w(K)$ of sub-graph. The root vertex has degree one is called root tree. The sub-graph contains all vertices with minimum number of edges are called a spanning tree. The Laplacian matrix $L = (l_{ij})$ of $G$ is defined as:

$$l_{ij} = \begin{cases} -a_{ij}, & i \neq j, \\ \sum_{i\neq k} a_{ik}, & i = k. \end{cases} \tag{15.3}$$

**Theorem 3.1** (Kirchhoff's matrix tree theorem) *For $n \geq 2$, assume that $c_i$ is the cofactor of $l_{ii}$ in L. Then, $c_i = \sum_{T \in T_i} \mathcal{W}(T), i = 1, 2, \ldots, n$, where $T_i$ is the set of all spanning tree T of graph $(G, A)$ that are rooted at vertex i. Moreover, if $(G, A)$ is strongly connected, then $c_i > 0$ for $1 \leq i \leq n$.*

**Lemma 3.1** *Let $c_i$ be as given in Kirchhoff's matrix tree theorem. If $a_{ij} > 0$ and $d^+(j) = 1$ for some $1 \leq i, j \leq n$, then $c_i a_{ij} = \sum_{k=1}^{n} c_j a_{jk}$.*

**Lemma 3.2** *Let $c_i$ be as given in the in Kirchhoff's matrix tree theorem. If $a_{ij} > 0$ and $d^-(i) = 1$ for some $1 \leq i, j \leq n$, then $c_i a_{ij} = \sum_{k=1}^{n} c_k a_{ki}$.*

**Theorem 3.2** *Suppose that the following assumptions are satisfied:*

1. *There exist functions $V_i : U \to \mathbb{R}$, $G_{ij} : U \to \mathbb{R}$ and constants $a_{ij} > 0$ such that for every $1 \leq i \leq n$, $V_i' \leq \sum_{j=1}^{n} G_{ij}(x)$ for $x \in U$.*
2. *For $M = [a_{ij}]$, each directed cycle C of $(G, M)$ has $\sum_{(s,r)\in\varepsilon(C)} G_{rs}(x) \leq 0$ for $x \in U$, where $\varepsilon(C)$ denotes the arc set of the directed cycle C.*

*Then, the function $V(x) = \sum_{i=1}^{n} c_i V_i(x)$, with constants $c_i \geq 0$ as given in* Theorem 3.1 *which satisfies $V' \leq 0$. Hence, V is a Lyapunov function for the system.*

**Theorem 3.3** *Endemic equilibrium point $Y^*\left(S_I^*, E^*, I_P^*, I_N^*, H_I^*, M^*\right)$ is globally asymptotically stable.*

***Proof*** To analyze the stability at endemic equilibrium point, we create Lyapunov function $V(t)$ using graph theoretical results. Therefore, we take $V_i$

$$V_1 = S_I - S_I^* - S_I^* \ln \frac{S_I}{S_I^*}, \quad V_2 = E - E^* - E^* \ln \frac{E}{E^*},$$
$$V_3 = I_P - I_P^* - I_P^* \ln \frac{I_P}{I_P^*}, \quad V_4 = I_N - I_N^* - I_N^* \ln \frac{I_N}{I_N^*},$$
$$V_5 = H_I - H_I^* - H_I^* \ln \frac{H_I}{H_I^*} \quad \text{and} \quad V_6 = M - M^* - M^* \ln \frac{M}{M^*}$$

Now, each $V_i$, for $1 \leq i \leq 6$ differentiating with respect to $t$ we have,

$$\begin{aligned} V_1' &= \left(1 - \frac{S_I^*}{S_I}\right) S_I' \\ &= \left(1 - \frac{S_I^*}{S_I}\right)(B - \alpha_1 S_I - \mu S_I) \\ &\leq \left(1 - \frac{S_I^*}{S_I}\right)(\alpha_1(S_I^* - S_I) + \mu(S_I^* - S_I)) \\ &= (\alpha_1 + \mu) S_I^* \left(1 - \frac{S_I^*}{S_I}\right)\left(1 - \frac{S_I}{S_I^*}\right) \\ &= a_{11} G_{11} \end{aligned}$$

Similarly,

$$
\begin{aligned}
V_2' &= \left(1 - \frac{E^*}{E}\right)(\alpha_1 S_I - \alpha_2 E I_P - \alpha_3 E I_N + \alpha_9 M - \mu E) \\
&\le \alpha_1 S_I^* \left(1 - \frac{E^*}{E}\right)\left(1 - \frac{S_I}{S_I^*}\right) + \alpha_9 M^* \left(1 - \frac{E^*}{E}\right)\left(1 - \frac{M}{M^*}\right) \\
&= a_{21} G_{21} + a_{26} G_{26}
\end{aligned}
$$

$$
\begin{aligned}
V_3' &= \left(1 - \frac{I_P^*}{I_P}\right)(\alpha_2 E I_P - \alpha_4 I_P + \alpha_5 I_N - \alpha_6 I_P - \mu I_P) \\
&\le \alpha_2 E^* I_P^* \left(1 - \frac{I_P^*}{I_P}\right)\left(1 - \frac{E I_P}{E^* I_P^*}\right) + \alpha_5 I_N^* \left(1 - \frac{I_P^*}{I_P}\right)\left(1 - \frac{I_N}{I_N^*}\right) \\
&= a_{32} G_{32} + a_{34} G_{34}
\end{aligned}
$$

$$
\begin{aligned}
V_4' &= \left(1 - \frac{I_N^*}{I_N}\right)(\alpha_3 E I_N + \alpha_4 I_P - \alpha_5 I_N - \alpha_7 I_N - \mu I_N) \\
&\le \alpha_3 E^* I_N^* \left(1 - \frac{I_N^*}{I_N}\right)\left(1 - \frac{E I_N}{E^* I_N^*}\right) + \alpha_4 I_P^* \left(1 - \frac{I_N^*}{I_N}\right)\left(1 - \frac{I_P}{I_P^*}\right) \\
&= a_{42} G_{42} + a_{43} G_{43}
\end{aligned}
$$

$$
\begin{aligned}
V_5' &= \left(1 - \frac{H_I^*}{H_I}\right)(\alpha_6 I_P + \alpha_7 I_N - \alpha_8 H_I - \mu H_I) \\
&\le \alpha_6 I_P^* \left(1 - \frac{H_I^*}{H_I}\right)\left(1 - \frac{I_P}{I_P^*}\right) + \alpha_7 I_N^* \left(1 - \frac{H_I^*}{H_I}\right)\left(1 - \frac{I_N}{I_N^*}\right) \\
&= a_{53} G_{53} + a_{54} G_{54}
\end{aligned}
$$

and

$$
\begin{aligned}
V_6' &= \left(1 - \frac{M^*}{M}\right)(\alpha_8 H_I - \alpha_9 M - \mu M) \\
&\le \alpha_8 H_I^* \left(1 - \frac{M^*}{M}\right)\left(1 - \frac{H_I}{H_I^*}\right) \\
&= a_{65} G_{65}
\end{aligned}
$$

where $a_{21} = \alpha_1 S_I^*$, $a_{26} = \alpha_9 M^*$, $a_{32} = \alpha_2 E^* I_P^*$, $a_{34} = \alpha_4 I_N^*$, $a_{42} = \alpha_3 E^* I_N^*$, $a_{43} = \alpha_5 I_P^*$, $a_{53} = \alpha_6 I_P^*$, $a_{54} = \alpha_7 I_N^*$ and $a_{65} = \alpha_8 H_I^*$.

Thus, using above calculations, we construct a weighted graph as shown in Fig. 15.2.

The weighted graph of this model $(G, A)$ has six vertices and cycles such as

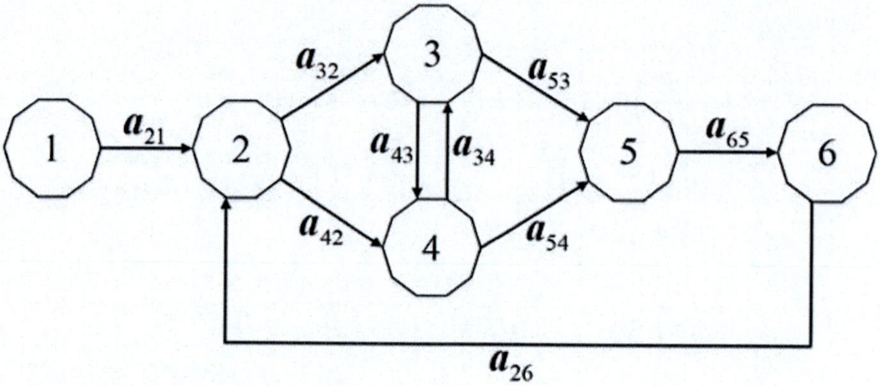


**Fig. 15.2** Weighted graph. *Source* Own

$G_{34} + G_{43} = 0, G_{32} + G_{53} + G_{65} + G_{26} = 0, G_{42} + G_{54} + G_{65} + G_{26} = 0,$
$G_{32} + G_{43} + G_{54} + G_{65} + G_{26} = 0, G_{42} + G_{34} + G_{53} + G_{65} + G_{26} = 0$ etc.

From Theorem 3.2 constants $c_i^{;}s$ for $1 \le i \le 6$ such that $V(x) = \sum_{i=1}^{n} c_i V_i(x)$ be a Lyapunov function.

Now, using Lemma 3.2, the relation between the constants $c_i^{;}s$ is

$$d^{+}(1) = 1 \Rightarrow c_1 = 0 d^{+}(5) = 1 \Rightarrow c_6 = c_5 \frac{(a_{53} + a_{54})}{a_{65}} d^{+}(6) = 1 \Rightarrow c_2 = c_5 \frac{(a_{53} + a_{54})}{a_{26}}$$

Let us take $c_3 = k_1, c_4 = k_2, c_5 = k_3$

$$\Rightarrow c_2 = \frac{k_3(\alpha_6 I_P^* + \alpha_7 I_N^*)}{\alpha_9 M^*}$$

Since

$$V(t) = \sum_{i=1}^{6} c_i V_i(t) = c_1 v_1 + c_2 v_2 + c_3 v_3 + c_4 v_4 + c_5 v_5 + c_6 v_6$$
$$= \frac{k_3(\alpha_6 I_P^* + \alpha_7 I_N^*)}{\alpha_9 M^*} v_2 + k_1 v_3 + k_2 v_4 + k_3 v_5 + \frac{k_3(\alpha_6 I_P^* + \alpha_7 I_N^*)}{\alpha_8 H_I^*} v_3$$

where $k_i$'s for $1 \le i \le 3$ are arbitrary constants.

This proves that $\{Y^*\}$ is the only invariant set in int $(\Lambda)$ as $V' = 0$ which proves that $Y^*$ is globally asymptotically stable.

In next section, numerical simulation is carried out to corroborate the findings using data given in Table 15.1.

## Numerical Simulation

Nowadays, parent's handover their smartphones to their kids and use it as a tool to keep them engaged and avoid mischief. Even while fooding, they use mobile phones to show them videos so that they eat without creating trouble. Here, as shown in Fig. 15.3, we can see that at an early stage, i.e. at approximately 3 years of age, children easily get exposed and are negatively infected which leads to different health issues. Moreover, for some cases, it is noted that these negatively infected individuals, after a period of time may use smartphones positively and vice versa. Also, after procuring medication, approximate 55% individuals again get negatively infected.

In Fig. 15.4, the direction shows the behavioural changes of individuals in their respective compartments. As shown in Fig. 15.4a, the intensity of exposed individuals to smartphone use decreases if they procure medication in terms of limiting excessive phone usage and involving them more in other fruitful activities. Moreover, as observed in Fig. 15.4b, individuals suffering from health issues due to addiction stabilizes, if they procure proper medication.

Figure 15.5a suggests that for every 26% individuals susceptible to smartphone use, 19% gets exposed. Overall, 11% are positively infected, 21% are negatively infected, and 11% suffer from different health issues. On the whole, 12% procures medication. Furthermore, as observed in Fig. 15.5b, for every 25% positively infected and 49% negatively infected individuals, by and large 26% suffer from health issues.

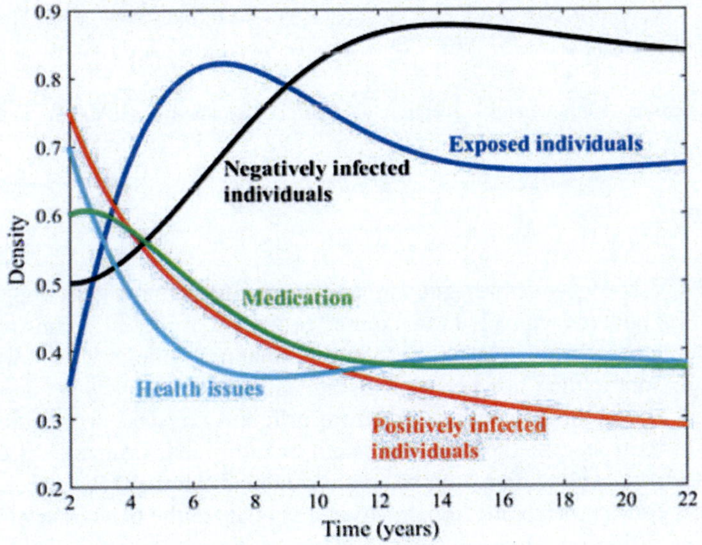


**Fig. 15.3** Density of individuals in different rooms. *Source* Own

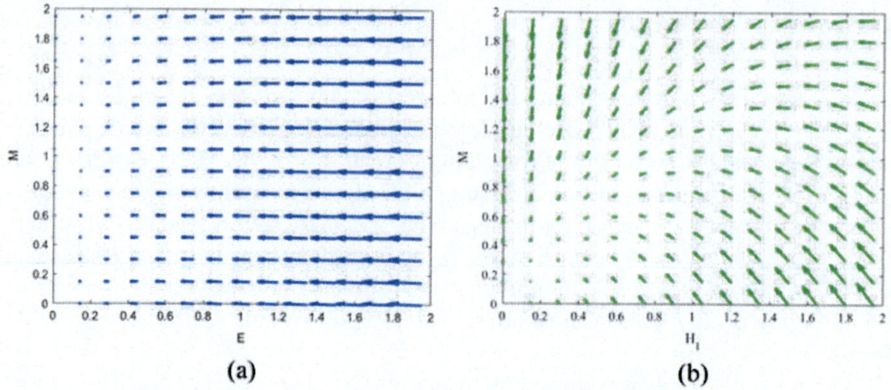


**Fig. 15.4** Behaviour of individuals in their respective compartments. *Source* Own

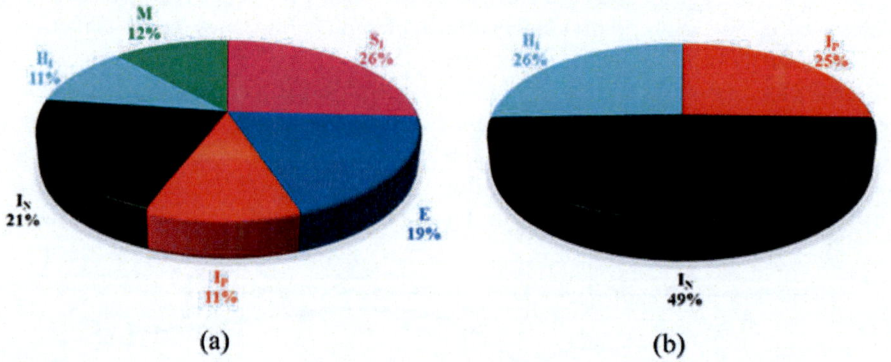


**Fig. 15.5** Percentage of individuals susceptible to smartphone addiction and its impacts. *Source* Own

## Conclusion

During the COVID-19 pandemic, 'quarantine' or 'work from home' was imposed for the safety of individuals which led to increased use of smartphones. In this research paper, a systematic pathway is used to design mathematical model, where individuals susceptible to smartphone use get exposed and may get positively or negatively infected. Due to which, they may suffer from different physical or mental health issues who further need medication. A system of non-linear ordinary differential equations is formulated using which basic reproduction number ($R_0$) is calculated using next generation matrix method, to study and analyze the behaviour of spread of smartphone addiction. Here, $R_0 = 3.49$ which connotes that an individual can expose or infect approximately three other individuals to smartphone use. Also, using graph theory, it is proved that the system is globally asymptotically stable, and for validation of data, numerical simulation is carried out. There are several ways that can help to control and minimize the negative impact of smartphone use in society by educating users on how to use smartphones smartly and engaging themselves in other activities which do not rely on technology.

**Acknowledgements** The authors thank reviewers for their constructive comments. The authors thank DST-FIST file # MSI-097 for technical support to the Department of Mathematics, Gujarat University. Second author (Purvi M. Pandya) would like to extend sincere thanks to the Education Department, Gujarat State for providing scholarship under ScHeme Of Developing High quality research (student AisheCode: 201901380135). Third author (Ekta N. Jayswal) is funded by the UGC granted National Fellowship for OBC (NFO-2018-19-OBC-GUJ-71790).

## References

1. Ballal, I. P., & Nadkarni, K. P. (2020). Data analysis and prediction of survey on effect of smart phones on society. In *Emerging trends in electrical, communications, and information technologies* (pp. 61–70). Singapore: Springer.
2. Diekmann, O., Heesterbeek, J. A. P., & Roberts, M. G. (2009). The construction of next-generation matrices for compartmental epidemic models. *Journal of the Royal Society, Interface, 7*(47), 873–885.
3. Enez Darcin, A., Kose, S., Noyan, C. O., Nurmedov, S., Yılmaz, O., & Dilbaz, N. (2016). Smartphone addiction and its relationship with social anxiety and loneliness. *Behaviour & Information Technology, 35*(7), 520–525.
4. Hale, J. K. (1980). *Ordinary differential equations*. RE Krieger Publication.
5. Harary, F. (1969). *Graph theory*. Boston, MA: Addison Wesley.
6. Hawi, N. S., & Samaha, M. (2017). Relationships among smartphone addiction, anxiety, and family relations. *Behaviour & Information Technology, 36*(10), 1046–1052.
7. Ibrahim, J., Ros, R. C., Sulaiman, N. F., Nordin, R. C., & Ze, L. (2014). Positive impact of Smartphone application: WhatsApp & Facebook for online business. *International Journal of Scientific and Research Publications, 4*(12), 1–4.
8. Iyengar, K., Upadhyaya, G. K., Vaishya, R., & Jain, V. (2020). COVID-19 and applications of smartphone technology in the current pandemic. *Diabetes & Metabolic Syndrome: Clinical Research & Reviews, 14*(5), 733–737.
9. Kashif, M., Aziz-Ur-Rehman, & Muhammad, K. J. (2020). Social media addiction due to coronavirus. *International Journal of Medical Science in Clinical Research and Review, 3*(04), 331–336.
10. Kim, E., Son, H., Choi, M. O., & Jeong, B. (2019). Development of a smartphone overdependence prevention group program in adolescents. *Asia-Pacific Journal of Convergent Research Interchange, 5*(4), 123–137.
11. Kim, H. (2013). Exercise rehabilitation for smartphone addiction. *Journal of exercise rehabilitation, 9*(6), 500.
12. King, D. L., Delfabbro, P. H., Billieux, J., & Potenza, M. N. (2020). Problematic online gaming and the COVID-19 pandemic. *Journal of Behavioral Addictions, 9*(2), 184–186.
13. Király, O., Potenza, M. N., Stein, D. J., King, D. L., Hodgins, D. C., Saunders, J. B. … Abbott, M. W. (2020). Preventing problematic internet use during the COVID-19 pandemic: Consensus guidance. *Comprehensive Psychiatry*, 152180.
14. Lee, J., Cho, B., Kim, Y., & Noh, J. (2015). Smartphone addiction in university students and its implication for learning. In *Emerging issues in smart learning* (pp. 297–305). Berlin: Springer.
15. Miakotko, L. (2017). *The impact of smartphones and mobile devices on human health and life*. New York University.

16. Mishra, M. (2017). Impact of smartphone on college students. *International Journal of Education and Management Studies, 7*(3), 310–313.
17. Park, C., & Park, Y. R. (2014). The conceptual model on smart phone addiction among early childhood. *International Journal of Social Science and Humanity, 4*(2), 147.
18. Samaha, M., & Hawi, N. S. (2016). Relationships among smartphone addiction, stress, academic performance, and satisfaction with life. *Computers in Human Behavior, 57*, 321–325.
19. Sañudo, B., Fennell, C., & Sánchez-Oliver, A. J. (2020). Objectively-assessed physical activity, sedentary behavior, smartphone use, and sleep patterns pre-and during-COVID-19 quarantine in young adults from spain. *Sustainability, 12*(15), 5890.
20. Sarwar, M., & Soomro, T. R. (2013). Impact of smartphone's on society. *European Journal of Scientific Research, 98*(2), 216–226.
21. Soni, R., Upadhyay, R., & Jain, M. (2017). Prevalence of smart phone addiction, sleep quality and associated behavior problems in adolescents. *International Journal of Research in Medical Sciences, 5*(2), 515–519.
22. Swar, B., & Hameed, T. (2017). Fear of missing out, social media engagement, smartphone addiction and distraction: Moderating role of self-help mobile apps-based interventions in the youth. *International Conference on Health Informatics, 6*, 139–146.
23. West, D. B. (1996). *Introduction to graph theory* (Vol. 2). Upper Saddle River, NJ: Prentice hall.
24. Yuniati, Y., & Yuningsih, A. (2017). Utilization of smartphone literacy in learning process. *Mimbar: Jurnal Sosial dan Pembangunan, 33*(1), 90–98.

# Chapter 16
# Modelling and Sensitivity Analysis of COVID-19 Under the Influence of Environmental Pollution

**Nitin K Kamboj, Sangeeta Sharma, and Sandeep Sharma**

**Abstract** The ongoing COVID-19 pandemic emerged as one of the biggest challenges of recent times. Efforts have been made from different corners of the research community to understand different dimensions of the disease. Some theoretical works have reported that disease becomes severe in the presence of environmental pollution. In this work, we propose a nonlinear mathematical model to study the influence of air pollution on the dynamics of the disease. The basic reproduction number plays a vital role in predicting the future of an epidemic. Therefore, we obtain the expression of the basic reproduction number and performed a detailed sensitivity and uncertainty analysis. The values of partial rank correlation coefficients (PRCC) have been calculated corresponding to six critical parameters. The positive values of PRCC for pollution-related parameters depicts that pollution enhances the chances of a rapid spread of COVID-19.

**Keywords** COVID-19 · Environmental pollution · Mathematical model · The basic reproduction number · Sensitivity analysis

## Introduction

The ongoing pandemic of COVID-19 was originated in Wuhan (China). Since then, it has invaded almost every country across the globe and created unprecedented stress on medical facilities and infrastructure. Due to its severity, the World Health Organization (WHO) first declared a Public Health Emergency of International Concern

---

N. K. Kamboj · S. Sharma (✉)
Department of Mathematics, DIT university, Dehradun, Uttarakhand 248009, India
e-mail: sampark81@gmail.com

N. K. Kamboj
e-mail: nkdrstats@gmail.com

S. Sharma
School of Life Sciences and Technology, IIMT University, Meerut, Uttar Pradesh 250001, India
e-mail: sangeeta4523@rediffmail.com



N. H. Shah and M. Mittal (eds.), *Mathematical Analysis for Transmission of COVID-19*, Mathematical Engineering, https://doi.org/10.1007/978-981-33-6264-2_16

on 30 January 2020, and subsequently a pandemic on 11 March 2020 [1]. Coronaviruses, responsible for COVID-19, belong to the Coronaviridae family in the Nidovirales order [2]. Due to the presence of crown-like spikes on its outer surface, it is known as coronavirus. The coronaviruses diameter ranges between 65–125 nm and contain a single-stranded RNA as a nucleic material. The coronaviruses family has four subgroups namely; alpha ($\alpha$), beta ($\beta$), gamma ($\gamma$) and delta ($\delta$) [2]. Before 2002, it was believed that coronaviruses infect only animals. The 2002 outbreak (in Guangdong, China) of severe acute respiratory syndrome (SARS) was the first event caused by SARS-CoV [3]. After 2002, the SARS outbreak, the world faced an outbreak of Middle East respiratory syndrome coronavirus (MERS-CoV) caused by the coronaviruses family [4].

The coronavirus (responsible for COVID-19) initially transmitted to humans from the Wuhan seafood market [5, 6]. Later, it was established that the disease can also spread through human to human transmission mode. As per scientific and clinical research, the transmission of the coronavirus from infected individuals to healthy individuals spreads due to physical contacts and droplets in the air due to sneezing and coughing of an infected person [7]. The primary symptoms of COVID-19 are approximately 95% similar to SARS coronavirus and include dry cough, abnormality in body temperature, breathing difficulty and bilateral lung infiltration [7, 8]. Moreover, the role of environmental pollution on the spread of waterborne diseases has also been studied [9–11]. The correlation between exposure to air pollution and COVID-19 mortality poses a serious question in front of the research community. Owning to this, attempts have been made to identify the possible impact of environmental pollution on the spread of COVID-19 and COVID-19 mortalities. In particular, the study carried out by Conticini et al. [12] explored the correlation between air pollution and COVID-19 mortality. They further observed that the regions with a high level of pollution (Lombardy and Emilia Romagna) are also registered the maximum number of deaths due to COVID-19. Further, based on the data, the study concludes that people living in areas with high pollutants concentration may easily fall prey to respiratory disease. The work carried out by Zoran et al. [13] investigates the correlation between high transmissibility and lethality COVID-19 and the surface air pollution in the Milan metropolitan area, Lombardy region, Italy. To achieve this goal, authors collected daily data (from 1 January to 30 April 2020) of average concentrations of $PM_{2.5}$ and $PM_{10}$ and maxima $PM_{10}$ ground-level atmospheric pollutants and air quality and climate variables (e.g., daily average temperature, wind speed, relative humidity, atmospheric pressure field, etc.). The study demonstrates the strong influence of daily averaged ground levels of particulate matter concentrations on COVID-19 cases outbreaks in Milan. The study, further, concludes that chronic or short-term exposure to particulate matter $PM_{2.5}$ or $PM_{10}$ carrying different viruses or bacteria has a major negative impact on the human immune system and thus makes people vulnerable towards COVID-19.

Mathematical modelling of infectious disease provides critical information about its transmission mechanism. The results obtained on the long-term dynamics of an epidemic through the analysis of the mathematical model are crucial in the planning intervention programs [14]. In case of COVID-19, many mathematical models have been deployed to study the difference dimensions of the disease (see [15–22, 22–26] and references cited therein).

In [21], the authors proposed a mathematical model to make short-term predictions about the future of COVID-19 in India, Argentina, Mexico and South Africa. Moreover, they also explore the conditions under which the proposed model exhibits backward bifurcation. In [23], a mathematical model is proposed to study the role of different nonpharmaceutical intervention policies (e.g. contact tracing, quarantine, social distancing, isolation, the use of face masks, etc.) in controlling the disease. With the help of the data of COVID-19 for USA (US), the authors demonstrate that these intervention strategies are successful in controlling the ongoing pandemic. In [24], the authors investigate the importance of the lockdown in controlling the wave of COVID-19 with the help of a mathematical model and simulating the same over the real data available for Florida, Arizona, New York and for the entire country. The study also underscores the importance of the identification of pre-symptomatic and asymptomatic patients. Further, authors also observe that control of COVID-19 will be significantly achieved if the implementation of lockdown is complemented by the use of face masks. In [17], the authors used a simple SIR epidemic model to study the COVID-19 scenario in France, China and Italy. Through the simulation of the model on the available data for the three countries, the authors identify that the recovery rate is the same for all the three countries. On the other hand, a very high variability has been observed among the three countries in terms of disease-induced death and disease transmission rate. In [22], authors extended the generic SEIR epidemic model by including a separate compartment for the super spreader, hospitalized and fatality class. The analysis of the model reveals the role of the super spreader in the dynamics of COVID-19 in Wuhan (China). In [20], authors proposed a mathematical model by including the bat population. Further, the authors extended the model to a fractional mathematical model to study the role of bats and the seafood market on the spread of COVID-19. In [26], authors proposed a mathematical model by incorporating some time dependent parameters to investigate the COVID-19 outbreak in Wuhan. Through the calculation of the effective daily reproduction ratio, authors demonstrate that delay in providing medical facilities play a key role in the increase in the size of the epidemic. In [18], authors formulated an eight stages SIDARTHE epidemic model to investigate the COVID-19 scenario in Italy. The model subsequently analysed to identify effective control strategies. The authors conclude that collective implementation of restrictive social distancing along with widespread testing and contact tracing may end the ongoing COVID-19 wave in Italy. In [19], a stochastic epidemic model has been used to study the COVID-19 pandemic in China. The model further used to estimate the reproduction number and gauge the success of implemented control measures. The authors also investigate the effect of the work timing on the disease dynamics. In [16], authors apply Monte–Carlo simulation on a stochastic SEIR compartmental model and predict the future of COVID-19 using the initial data of the reported cases in India. In [22], authors used a compartmental model to investigate the spread of the COVID-19 in Wuhan. Subsequently, the authors proposed a detailed dynamical study of the model and obtained the necessary conditions for the stability of the equilibrium solutions.

Sensitivity analysis also carried out to gauge the impact of the individual parameters on the dynamics of the disease. In [25], authors demonstrate the success of the SIR epidemic model over the SEIR epidemic model (in terms of the representation of confirmed cases) using the Akaike Information Criterion (AIC) in the term. In [15], authors used a SLIAR epidemic model to study the future of COVID-19 using the initial data of the disease. They demonstrate that during the initial phase of the infection simple models are more helpful and provides significant information about the disease.

From the above discussion, it can be concluded that mathematical models are vastly used to study different dimensions of the COVID-19. Moreover, the model also employed to identify robust control measures to reduce the size of the epidemic.

During the formulation of a disease model, one has to make different assumptions helping in representing the transmission mechanism of disease in the language of mathematics. But, the assumptions lead to the formulation of an epidemic model often introduce uncertainties in the estimation of parameters involved in the model. In many cases, this seriously affects the accuracy of the results obtained through the analysis of the model. The information obtained through the sensitivity analysis will provide key information about the impact of different parameters on the dynamics of the disease. Due to this, sensitivity and uncertainty analysis of epidemic models is an important area of research, and many researchers are performing sensitivity analysis of epidemic models [27–29]. A nice review of the methods and techniques of sensitivity analysis can be found in [30].

Recently, a number of studies are found an active role of the environmental pollution in the spread of a number of infections [31–35]. Some studies observed that regular exposure to environmental pollution results in diminishing of immunity [31, 36, 37]. This enhances the susceptibility of an individual towards an infection. In particular, Lafferty et al. [32] give a nice illustration of the impact of environmental stress on the dynamics of an infectious disease. The work carried out in [38] proposed an SIS type epidemic model by including a separate compartment of stressed individuals (those having regular exposure to environmental pollution) demonstrate the positive impact of environmental pollution on the spread of the disease. Despite this, to the best of our knowledge, no mathematical model available to study the correlation between environmental pollution and COVID-19. To fill this gap, in this work, we propose a new mathematical model to investigate the possible impact of environmental pollution on the spread of COVID-19. We obtain the expression of the basic reproduction number for the proposed model and then perform a rigorous sensitivity analysis of the same. The sensitivity analysis helps us to identify the impact of a particular parameter on the dynamics of the system. In particular, the current work demonstrates that environmental pollution may play a significant role and should be considered as one of the factors in the study of COVID-19.

## Mathematical Model

In this section, we present the mathematical model considered during the work. In the modelling process, we divide the total human population $N(t)$ at a particular time $t$ into seven mutually exclusive compartments; namely susceptible $S_1(t)$, stressed (those with weaken immune system due to regular or frequent exposure to environmental pollution) $S_2(t)$, exposed $E(t)$, quarantine $Q(t)$, infected $I(t)$, hospitalized $H(t)$ and recovered $R(t)$.

$$
\begin{aligned}
\frac{\mathrm{d}S_1}{\mathrm{d}t} &= \Lambda - \frac{\delta S_1(I+\delta_0 E)}{N} - dS_1 - \gamma S_1 \\
\frac{\mathrm{d}S_2}{\mathrm{d}t} &= \gamma S_1 - \frac{\delta(1+\epsilon\delta')S_2(I+\delta_0 E)}{N} - dS_2 \\
\frac{\mathrm{d}E}{\mathrm{d}t} &= \frac{\delta S(I+\delta_0 E)}{N} + \frac{\delta(1+\epsilon\delta')S_2(I+\delta_0 E)}{N} - (d+\theta)E \\
\frac{\mathrm{d}Q}{\mathrm{d}t} &= \theta(1-p)E - (d+\alpha)Q - \phi Q - \phi_0 Q \\
\frac{\mathrm{d}I}{\mathrm{d}t} &= \theta pE - (d+\alpha+r)I - \psi I \\
\frac{\mathrm{d}H}{\mathrm{d}t} &= \psi I - (d+\alpha\alpha')H - \xi H + \phi_0 Q \\
\frac{\mathrm{d}R}{\mathrm{d}t} &= \xi H + \phi Q - dR + rI
\end{aligned}
\tag{16.1}
$$

In the model system, $\Lambda$ is the constant recruitment rate. $\delta$ is the disease transmission rate for the individuals of $S_1$ class from the infected individuals. It is assumed that disease transmission rate from exposed individuals to susceptible individuals is less than $\delta$ and to incorporate the same we introduce $\delta_0$ as the reduction factor. Next, using the approach of Lafferty et al. [32], we modify the disease transmission rate for stressed individuals ($S_2$) as $\delta(1+\epsilon\delta')$, where $\epsilon$ measures the impact of environmental pollution on the transmission rate and the effect of pollution $\delta$ is represented by $\delta'$. The model also considers the natural death rate of $d$. $\theta$ is the rate at which exposed individuals leave the exposed class and join the infected class, out of which one fraction $p$ joins the infected class while the remaining $1-p$ joins the quarantine class. Parameter $\alpha$ represents the disease-induced death rate. It is also assumed that the disease-induced death rate for hospitalized individuals is less (due to availability of medical treatment) than that of infected individuals and to incorporate this we introduce $\alpha'$ as the reduction factor. $\psi$ is the rate at which infected individuals admitted to hospitals. The hospitalized individuals after initial treatment/checkup are allowed to move in self-quarantine at the rate $\phi_0$. Parameters $\phi$, $\xi$ and $r$ represent the recovery rates for individuals of quarantine, hospitalized and infected class, respectively.

## The Basic Reproduction Number

It is trivial to observe that proposed model system possesses a unique disease free equilibrium point $E_0 = \left(S_1^0, S_2^0, 0, 0, 0, 0, 0\right)$, where $S_1^0 = \frac{\Lambda}{(\gamma+d)}$ and $S_2^0 = \frac{\gamma\Lambda}{d(\gamma+d)}$. Next, to calculate the basic reproduction number for the proposed model system 16.1, we use the popular next generation matrix method [39]. Now, the matrix $F$ and $V$ can be obtained as

$$\mathscr{F} = \begin{bmatrix} \frac{\delta S(I+\delta_0 E)}{N} + \frac{\delta(1+\epsilon\delta')S_2(I+\delta_0 E)}{N} \\ 0 \\ 0 \end{bmatrix}, \quad \mathscr{V} = \begin{bmatrix} (\theta + d)E \\ -\theta(1-p)E + (\alpha + d + \phi + \phi_0)Q \\ -p\theta E + (\alpha + d + r + \psi)I \end{bmatrix}$$

Subsequently, we can obtained the Jacobians $F$ and $V$ of $\mathscr{F}$ and $\mathscr{V}$, respectively, at the disease free equilibrium point $E_0$ as

$$F = \begin{bmatrix} \frac{\delta\delta_0 S_1^0 + \delta\delta_0(1+\epsilon\delta')S_2^0}{N} & \frac{\delta S_1^0 + \delta(1+\epsilon\delta')S_2^0}{N} & 0 \\ 0 & 0 & 0 \\ 0 & 0 & 0 \end{bmatrix}$$

Similarly,

$$V = \begin{bmatrix} (\theta + d) & 0 & 0 \\ -\theta(1-p) & 0 & (\alpha + d + \phi + \phi_0) \\ -p\theta & (\alpha + d + r + \psi) & 0 \end{bmatrix}$$

Now, the spectral radius of the matrix $FV^{-1}$ will provide us the expression of the basic reproduction number ($R_0$) as

$$R_0 = \frac{\delta\delta_0 d + \delta\delta_0(1+\epsilon\delta')\gamma}{(\gamma + d)(\theta + d)} + \frac{p\theta\delta d + \delta(1+\epsilon\delta')}{(\theta + d)(\gamma + d)(\alpha + d + r + \psi)} \tag{16.2}$$

## Sensitivity Analysis

Mathematical models pertaining to infectious disease are considered as an exciting field of research. The availability of a variety of mathematical models helps agencies to reduce the burden of a number of infectious diseases. In the study of disease dynamics using mathematical models, basic reproduction plays a very important role as it provides the threshold for the disease elimination and persistence. The mathematical epidemiological model may acquire a complex structure depending upon the complexity involved in the transmission mechanism of a particular infection. Therefore, in many cases, it will be difficult to identify the impact of an individual parameter on the dynamics of the disease. Sensitivity analysis of the epidemic model plays an important role to overcome this problem. The sensitivity analysis corresponding to some key parameters of an epidemic model provides crucial information that helps the authorities to frame some robust policies to combat infectious disease.

In this section, the sensitivity analysis of $R_0$ using two sampling schemes (namely random and Latin hypercube) to investigate how sensitive $R_0$ to an input parameter involved in the mathematical model. Sensitivity analysis generally constitutes a series of tests involving different sets of input parameters. This helps us to observe how a change in the predictor parameter values changes the dynamical behaviour of the system. It also provides information on how closely input parameters are related to a particular predictor parameter. The results obtained through sensitivity analysis helps in determining the level of change necessary for an input parameter to obtain the desired value of a predictor parameter.

In the study of disease models, sensitivity analysis is used to identify the key parameters, among the parameters involved in the model, which have a significant impact on the outcome of $R_0$ depending on the uncertainty involved in their estimation. Partial rank correlation coefficients (PRCCs) are a popular technique used to determine the statistical influence of any parameter on the $R_0$. In the present work, we first performed uncertainty analysis for $R_0$, subsequently, we obtain partial rank correlation coefficients corresponding to all uncertain parameters.

There are 11 parameters involved in the model system (16.1) and the estimated values of the same have been given in Table 16.1. Out of 11 parameters, $\delta$, $\theta$, $p$, $r$, $\epsilon$ and $\delta'$ have been identified to conduct uncertainty analysis due to uncertainties involved in their estimation. To carry out the uncertainty analysis, we consider each of these parameters as a random variable with an appropriate probability density function. The remaining five parameters are kept fix and their values have been taken from the literature (given in Table 16.1).

Following are the distributions selected for six parameters

1. Four parameters consider to follow uniform distribution as discussed in [15]

   (a) $\delta$ with minimum $5 \times (10)^{-5}$ and maximum $3 \times (10)^{-4}$.
   (b) $\theta$ with minimum 1 and maximum 14.
   (c) $p$ with minimum 0 and maximum 1.
   (d) $r$ with minimum 2 and maximum 14.

2. $\epsilon$ follows Weibull distribution with parameters 0.6386 (scale) and 12.766 (shape).
3. $\delta'$ follows exponential distribution with rate 1.

Two sampling methods random sampling (RS) and Latin hypercube sampling (LHS) have been considered to generate the values of these six uncertain parameters. A set of 1000 parameter values have been sampled using RS and LHS for six parameters from different types of parameter distribution. Histograms of the parameters considered for the study are given in Figs. 16.1 and 16.2. Histograms and box plots for the distributions of $R_0$ are shown in Fig. 16.5. These histograms and box plots have been generated from Eq. 16.2 using RS and LHS.

**Table 16.1** Model parameters with interpretation and values/distribution

| Parameter | Description | Value/distribution | References |
|---|---|---|---|
| $\delta$ | Disease transmission rate from infected to susceptible individuals | Uniform | [15] |
| $\delta_0$ | Reduction in disease transmission for exposed individuals | 0.2 | [40] |
| $d$ | Natural death rate | 0.00003961 | [41] |
| $\theta$ | Rate at which exposed individuals leaving exposed class to infected class | $\frac{1}{5.1}$ | [16] |
| $p$ | Fraction of exposed class joining infected class | Uniform | – |
| $\alpha$ | Disease related death rate | 0.0175 | [40] |
| $\alpha'$ | Reduction factor and lies between 0 and 1 | 0.1 | Assumed |
| $r$ | Recovery rate for infected class | Uniform | [15] |
| $\psi$ | Rate at which infected individuals admitted to hospitals | 0.2174 | [42] |
| $\rho$ | Recovery rate of hospitalized individuals | $\frac{1}{14}$ | [40] |
| $\phi$ | Recovery rate of quarantine individuals | 0.1162 | [40] |
| $\epsilon$ | Amount by which environmental pollution affects the transmission rate | Weibull | [43] |
| $\delta'$ | Effect of pollution on $\delta$ | Exponential | [32] |
| $\gamma$ | Rate at which individuals of $S_1$ class join $S_2$ class | 0.004 | Assumed |

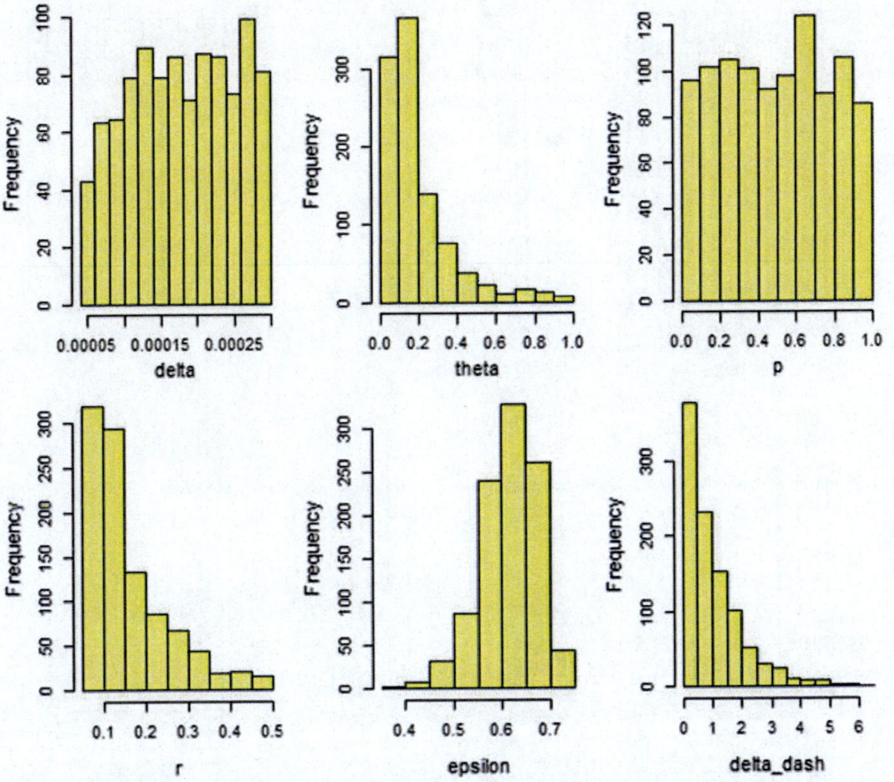


**Fig. 16.1** Histograms of the values obtained through random sampling with a sample size of 1000 (*Source* own)

## Results and Discussion

The partial rank correlation coefficients (PRCC) indicate the degree of effect of a particular parameter on the outcome. Scatter plots comparing the basic reproduction number for each of the parameters; $\delta$, $\theta$, $p$, $r$, $\epsilon$ and $\delta'$ are shown in Figs. 16.3 and 16.4 for RS and LHS, respectively. These scatter plots clearly show the linear relationships between input parameters and $R_0$.

The sign of the PRCC depicts the qualitative relationship between the input parameter and the related output variable. The positive sign of the PRCC of the variables corresponds to the situation that an increase in the value of the input parameter, the value of $R_0$ also increases. On the other hand, the negative sign of the PRCC signifies that any increase in the corresponding parameter results in the decrease of the basic reproduction number, and hence such parameters have the potential to reduce the size of the epidemic.

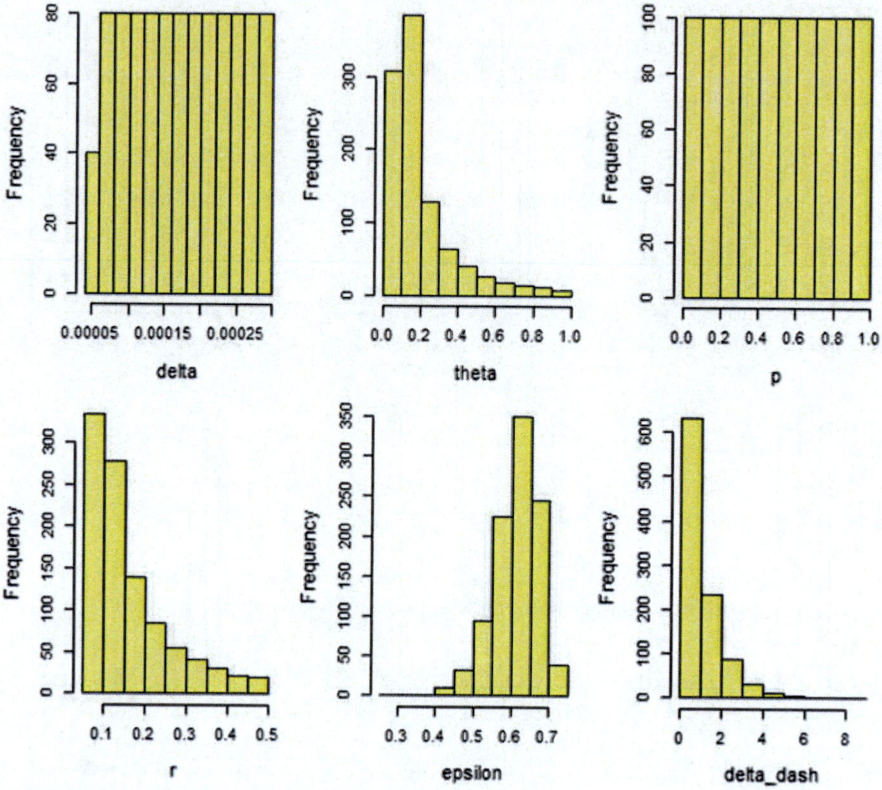


**Fig. 16.2** Histograms of the values obtained through Latin hypercube sampling with a sample size of 1000 (*Source*: own)

The results obtained through sensitivity analysis are given in Table 16.2. On the basis of random sampling, we can conclude that parameters $\theta$, $\delta$ and $\delta'$ have higher degree of relationship with $R_0$ with PRCC values 0.9077, 0.8591 and 0.749, respectively. From the values of the parameters, it is easy to observe that $\delta$ and $\delta'$ have positive association while the parameter $\theta$ has a negative association. Further, the other pollution-related parameters $\epsilon$ also recorded the positive value of PRCC.

Similar results have been obtained for the Latin hypercube sampling as $\theta$ emerged as the most sensitive parameter with the highest value (−0.924458593) of PRCC. $\delta$ and $\delta'$ are observed as the next two most sensitive parameters with PRCC values 0.879221188 and 0.763274350, respectively. The PRCC value for $\epsilon$ is 0.096838976, which clearly reflects the positive association of the pollution on the spread of the disease.

The positive values of PRCC clearly depict that pollution plays a supportive role in the spread of the disease and can increase the size of the epidemics. Since pollution is a global problem and many developing countries are suffering severely with the menace

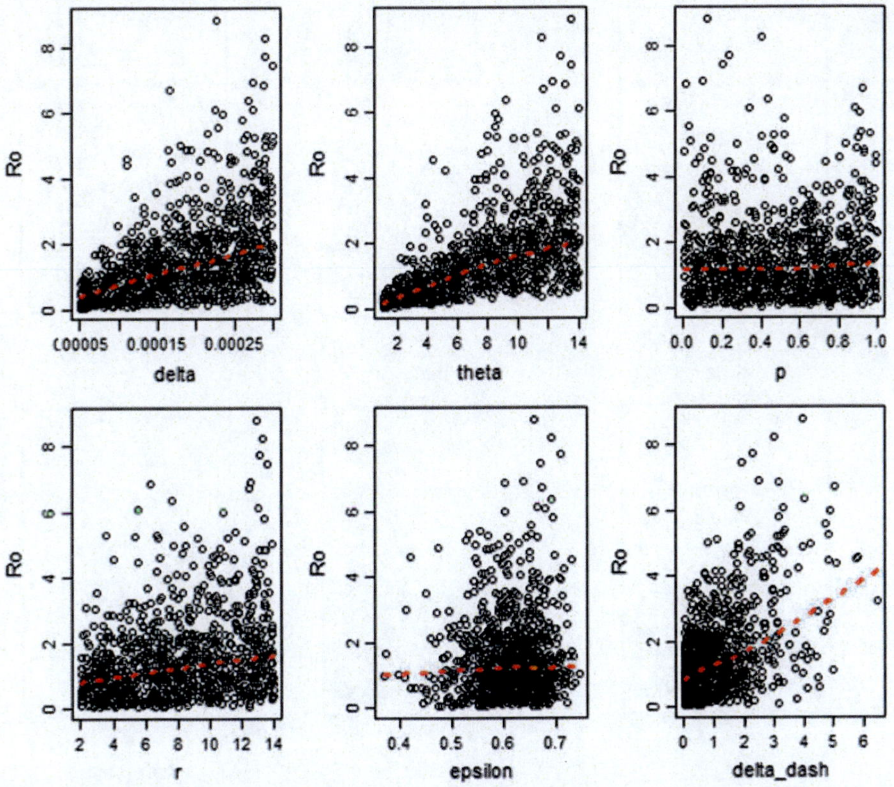


**Fig. 16.3** Scatter plots for $R_0$ obtained from random sampling (*Source*: own)

of pollution. These countries also lack a robust medical infrastructure. Therefore, the results obtained in this work clearly highlights the need for some appropriate steps from the different research community in order to control the ongoing COVID-19 pandemic.

In short, the following points can be concluded from the current study

1. $\theta$ is the most sensitive parameter followed by $\delta$ and $\delta'$.
2. The positive values of PRCC for pollution-related parameters ($\delta'$ and $\epsilon$) are positive. This clearly reflects the positive association of pollution on the spread of the disease.
3. The PRCC value for $\delta'$ is significantly high, which shows that disease spread more rapidly in the presence of pollution.
4. Disease eradication needs more effort in the presence of pollution.

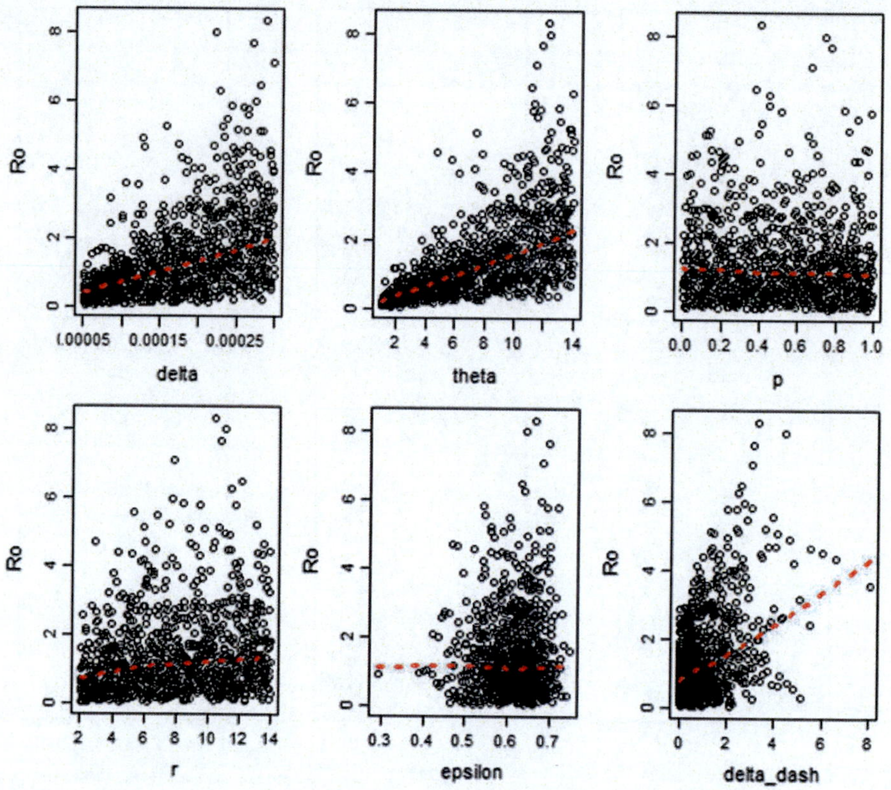


**Fig. 16.4** Scatter plots for $R_0$ obtained from Latin hypercube sampling (*Source*: own)

**Table 16.2** PRCC values for the basic reproduction number $R_0$

| Parameter | Sampling | PRCCs |
|---|---|---|
| $\delta$ | RS | 0.85912837 |
| | LHS | 0.879221188 |
| $\theta$ | RS | −0.90773120 |
| | LHS | −0.924458593 |
| $p$ | RS | 0.02250202 |
| | LHS | 0.009476424 |
| $r$ | RS | −0.57183391 |
| | LHS | −0.617232256 |
| $\epsilon$ | RS | 0.07814241 |
| | LHS | 0.096838976 |
| $\delta'$ | RS | 0.74904911 |
| | LHS | 0.763274350 |

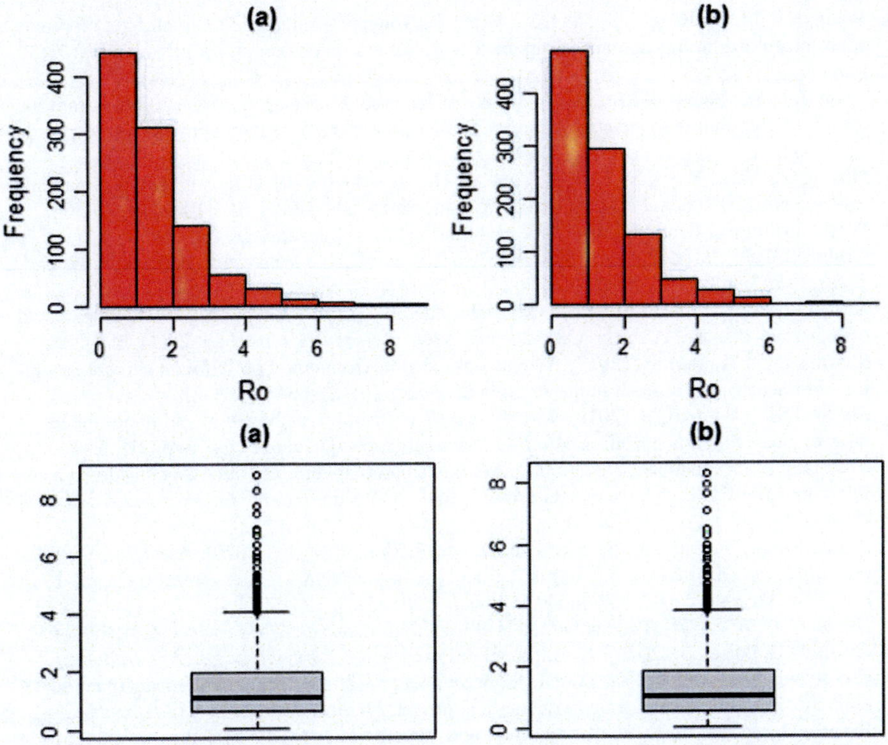


Fig. 16.5 Histogram and box plot of the basic reproduction number, **a** obtained from random sampling and **b** obtained from Latin hypercube sampling (*Source*: own)

Accounts to the rapid growth in the industry and automobiles, the level of pollution will rise significantly in the near future. And there is a high chance that many new chemical pollutants will come into existence. Therefore, the current study can be extended, in the future, by incorporating more variables or parameters as per the available field data.

## References

1. https://www.who.int/emergencies/diseases/novel-coronavirus-2019/events-as-they-happen .
2. Shereen, M. A., Khan, S., Kazmi, A., Bashir, N., & Siddique, R. (2020). Covid-19 infection: Origin, transmission, and characteristics of human coronaviruses. *Journal of Advanced Research*.
3. Zhao, Z., Zhang, F., Xu, M., Huang, K., Zhong, W., Cai, W., et al. (2003). Description and clinical treatment of an early outbreak of severe acute respiratory syndrome (sars) in Guangzhou, PR china. *Journal of Medical Microbiology*, *52*(8), 715–720.

4. Wang, N., Shi, X., Jiang, L., Zhang, S., Wang, D., Tong, P., et al. (2013). Structure of mers-cov spike receptor-binding domain complexed with human receptor dpp4. *Cell research, 23*(8), 986–993.
5. Chan, J. F. W., Yuan, S., Kok, K. H., To, K. K. W., Chu, H., Yang, J., et al. (2020). A familial cluster of pneumonia associated with the 2019 novel coronavirus indicating person-to-person transmission: A study of a family cluster. *The Lancet, 395*(10223), 514–523.
6. Huang, C., Wang, Y., Li, X., Ren, L., Zhao, J., Hu, Y., et al. (2020). Clinical features of patients infected with 2019 novel coronavirus in Wuhan, China. *The lancet, 395*(10223), 497–506.
7. World health organization. https://www.who.int/health-topics/coronavirus.
8. Gralinski, L. E., & Menachery, V. D. (2020). Return of the coronavirus: 2019-ncov. *Viruses, 12*(2), 135.
9. Kumari, N., & Sharma, S. (2016). Does water disinfectant play a supportive role in the spread of infectious disease? A mathematical study. *Natural Resource Modeling, 29*(2), 259–288.
10. Sharma, S., & Kumari, N. (2017). Why to consider environmental pollution in cholera modeling? *Mathematical Methods in the Applied Sciences, 40*(18), 6348–6370.
11. Sharma, S., & Kumari, N. (2019). Dynamics of a waterborne pathogen model under the influence of environmental pollution. *Applied Mathematics and Computation, 346*, 219–243.
12. Conticini, E., Frediani, B., & Caro, D. (2020). Can atmospheric pollution be considered a co-factor in extremely high level of sars-cov-2 lethality in northern italy? *Environmental pollution* 114465.
13. Zoran, M. A., Savastru, R. S., Savastru, D. M., & Tautan, M. N. (2020). Assessing the relationship between surface levels of pm 2. 5 and pm10 particulate matter impact on covid-19 in Milan, Italy. *Science of The Total Environment*, 139825.
14. Hethcote, H. W. (2000). The mathematics of infectious diseases. *SIAM review, 42*(4), 599–653.
15. Arino, J., & Portet, S. (2020). A simple model for covid-19. *Infectious Disease Modelling*.
16. Chatterjee, K., Chatterjee, K., Kumar, A., & Shankar, S. (2020). Healthcare impact of covid-19 epidemic in India: A stochastic mathematical model. *Medical Journal Armed Forces India*.
17. Fanelli, D., & Piazza, F. (2020). Analysis and forecast of covid-19 spreading in china, Italy and France. *Chaos, Solitons & Fractals, 134*, 109761.
18. Giordano, G., Blanchini, F., Bruno, R., Colaneri, P., Di Filippo, A., Di Matteo, A., & Colaneri, M. (2020). Modelling the covid-19 epidemic and implementation of population-wide interventions in Italy. *Nature Medicine*, 1–6.
19. He, S., Tang, S., & Rong, L. (2020). A discrete stochastic model of the covid-19 outbreak: Forecast and control. *Mathematical Biosciences and Engineering, 17*, 2792–2804.
20. Khan, M. A., & Atangana, A. (2020). Modeling the dynamics of novel coronavirus (2019-ncov) with fractional derivative. *Alexandria Engineering Journal*.
21. Nadim, S. S., & Chattopadhyay, J. (2020). *Occurrence of backward bifurcation and prediction of disease transmission with imperfect lockdown: A case study on covid-19* (p. 110163). Solitons & Fractals: Chaos.
22. Ndairou, F., Area, I., Nieto, J. J., & Torres, D. F. (2020). *Mathematical modeling of covid-19 transmission dynamics with a case study of wuhan* (p. 109846). Solitons & Fractals: Chaos.
23. Ngonghala, C. N., Iboi, E., Eikenberry, S., Scotch, M., MacIntyre, C. R., Bonds, M. H., & Gumel, A. B. (2020) Mathematical assessment of the impact of non-pharmaceutical interventions on curtailing the 2019 novel coronavirus. *Mathematical Biosciences*, 108364
24. Ngonghala, C. N., Iboi, E., & Gumel, A. B. (2020) Could masks curtail the post-lockdown resurgence of covid-19 in the us? *Mathematical Biosciences*, 108452
25. Roda, W. C., Varughese, M. B., Han, D., & Li, M. Y. (2020). Why is it difficult to accurately predict the covid-19 epidemic? *Infectious Disease Modelling*.
26. Wang, L., Wang, J., Zhao, H., Shi, Y., Wang, K., Wu, P., & Shi, L. (2020) Modelling and assessing the effects of medical resources on transmission of novel coronavirus (covid-19) in Wuhan, China.
27. Arino, J., Brauer, F., Van Den Driessche, P., Watmough, J., & Wu, J. (2008). A model for influenza with vaccination and antiviral treatment. *Journal of Theoretical Biology, 253*(1), 118–130.

28. Matsuyama, R., Akhmetzhanov, A. R., Endo, A., Lee, H., Yamaguchi, T., Tsuzuki, S., et al. (2018). Uncertainty and sensitivity analysis of the basic reproduction number of diphtheria: A case study of a Rohingya refugee camp in Bangladesh, November-December 2017. *PeerJ*, *6*, e4583.
29. Samsuzzoha, M., Singh, M., & Lucy, D. (2013). Uncertainty and sensitivity analysis of the basic reproduction number of a vaccinated epidemic model of influenza. *Applied Mathematical Modelling*, *37*(3), 903–915.
30. Marino, S., Hogue, I. B., Ray, C. J., & Kirschner, D. E. (2008). A methodology for performing global uncertainty and sensitivity analysis in systems biology. *Journal of Theoretical Biology*, *254*(1), 178–196.
31. Dobson, A. (2009). Climate variability, global change, immunity, and the dynamics of infectious diseases. *Ecology*, *90*(4), 920–927.
32. Lafferty, K. D., & Holt, R. D. (2003). How should environmental stress affect the population dynamics of disease? *Ecology Letters*, *6*(7), 654–664.
33. Lipp, E. K., Huq, A., & Colwell, R. R. (2002). Effects of global climate on infectious disease: The cholera model. *Clinical Microbiology Reviews*, *15*(4), 757–770.
34. McMichael, A. J., & Woodruff, R. E. (2005). Detecting the health effects of environmental change: Scientific and political challenge.
35. Schwarzenbach, R. P., Egli, T., Hofstetter, T. B., Von Gunten, U., & Wehrli, B. (2010). Global water pollution and human health. *Annual Review of Environment and Resources*, *35*, 109–136.
36. Gavrilescu, M., Demnerová, K., Aamand, J., Agathos, S., & Fava, F. (2015). Emerging pollutants in the environment: Present and future challenges in biomonitoring, ecological risks and bioremediation. *New biotechnology*, *32*(1), 147–156.
37. Huntingford, C., Hemming, D., Gash, J., Gedney, N., & Nuttall, P. (2007). Impact of climate change on health: What is required of climate modellers? *Transactions of the Royal Society of Tropical Medicine and Hygiene*, *101*(2), 97–103.
38. Kumari, N., & Sharma, S. (2018). Modeling the dynamics of infectious disease under the influence of environmental pollution. *International Journal of Applied and Computational Mathematics*, *4*(3), 84.
39. Van den Driessche, P., & Watmough, J. (2002). Reproduction numbers and sub-threshold endemic equilibria for compartmental models of disease transmission. *Mathematical biosciences*, *180*(1–2), 29–48.
40. Senapati, A., Rana, S., Das, T., & Chattopadhyay, J. (2020). Impact of intervention on the spread of covid-19 in India: A model based study. arXiv preprint arXiv: 2004.04950.
41. https://data.worldbank.org/indicator/SP.DYN.LE00.IN?locations=IN .
42. Li, Q., Guan, X., Wu, P., Wang, X., Zhou, L., Tong, Y., et al. (2020). Early transmission dynamics in Wuhan, China, of novel coronavirus-infected pneumonia. *New England Journal of Medicine*.
43. Oguntunde, P., Odetunmibi, O., & Adejumo, A. (2014). A study of probability models in monitoring environmental pollution in Nigeria. *Journal of Probability and Statistics*.

# Chapter 17
# Bio-waste Management During COVID-19

**Nita H. Shah, Ekta N. Jayswal, and Purvi M. Pandya**

**Abstract** Ever since the transmission of novel coronavirus through human-to-human hit the world. As this disease is spreading every day, hospitalisation of individuals increased. Consequence of this, there is a sudden surge of millions of gloves, masks, hand sanitizers and the other essential equipment in each month. Disposal of these commodities is a big challenge for hospitals and COVID-centre, as they may became the reason of creating pollution and infect the surroundings. Increasing hospitalisation cases of COVID-19 results in raising bio-waste which creates pollution. Observing the scenario, a mathematical model with four compartments is constructed in this article. The threshold value indicates the intensity of pollution that emerged from bio-waste. Stability of the equilibrium point gave the necessary condition. Optimal control theory is outlined to achieve the purpose of this chapter by reducing pollution. Outcomes are analytically proven and also numerically simulated.

**Keywords** COVID-19 outbreak · Bio-waste · Hospitalisation · Optimal control · Pollution

**Mathematics Subject Classification** 37NXX · 97NXX

## Introduction

In February 2020, COVID 19 as pandemic globally, the World Health Organization (WHO) stated that this outbreak is highly infectious [1]. Such a contagious and contaminate disease, plastic plays a vital role in making confident social worker to deal with it. As per the latest (2020c) WHO evaluations, the world has consumed

N. H. Shah · E. N. Jayswal (✉) · P. M. Pandya
Department of Mathematics, Gujarat University, Ahmedabad 380009, Gujarat, India
e-mail: jayswal.ekta1993@gmail.com

N. H. Shah
e-mail: nitahshah@gmail.com

P. M. Pandya
e-mail: pandya091@gmail.com

 
N. H. Shah and M. Mittal (eds.), *Mathematical Analysis for Transmission of COVID-19*, Mathematical Engineering, https://doi.org/10.1007/978-981-33-6264-2_17

89 million masks, 30 million gowns, 1.59 million goggles, and 76 million gloves per month alone due to the COVID-19 [2]. As these things cannot be reused, it can only be dumped or recycled becomes waste. Thus, along with Personal Protective Equipment (PPEs) and sanitization, this biomedical waste is also a big challenge for any hospital having COVID patients. The waste produced by healthcare taken at home during COVID-19 pandemic time is also classified as bio-waste but this research only includes bio-waste produced due to hospitalisation. As time passed, the disposal of bio-waste creates pollution.

In March 2020, the government of India has released guidelines for handling safe disposal of bio-waste generated during the treatment of a patient having novel coronavirus [3]. In a very short time, the COVID-19 outbreak becomes epidemic in some highly populated regions and it becomes more complicated for doctors to handle the large scale of infected population henceforth, rigorous monitoring may not be possible. If medication guidelines could be followed then transmission of COVID-19 and pollution through bio-waste can also be controlled [4].

Observing each situation, a mathematical model with a system of nonlinear differential equations is illustrated in Sect. Mathematical Modelling. The data collection is done precisely. The data which are not available taken as hypothetically. The formulation of the basic reproduction number gives a significant result of the model that resolves the amount of pollution created from bio-waste. Section Stability includes the stability of equilibrium points of the system. To control pollution, optimal control theory is used and three controls are applied to the system in Sect. Optimal Control. Section Numerical Simulation concludes all results graphically with validated data.

## Mathematical Modelling

Transmission of COVID infection is spreading frequently all-around the world. A mathematical model consists four compartments. The first compartment is infected individuals by the coronavirus $(C_I)$ who are getting hospitalised $(H)$. Due to hospitalisation, all necessary equipment are used, which increase the density of bio-waste $(B_W)$ and day by day increment contaminates the environment called pollution $(P)$.

Dynamics of the model are shown in Fig. 17.1. Parameters that connect each compartment called rates show the flow of the model.

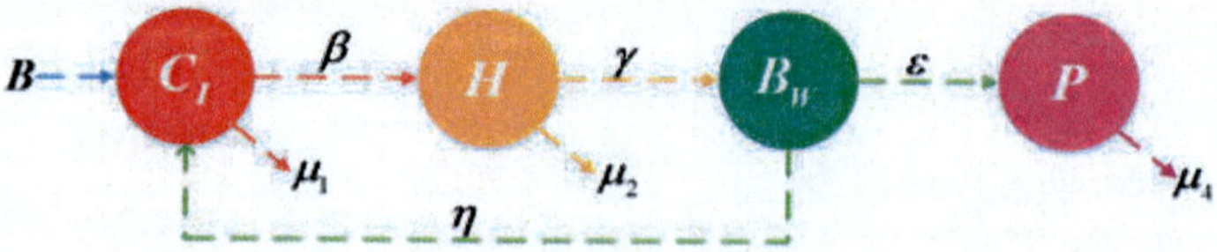


**Fig. 17.1** Transmission of model. *Source* Own

**Description and parametric values:**

Data is taken on July 23, 2020, 07:55 GMT.

$B = 0.22$, growth rate is the ratio of active cases to total cases, i.e. 3,474,060/15,642,265 [5].

$\beta = 0.5760$, the rate at which COVID infected individuals are getting hospitalised. Centers for Disease Control and Prevention (CDC) reported that the overall cumulative COVID-19 hospitalisation rate is 113.6 per 1,00,000 [6]. Hence, from this rate, total hospitalised infected individuals are 88,60,741; hence, the rate at which infected individuals are hospitalised is the ratio of the number of hospitalised cases to total infected cases.

$\mu_1 = 0.6537$, is death rate of coronavirus infected individuals. This is the ratio of death cases to total infected cases, i.e. 6,30,370/15,382,854 [5].

$\mu_2 = 0.0711$, is death rate of hospitalised individuals. This is the ratio of death cases to total hospitalisation cases, i.e. 6,30,370/88,60,741 [6].

$\gamma = 0.75$, is the rate bio-waste created due to hospitalisation. This data is from different channels suggests that world could generate an entire year's worth of medical waste in two or three months because of the impact of COVID-19 [7]. It means we can say that there is approximately 0.75% raise is seen in biomedical waste.

$\varepsilon = 0.6$ is increasing rate of pollution due to produced bio-waste (data is taken from [8]).

$\eta = 0.3$ is the rate at which bio-waste infect the individual.

$\mu_4 = 0.4$ is the escape rate.

Note that $\mu_4$ and $\eta$ are taken hypothetically considering by current situation of world due to novel corona virus.

By using four compartments which are connected with these parameters mentioned in Sect. Mathematical Modelling, the following system of equations can be generated,

$$\begin{aligned}
\frac{dC_I}{dt} &= B - \beta H C_I + \eta B_W - \mu_1 C_I \\
\frac{dH}{dt} &= \beta H C_I - \gamma H - \mu_2 H \\
\frac{dB_W}{dt} &= \gamma H - \varepsilon B_W - \eta B_W \\
\frac{dP}{dt} &= \varepsilon B_W - \mu_4 P
\end{aligned} \tag{17.1}$$

with $C_I > 0$ and $H, B_W, P \geq 0$.

By accumulating and this initial condition gives the feasible region of the model, $\Lambda = \left\{(C_I, H, B_W, P) \in R^4 : C_I + H + B_W + P \leq \frac{B}{\mu_1}\right\}$.

Solving system (17.1), two equilibrium points are evaluated. First is pollution-free and another is endemic equilibrium point having optimum value.

(i) $E_0\left(\frac{B}{\mu_1}, 0, 0, 0\right)$
(ii) $E^*\left(C_I^*, H^*, B_W^*, P^*\right)$

where $C_I^* = \frac{\gamma_1+\mu_2}{\beta}$, $H^* = \frac{(B\beta-\mu_1(\gamma_1+\mu_2))(\varepsilon+\eta)}{\beta(\varepsilon\gamma_1+\mu_2(\varepsilon+\eta))}$, $B_W^* = \frac{\gamma_1(B\beta-\mu_1(\gamma_1+\mu_2))}{\beta(\varepsilon\gamma_1+\mu_2(\varepsilon+\eta))}$ and $P^* = \frac{\varepsilon\gamma_1(B\beta-\mu_1(\gamma_1+\mu_2))}{\beta\mu_4(\varepsilon\gamma_1+\mu_2(\varepsilon+\eta))}$.

The threshold value is calculated by using next-generation matrix method called basic reproduction number ($R_0$) [9].

$$F = \begin{bmatrix} \beta H C_I \\ 0 \\ 0 \\ 0 \end{bmatrix} \text{ and } V = \begin{bmatrix} \gamma H + \mu_2 H \\ -\gamma H + \varepsilon B_W + \eta B_W \\ -\varepsilon B_W + \mu_4 P \\ -B + \beta H C_I - \eta B_W + \mu_1 C_I \end{bmatrix}.$$

Calculating Jacobian matrix of $F$ and $V$, the mathematical expression of $R_0$ is, $\frac{B\beta}{\mu_1(\gamma_1+\mu_2)}$.

$R_0$ is very important mathematical term for the dynamical model. After substitution of the parameter threshold value suggest that, 23.64% pollution is observed.

## Stability

### *Local Stability*

**Theorem 1** The pollution-free equilibrium point $E_0$ is locally asymptotically stable if $B\beta < \mu_1(\gamma_1 + \mu_2)$.

***Proof*** The Jacobian matrix of the system at $E_0$ is derived and eigenvalues of this matrix are:

(i) $-\mu_4$, (ii) $-\mu_1$, (iii) $-(\varepsilon + \eta)$, (iv) $\frac{B\beta-\mu_1(\gamma_1+\mu_2)}{\mu_1}$.

Equilibrium point is stable if each eigenvalue of Jacobian matrix at $E_0$ is negative and here it seems that, fourth eigenvalue is positive. Hence, equilibrium point is locally asymptotically stable if $B\beta < \mu_1(\gamma_1 + \mu_2)$.

**Theorem 2** The optimum issue point $E^*$ is locally asymptotically stable if $C_I\beta < \min\{1, \varepsilon + \eta, \varepsilon + \eta + \mu_1 + \mu_4\} + (\gamma_1 + \mu_2)$.

***Proof*** The characteristic equation of the Jacobian matrix at endemic equilibrium point,

$$\lambda^4 + a_1\lambda^3 + a_2\lambda^2 + a_3\lambda + a_4 = 0$$

where

$$a_1 = -C_I\beta + H\beta + \varepsilon + \eta + \gamma + \mu_1 + \mu_2 + \mu_4$$

$$a_2 = (\mu_1 + \mu_4)(-C_I\beta + \varepsilon + \eta + \gamma + \mu_2) + \mu_1\mu_4 + (\varepsilon + \eta)(-C_I\beta + H\beta + \gamma + \mu_2)$$
$$a_3 = (\varepsilon + \eta)(\mu_1 + \mu_4)(-C_I\beta + \gamma + \mu_2) + \mu_1\mu_4(-C_I\beta + \varepsilon + \eta + \gamma + \mu_2) + H\beta((\varepsilon + \eta)(\mu_1 + \mu_4) + \gamma(\varepsilon + \mu_4) + \mu_2\mu_4)$$
$$a_4 = \mu_4((\varepsilon + \eta)(\mu_1(-C_I\beta + \gamma + \mu_2) + H\beta\mu_2) + H\beta\varepsilon\gamma)$$

Using Routh-Hurwitz (1877) [10] criteria, this optimum issue point is globally asymptotically stable if $C_I\beta < \min\{1, \varepsilon+\eta, \varepsilon+\eta+\mu_1+\mu_4\}+(\gamma_1 + \mu_2)$ condition is satisfied.

Stability analysis advices that if $\beta$ that is the rate at which infected people getting hospitalised is lesser then model will be stable.

## *Global Stability*

**Theorem 3** The pollution-free equilibrium point $E_0$ is globally asymptotically stable whenever $R_0 \leq 1$.

***Proof*** Consider Lyapunov's function, $L_0(t) = A_1C_I(t) + A_2H(t)$.

$\therefore L_0'(t) = A_1(B - \beta HC_I + \eta B_W - \mu_1 C_I) + A_2(\beta HC_I - \gamma H - \mu_2 H)$

$\therefore L_0'(t) \leq \left(A_2(\gamma + \mu_2)\left(\frac{B\beta}{(\gamma+\mu_2)\mu_1} - 1\right) - \frac{A_1B\beta}{\mu_1}\right)H^0$

$\therefore L_0'(t) = A_1C_I'(t) + A_2H'(t)$

$\therefore L_0'(t) \leq \left(A_2(\gamma + \mu_2)(R_0 - 1) - \frac{A_1B\beta}{\mu_1}\right)H^0$

$\therefore L_0'(t) \leq 0$ if $R_0 \leq 1$.

Hence, $E_0$ is globally asymptotically stable if $R_0 \leq 1$.

**Theorem 4** The endemic equilibrium point $E^*$ is globally asymptotically stable if $\frac{C_I}{C_I^*} = \frac{H}{H^*} = \frac{B_W}{B_W^*} = \frac{P}{P^*} < 1$.

***Proof*** Let Lyapunov's function be, $L_1(t) = \varphi\left(\frac{C_I}{C_I^*}\right) + \varphi\left(\frac{H}{H^*}\right) + \varphi\left(\frac{B_W}{B_W^*}\right) + \varphi\left(\frac{P}{P^*}\right)$.

Such that, $\varphi(x) = x - 1 - \ln(x)$, $x > 0$ is an increasing function.

$\therefore L_1'(t) = \left(1 - \frac{C_I^*}{C_I}\right)C_I' + \left(1 - \frac{H^*}{H}\right)H' + \left(1 - \frac{B_W^*}{B_W}\right)B_W' + \left(1 - \frac{P^*}{P}\right)P'$

$$\therefore L_1'(t) = (B - \mu_1 C_I^*)\left(1 - \frac{C_I^*}{C_I}\right) - \mu_2 H^*\left(1 - \frac{H^*}{H}\right) - \mu_4 P^*\left(1 - \frac{P^*}{P}\right) - \beta H^* C_I^*\left(\frac{H^*}{H} - \frac{C_I^*}{C_I}\right) - \eta B_W^*\left(\frac{C_I^*}{C_I} - \frac{B_W^*}{B_W}\right) - \varepsilon B_W^*\left(\frac{P^*}{P} - \frac{B_W^*}{B_W}\right) - \gamma_1 H^*\left(\frac{B_W^*}{B_W} - \frac{H^*}{H}\right)$$

Here, $L_1'(t) \leq 0$ at $E^*$ that is $C_I = C_I^*$, $H = H^*$, $B_W = B_W^*$, $P = P^*$ and $\frac{C_I^*}{C_I} = \frac{H^*}{H} = \frac{B_W^*}{B_W} = \frac{P^*}{P} < 1$. Hence, $E^*$ is globally asymptotically stable.

## Optimal Control

Optimal control helps to optimise objective function over a few periods. Three controls are applied to the system. The first control ($u_1$) is on COVID-19 infected people to optimise hospitalisation. The second control ($u_2$) is treated on hospitalised individuals which decreases bio-waste and the third control ($u_3$) is a treatment applied to the bio-waste material that can optimise pollution. Therefore, the system seems,

$$\begin{aligned}
\frac{dC_I}{dt} &= B - \beta H C_I + \eta B_W - \mu_1 C_I + u_1 H \\
\frac{dH}{dt} &= \beta H C_I - \gamma H - \mu_2 H - u_1 H + u_2 B_W \\
\frac{dB_W}{dt} &= \gamma H - \varepsilon B_W - \eta B_W - u_2 B_W + u_3 P \\
\frac{dP}{dt} &= \varepsilon B_W - \mu_4 P - u_3 P
\end{aligned} \tag{17.2}$$

The objective function is,

$$\begin{aligned}
J(c_i, \Lambda) = \int_0^T & (A_1 C_I^2 + A_2 H^2 + A_3 B_W^2 + A_4 P^2 \\
& + w_1 u_1^2 + w_2 u_2^2 + w_3 u_3^2))\mathrm{d}t
\end{aligned} \tag{17.3}$$

where, $\Lambda$ denotes set of all compartmental variables, $A_1, A_2, A_3, A_4$ denote non-negative weight constants for compartments $C_I, H, B_W, P$ respectively. $w_1$, $w_2$ and $w_3$ are the weight constants for each control $u_i$ where $i = 1, 2, 3$, respectively.

Compute every values of control variables from $t = 0$ to $t = T$ such that,

$$J(u_i(t)) = \min\{J(u_i^*, \Lambda)/(u_i) \in \phi\}, \quad i = 1, 2, 3$$

where $\phi$ is a smooth function on the interval [0, 1].

The optimal control denoted by $u_1, u_2$ and $u_3$ are found by computing all the integrands of Eq. (17.3) using the lower bounds and upper bounds respectively with the results of Fleming and Rishel [11].

For minimising the cost function in Eq. (17.3), using the Pontryagin principle [12] by constructing Langrangian function consisting of state equations and adjoint variables $\lambda_1, \lambda_2, \lambda_3, \lambda_4, \lambda_5$ as follows:

$$\begin{aligned}
L(\Lambda, A_i) = {} & A_1 C_I^2 + A_2 H^2 + A_3 B_W^2 + A_4 P^2 + w_1 u_1^2 + w_2 u_2^2 \\
& + w_3 u_3^2 + \lambda_1 (B - \beta H C_I + \eta B_W - \mu_1 C_I \\
& + u_1 H) + \lambda_2 (\beta H C_I - \gamma H - \mu_2 H - u_1 H + u_2 B_W) \\
& + \lambda_3 (\gamma H - \varepsilon B_W - \eta B_W - u_2 B_W + u_3 P)
\end{aligned}$$

$$+\lambda_4(\varepsilon B_W - \mu_4 P - u_3 P)$$

The adjoint equation variables, $\lambda_i = (\lambda_1, \lambda_2, \lambda_3, \lambda_4)$ for the system (2) is calculated by taking partial derivatives of the Langrangian function with respect to each compartment variable.

$$\begin{aligned}
\overset{\bullet}{\lambda_1} &= -\frac{\partial L}{\partial C_I} = -2A_1C_I + (\lambda_1 - \lambda_2)\beta H + \lambda_1\mu_1 \\
\overset{\bullet}{\lambda_2} &= -\frac{\partial L}{\partial H} = -2A_2H + (\lambda_1 - \lambda_2)\beta C_I + (\lambda_2 - \lambda_3)\gamma + (\lambda_2 - \lambda_1)u_1 + \lambda_2\mu_2 \\
\overset{\bullet}{\lambda_3} &= -\frac{\partial L}{\partial B_W} = -2A_3B_W + (\lambda_3 - \lambda_1)\eta + (\lambda_3 - \lambda_4)\varepsilon + (\lambda_3 - \lambda_2)u_2 \\
\overset{\bullet}{\lambda_4} &= -\frac{\partial L}{\partial P} = -2A_3P + (\lambda_4 - \lambda_3)u_3 + \mu_4\lambda_4
\end{aligned}$$

The necessary conditions for Langrangian function $L$ to be optimal are

$$\begin{aligned}
\overset{\bullet}{u_1} &= -\frac{\partial L}{\partial u_1} = -2w_1u_1 + (\lambda_2 - \lambda_1)H = 0 \\
\overset{\bullet}{u_2} &= -\frac{\partial L}{\partial u_2} = -2w_2u_2 + (\lambda_3 - \lambda_2)B_W = 0 \\
\overset{\bullet}{u_3} &= -\frac{\partial L}{\partial u_3} = -2w_3u_3 + (\lambda_4 - \lambda_3)P = 0
\end{aligned}$$

Hence, one can simplify as,

$$u_1 = \frac{(\lambda_2 - \lambda_1)H}{2w_1},\ u_2 = \frac{(\lambda_3 - \lambda_2)B_W}{2w_2} \text{ and } u_3 = \frac{(\lambda_4 - \lambda_3)P}{2w_3}.$$

Using Pontryagin max–min principle, the optimum value of each control is given by,

$$u_1^* = \max\left(a_1, \min\left(b_1, \frac{(\lambda_2 - \lambda_1)H}{2w_1}\right)\right),\ u_2^* = \max\left(a_2, \min\left(b_2, \frac{(\lambda_3 - \lambda_2)B_W}{2w_2}\right)\right)$$

$$\text{and } u_3^* = \max\left(a_3, \min\left(b_3, \frac{(\lambda_4 - \lambda_3)P}{2w_3}\right)\right)$$

For simulating result, this computation is done in this section analytically.

## Numerical Simulation

In this section, results are simulated numerically wherein the parametric values are considered from Sect. Mathematical Modelling.

Figure 17.2 shows the compartmental behaviour of the model. COVID-19 is spreading day by day. COVID-infected individuals are getting hospitalised, and this interprets that, after approximately one-week bio-waste generated during the pandemic. This collection of bio-waste creates pollution after three and a half weeks if there is not taken any solution to prevent this thing by waste management.

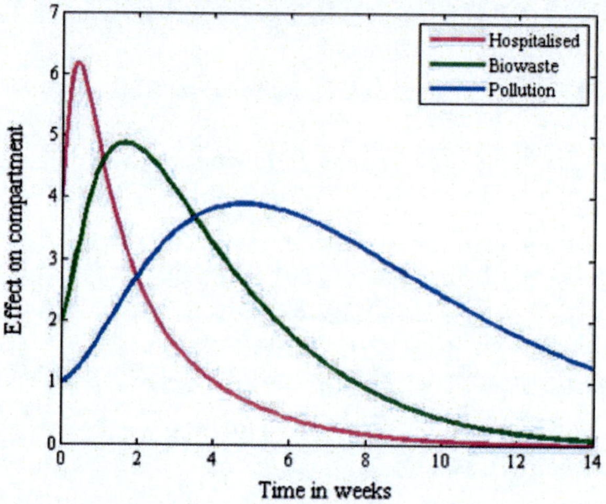


**Fig. 17.2** Transmission in compartments. *Source* Own

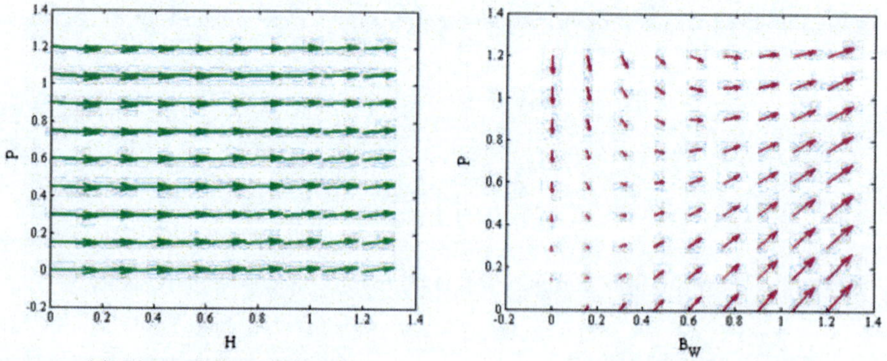


(a) Through hospitalisation (b) Through bio-waste

**Fig. 17.3** Intensity towards pollution. *Source* Own

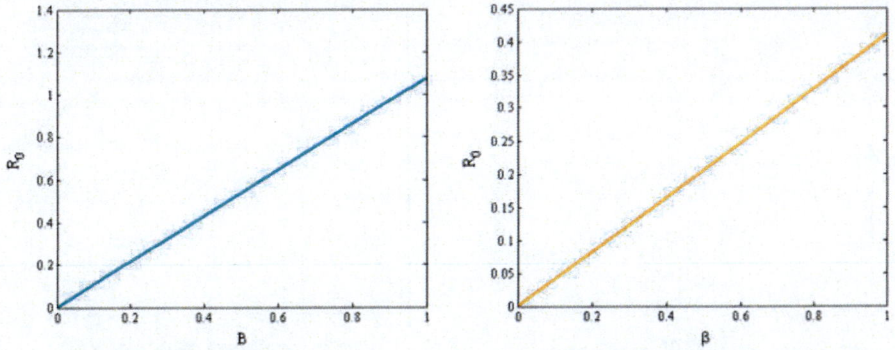


(a) On growth rate $B$ (b) On hospitalisation rate $\beta$

**Fig. 17.4** Effect of $R_0$. *Source* Own

The figure depicts the directional behaviour of bio-waste (Fig. 17.3a) and hospitalised individuals (Fig. 17.3b) towards emerging pollution. As hospitalisation increases, medical staff has to use safety things for protection and hence tons of bio-waste have been produced in each month which increases pollution.

The direction of growth rate (Fig. 17.4a) and hospitalisation rate (Fig. 17.4b) with the basic reproduction number are plotted. Both figures illustrate that the threshold value increases with an increase in both the parametric values. These parameters are more sensitive to the threshold value and hence the model. Other parameters have negligible effect. Therefore, if both parameters have been undertaken care which means that infected people are getting cured by only home isolation then hospitalisation reduce which causes a decrease in pollution.

Three controls are applied to the system as mentioned in Sect. Optimal Control. One can inspects that if 21.7% hospitalisation (Fig. 17.5a) of COVID-infected people decreases then 15.56% of bio-waste (Fig. 17.5b) reduces from the medical sector which results in a reduction of 35.07% in pollution (Fig. 17.5c). This study advocates that control $u_1$ can be performed if individuals follow basic rules and regulation during this pandemic situation. During hospitalisation control $u_2$ is applicable if some changes have been taken care of COVID-19-infected people. General waste does not contaminate the environmental things hence control $u_3$ is beneficial when bio-waste is analysed and also obeys government guidelines.

Figure 17.6 interprets the area-wise distribution of three compartments; hospitalised individuals, bio-waste and pollution created by bio-waste, where the area of respective compartment shows its intensity. Analysis of intensity distribution of the respective compartment shows that, increasing hospitalised cases leads to an increase in bio-waste which accelerates the intensity of pollution.

Result of Fig. 17.7 illustrates that, after getting hospitalisation of infected by coronavirus produces bio-waste which creates pollution. Here, black colour shows that, the everything becomes chaos and this scenario definitely a part of causing pollution.

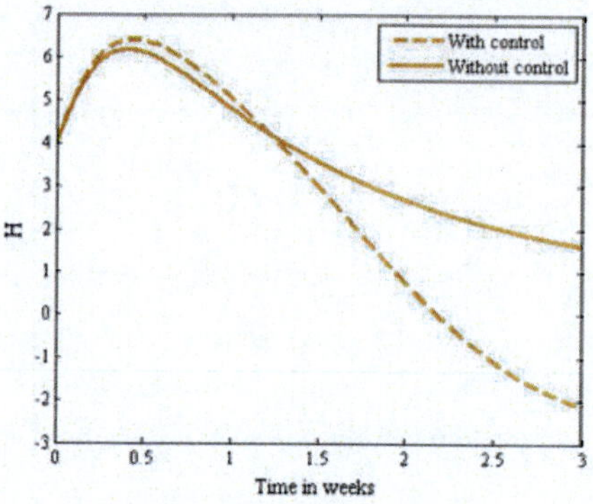


(a) Hospitalised individuals

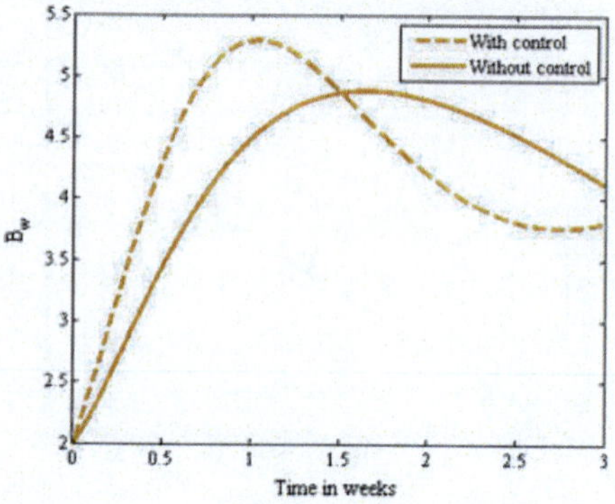


(b) Bio-waste

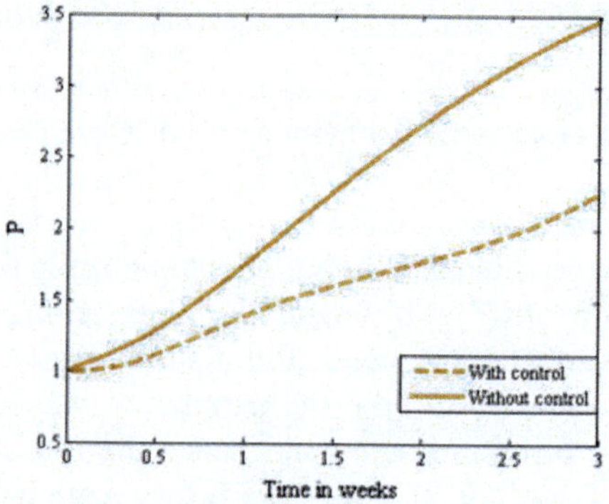


(c) Pollution

**Fig. 17.5** After applying control. *Source* Own

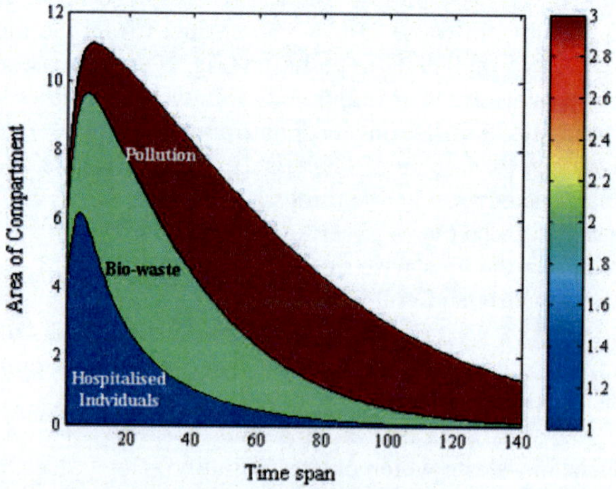


**Fig. 17.6** Area plot. *Source* Own

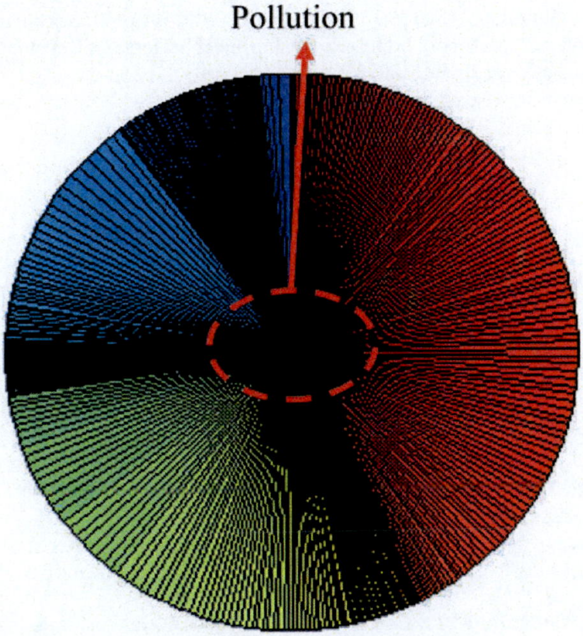


**Fig. 17.7** Intensity pie-chart. *Source* Own

## Conclusion

This study highlights how pollution due to bio-waste is regulated. Due to its adverse effect on human health, many organisations around the world are working in this direction during the current pandemic time. In an account of that, the compartmental model has been developed using some affected parameters and the data for simulation is taken from a recent ongoing COVID-19 outbreak. By solving the system, two equilibrium points are evaluated and both of them are stabilised locally and globally. The calculated threshold value suggests that 23.64% of pollution is created from biomedical waste from hospitalisation during the COVID-19 outbreak. This analysis recommends that if the rate at which infected people getting hospitalised is lesser, then this bio-waste is controlled and that reduces the pollution. Optimal control theory advices that, this bio-waste is started with having an infection coronavirus. Everyone has to get over it. Simulation suggests the reduction of pollution by 35.07%. Heartfelt thanks to all corona warriors. Its our prime duty to follow rules to fight again this contagious coronavirus and can contribute to make planet pollution-free and healthier.

**Acknowledgements** The authors thank reviewers for their constructive comments. All authors are thankful to DST-FIST file # MSI-097 for technical support to the department of Gujarat University. Second author (ENJ) is funded by UGC granted National Fellowship for Other Backward

Classes (NFO-2018-19-OBC-GUJ-71790). Third author (PMP) is funded a scholarship by Education Department, Gujarat State for providing scholarship under ScHeme Of Developing High quality research (student AisheCode: 201901380135).

## References

1. Sohrabi, C., Alsafi, Z., O'Neill, N., Khan, M., Kerwan, A., Al-Jabir, A., et al. (2020). World Health Organization declares global emergency: A review of the 2019 novel coronavirus (COVID-19). *International Journal of Surgery*, 71–76.
2. Mahanwar, P. A., & Bhatnagar, M. P. (2020). Medical Plastics Waste.
3. https://qz.com/india/1824884/india-frames-rules-for-disposing-coronavirus-waste-from-hospitals/.
4. Ranjan, M. R., Tripathi, A., & Sharma, G. (2020). Medical waste generation during COVID-19 (SARS-CoV-2) pandemic and Its management: An Indian perspective. *Asian Journal of Environment & Ecology*, 10–15.
5. https://www.worldometers.info/coronavirus/.
6. https://www.cdc.gov/coronavirus/2019-ncov/covid-data/covidview/index.html.
7. https://ww2.frost.com/frost-perspectives/managing-the-growing-threat-of-covid-19-generated-medical-waste/.
8. Sharma, H. B., Vanapalli, K. R., Cheela, V. S., Ranjan, V. P., Jaglan, A. K., Dubey, B., et al. (2020). Challenges, opportunities, and innovations for effective solid waste management during and post COVID-19 pandemic. *Resources, Conservation and Recycling*, 105052.
9. Diekmann, O., Heesterbeek, J. A. P., & Metz, J. A. (1990). On the definition and the computation of the basic reproduction ratio $R_0$ in models for infectious diseases in heterogeneous populations. *Journal of Mathematical Biology, 28*(4), 365–382.
10. Routh, E. J. (1877). *A treatise on the stability of a given state of motion: Particularly steady motion.* Macmillan and Company.
11. Fleming, W., & Lions, P. L. (Eds.). (2012). Stochastic differential systems, stochastic control theory and applications. In *Proceedings of a Workshop, held at IMA*, June 9–19, 1986 (Vol. 10). Springer Science & Business Media.
12. Pontryagin, L. S., Boltyanskii, V. G., Gamkrelidze, R. V., & Mishchenko, E. F. (1986). *The mathematical theory of optimal process* (pp. 4–5). New York: Gordon and Breach Science Publishers.

# Chapter 18
# Mathematical Modelling of COVID-19 in Pregnant Women and Newly Borns

**Navneet Kumar Lamba, Shrikant D. Warbhe, and Kishor C. Deshmukh**

**Abstract** Enlightened by the Coronavirus, the present paper deals with a mathematical model of COVID-19 to investigate the impact of S-I-R-M model on the pregnant women and the newly borns due to the influence of availability of suitable conditions. The rates of infection, rate of recovery, rate of mortality for pregnant women before and after delivery and for newly born babies due to the transmission rate have been discussed for the present observed data. The numerical illustrations have been carried out for the parameters, functions and represented graphically by MATHEMATICA Software. Moreover some comparisons have been shown in the figure to estimate the impact of susceptible conditions and represent the particular cases of S-I-R-M model.

**Keywords** S-I-R-M model · COVID-19 · Pregnant women · New born babies

**Mathematics Subject Classification** 00A71 · 93A30 · 03C99

## Introduction

The WHO announce pandemic situation worldwide due to the outspread of deadly coronavirus which is originated from the Wuhan city (China) in Dec. 2019. Further this deadly virus spread rapidly with a highest rate more than expectation in a short time period across globally. The virus is medically referred to as SARS-CoV-2, while the associated disease is named by the WHO as Covid-19.

N. K. Lamba (✉)
Department of Mathematics, S.L.P.M. Mandhal, R.T.M. Nagpur University, Maharashtra, India
e-mail: navneetkumarlamba@gmail.com

S. D. Warbhe
Department of Mathematics, Laxminarayan Institute of Technology, R.T.M. Nagpur University, Maharashtra, India
e-mail: sdwarbhe@rediffmail.com

K. C. Deshmukh
Department of Mathematics, R.T.M. Nagpur University, Nagpur, India
e-mail: kcdeshmukh2000@rediffmail.com



N. H. Shah and M. Mittal (eds.), *Mathematical Analysis for Transmission of COVID-19*, Mathematical Engineering, https://doi.org/10.1007/978-981-33-6264-2_18

In pandemic situations, pregnancy becomes more stressful and the anxiety level increases in exponential rate. According to senior Gynaecologist-obstetrician of All India Institute of Medical Sciences (AIIMS), New Delhi, there is a possibility that the number of infected pregnant women may increases [1]. The significant changes occur in the body during pregnancy including the immune system and to avoid infection catching it is very much important to take extra care and precautions. Also because of transmission rate, there is an increase of infection rate, it became dangerous for those women who are ready to deliver the new born baby.

Deshmukh et al. [2] prepared S-I-R-M model for the old and young age groups and discuss the rate of infection, recovery, mortality and susceptibles in the present scenario. Deshmukh et al. [3] presented a novel mathematical model of COVID-19 to investigate the impact of S-I-R-M model due to the influence of prevention factor and suitability conditions. Also they have been investigated and discussed the numerical calculation for infectious rate, recovery rate and mortality rate and displayed graphically. Ndairou et al. [4] proposed a compartmental mathematical model deals with the spread of the COVID-19 by focus on the transmissibility of super-spreaders individuals and computed the basic reproduction number. The suitability of the model had shown numerical for the outbreak occurred in Wuhan, China. Ivorra et al. [5] studied a new O-SEIHRD Mathematical modeling of the spread of the Coronavirus disease 2019 by taking into account the undetected infections. The above study done by taking on the particular case of China, reported data was collected and identified, which can be helpful for estimating the spread of COVID-19 in other countries. Also proposed model is able in estimating numbers of beds needed in hospitals. Li [6] studied Mathematical modeling related to Epidemic Prediction of COVID-19 and discussed its significance with prevention and control measures. The dynamic models of the six chambers was established, which based on the transmission mechanism of COVID-19 in the population. Also the time series models with different mathematical formulas established according to the variation law of the original data. Bhola et al. [7] discussed a predictive mathematical model on Corona epidemic in India context. The above model gives some idea about the fate of the virus and helpful in understanding and studying its future projections. The provided data in this model may be useful for the health care and other agencies to make suitable arrangements to fight the pandemic. Cakir and Savas [8] investigated the course of pandemic by mathematical modelling based on time dependent spreading of Novel 2019 coronavirus SARS-CoV-2(COVID-19). Here spreading rate of contagious disease assumed proportional to multiplication of those numbers who have caught the disease and those who have not.

Mbae [9] studied COVID-19 model for Kenya. In this model, author discussed about COVID-19 cases in regards of infection, recovery and mortality especially in Kenya and focused that each department and sector work together which helps fight the virus. Santacroce [10] pointed out the current state of the COVID-19 pandemics in Italy and presented an overview of Italian health services. Also they discussed how Italy facing crisis after two months from the first internal reported cases as well study concern about the non-sanitary issues caused its outbreak. Shaikh et al. [11] constructed a mathematical model of COVID-19 using fractional derivative based on Bats-Hosts-Reservoir-People transmission outbreak in India. Iterative Laplace transform method utilized for numerical computations and estimation of effective preventive measure, future prediction outbreaks and control strategies are analyzed. Thevarajan et al. [12] through their study presented the symptoms related to COVID-19 which majorly include high fever, cough and difficulties in inhalation. Riou and Althaus [13] constructed a mathematical model in which they shown that how people may get infected while breathing in an infected environment or touching any infected surface or by coming in contact of an infected person. Researcher like Sahin et al. [14], Cheng and Shan [15] involved the risk factor in their study as people are unaware of reason behind the spreading the virus.

In USA, the first known death of an infant was noticed when a positive-tested Covid-19 baby has died due to Coronavirus [16]. There was no coronavirus spread observed from breast feeding, amniotic fluid and cord blood in infected Covid-19 pregnant women at Wuhan City (China) [17]. As per the report published in the month of April 2020, there were not so significant symptoms noticed in 43 Covid-19 infected pregnant women in New York City [17]. In, Mumbai City of India, contaminated mothers due to the Novel Coronavirus gives birth to more than hundred of sound infants [18]. In April, just three out of the 115 babies born from contaminated moms who results positive for Covid-19 at L.T.M.G. Mumbai (Lokmanya Tilak Municipal General Hospital). Other two infected pregnant ladies passed at the hospital including one who died before delivery. In India, about 24,000 contaminated cases revealed in May 2020 and more than 840 deaths so far at the epicenter of Covid-19. As per the research letter by JAMA published in the period of May, it is investigated that [17]; Out of 782 pregnant ladies, 1.5 per cent had previously tested positive for SARS-Cov-2. In the remaining of the ladies 3.9 per cent tested positive and left of them have no symptoms. Working of immune system found less aggressive during pregnancy due to it mother become more susceptible to infection. At the end of last three months, pregnancy due to decrease in size of chest cavity, lungs left with less space to work which makes breathing difficult. It could also make COVID-19 more severe.

In present situations, entire world stand together to battle against the Novel Coronavirus, this situation is very critical for those who are in planning to welcome a newborn at their homes. Due to unavailability of proper literature, the preparation of mathematical model becomes more difficult; therefore, in this modelling, a hypothetical data is considered to investigate the impact of COVID-19 on pregnant women.

Hence, in this article, we prepare a mathematical model S-I-R-M to compare the study of coronavirus in pregnant women before and after delivery by using the influence of incubation, social distancing and medicines which contain vitamins to strength the immunity. Also, discuss the recovery, infection and mortality rate of the newly born babies. This model is prepared physically in nature and illustrated numerically by Mathematica Software.

## Proposed Mathematical Model Formulation in the Case of Pregnant Women

We separate the entire pregnant women population $N$(assumed constant) into four categories: the susceptible (denoted by $S$), the infected (denoted by $I$), and the recovered (denoted by $R$) and mortality (denoted by M), i.e., $S(t) + I(t) + R(t) + M(t) = N$.

Every infected person has some probability of passing the infection to other susceptible women. The probability of passing infection varies from person to person is directly depend upon the rate of transmission of virus. It is well known that with increase in $I$ the rate of new infection increases constantly with respect of time (measured in weeks).

To investigate the physical impact of COVID-19 modelling, in the case of pregnant women, the proposed model is split-up into two major portions: in first portion, we determine the infection rate, recovery rate, mortality rate in both before delivery and after delivery cases. While in second portion, we studied the rate of infection, recovery, mortality in newly born baby's in correlation (infected women to newly borns) with pregnant women.

Formulate the following model to describe the transmission dynamics of the COVID-19 epidemic in pregnant women before delivery:

$$\frac{\mathrm{d}I_{\mathrm{bd}}}{\mathrm{d}t} = cS_{\mathrm{bd}}I - r_{\mathrm{bd}}I_{\mathrm{bd}} - m_{\mathrm{bd}}I_{\mathrm{bd}} \tag{18.1}$$

$$\frac{\mathrm{d}S_{\mathrm{bd}}}{\mathrm{d}t} = -cS_{\mathrm{bd}}I \tag{18.2}$$

$$\frac{\mathrm{d}R_{\mathrm{bd}}}{\mathrm{d}t} = r_{\mathrm{bd}}I_{\mathrm{bd}} \tag{18.3}$$

$$\frac{\mathrm{d}M_{\mathrm{bd}}}{\mathrm{d}t} = m_{\mathrm{bd}}I_{\mathrm{bd}} \tag{18.4}$$

Here, pregnant population before delivery $N_{b\,d}$, can be further classified into major four broad classes as:

| | |
|---|---|
| $I_{\mathrm{bd}}(t)$ | Denotes the number of infectives pregnant women before delivery |
| $S_{\mathrm{bd}}(t)$ | Denotes the number of susceptible pregnant women before delivery |
| $R_{\mathrm{bd}}(t)$ | Denotes the number recovered pregnant women before delivery |
| $M_{\mathrm{bd}}(t)$ | Denotes the number of mortality in pregnant women before delivery |
| $r_{\mathrm{bd}}$ | Rate of recovery in pregnant women before delivery |
| $m_{\mathrm{bd}}$ | Rate of mortality in pregnant women before delivery |
| $c$ | Denotes the transmission rate. |

Clearly,

$$I_{bd}(t) + S_{bd}(t) + R_{bd}(t) + M_{bd}(t) = N_{bd} \tag{18.5}$$

Next, we describe the transmission dynamics of the COVID-19 epidemic in pregnant women after delivery:

$$\frac{dI_{ad}}{dt} = cS_{ad}I - r_{ad}I_{ad} - m_{ad}I_{ad} \tag{18.6}$$

$$\frac{dS_{ad}}{dt} = -cS_{ad}I \tag{18.7}$$

$$\frac{dR_{ad}}{dt} = r_{ad}I_{ad} \tag{18.8}$$

$$\frac{dM_{ad}}{dt} = m_{ad}I_{ad} \tag{18.9}$$

Here, pregnant women population after delivery $N_{ad}$ is defined as $N_{ad} = N_{bd} - M_{bd}$, further it can be further classified into major four broad classes as:

| | |
|---|---|
| $I_{ad}(t)$ | the number of infectives pregnant women after delivery |
| $S_{ad}(t)$ | the number of susceptible pregnant women after delivery |
| $R_{ad}(t)$ | the number recovered pregnant women after delivery |
| $M_{ad}(t)$ | the number of mortality in pregnant women after delivery |
| $r_{ad}$ | Rate of recovery in pregnant women after delivery |
| $m_{ad}$ | Rate of mortality in pregnant women after delivery. |

Clearly,

$$I_{ad}(t) + S_{ad}(t) + R_{ad}(t) + M_{ad}(t) = N_{bd} - M_{bd} \tag{18.10}$$

Further, we describe the transmission dynamics of the COVID-19 epidemic in newly baby birth given by pregnant women after delivery:

$$\frac{dI_{bb}}{dt} = cS_{bb}I - r_{bb}I_{bb} - m_{bb}I_{bb} \tag{18.11}$$

$$\frac{dS_{bb}}{dt} = -cS_{bb}I \tag{18.12}$$

$$\frac{dR_{bb}}{dt} = r_{bb}I_{bb} \tag{18.13}$$

$$\frac{dM_{bb}}{dt} = m_{bb}I_{bb} \tag{18.14}$$

Here, newly baby birth population is denoted as $N_{\text{bb}}$, which is further can be further classified into major four broad classes as:

| | |
|---|---|
| $I_{\text{bb}}(t)$ | the number of infectives baby |
| $S_{\text{bb}}(t)$ | the number of susceptible baby |
| $R_{\text{bb}}(t)$ | the number recovered baby |
| $M_{\text{bb}}(t)$ | the number of mortality in baby |
| $r_{bb}$ | Rate of recovery in baby |
| $m_{bb}$ | Rate of mortality in baby. |

Clearly,

$$I_{\text{bb}}(t) + S_{\text{bb}}(t) + R_{\text{bb}}(t) + M_{\text{bb}}(t) = N_{\text{bb}} \tag{18.15}$$

## Numerical Illustration

As a special case, we prepared the mathematical model S-I-R-M for the pregnant women for different parameters, functions, assume data and illustrated with Mathematica Software.

**Assumed data:**

**Before delivery**: $N = 100$, $S_{\text{bd}} = 85$, $I_{\text{bd}} = 10$,$R_{\text{bd}} = 4$, $M_{\text{bd}} = 1$, $c_{\text{bd}} = 0.117$, $r_{\text{bd}} = 0.40$, $m_{\text{bd}} = 0.1$
**After delivery**: $N = 99$, $S_{\text{ad}} = 75$, $I_{\text{ad}} = 16$,$R_{\text{ad}} = 6$, $M_{\text{ad}} = 2$, $c_{\text{ad}} = 0.213$, $r_{\text{ad}} = 0.375$, $m_{\text{ad}} = 0.125$.
**Baby birth**: $N = 99$, $S_{\text{bb}} = 89$, $I_{\text{bb}} = 6$, $R_{\text{bb}} = 4$, $M_{\text{bb}} = 2$, $c_{\text{bb}} = 0.067$, $r_{\text{bb}} = 0.66$, $m_{\text{bb}} = 0.33$.

For the above data choose time $t = 2$ weeks.

## Observation and Discussion

From the graphical representation, the authors hereby presented the entire observations related to pregnant women and their newly born babies due to the influence of Covid-19.

From Figs. 18.1, 18.2, 18.3, 18.4 and 18.5: It is observed that the rate of infection before delivery and after delivery in pregnant women is significantly equal or we can say that there is a minimum discrepancy between infection rates is found. After delivery, the rate of recovery is noted high and mortality is low because of special care of diagnosed women with Covid-19 was monitored closely by keeping her in incubation or isolation but in case of before delivery recovery and mortality is slight less.

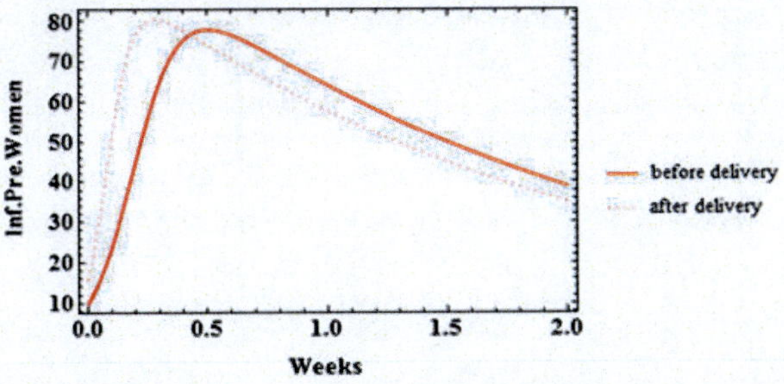


**Fig. 18.1** Variation of infection in pregnant women before and after delivery versus time in weeks. *Source* Own

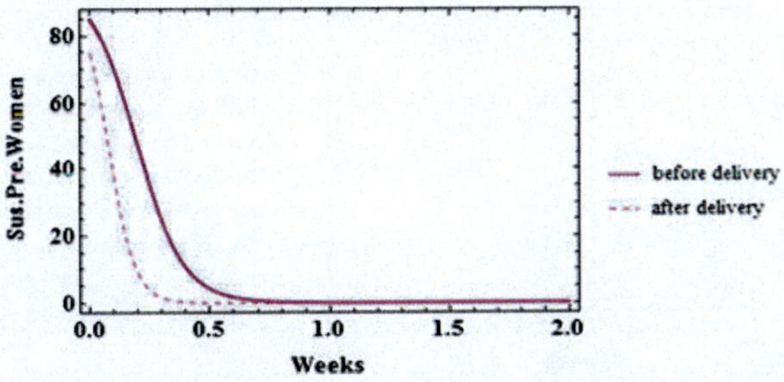


**Fig. 18.2** Variation of suspectibles in pregnant women before and after delivery versus time in weeks. *Source* Own

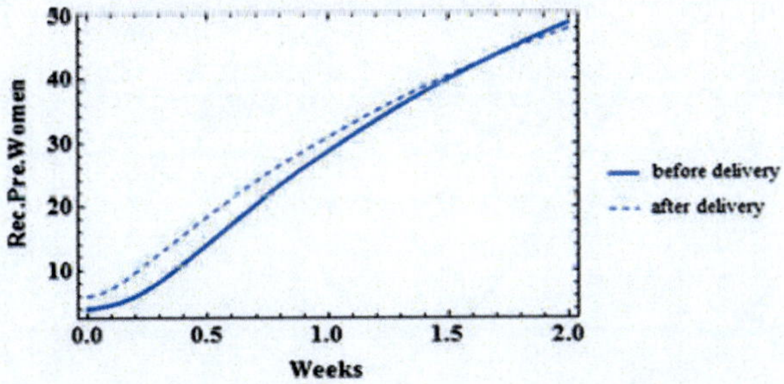


**Fig. 18.3** Variation of recovery in pregnant women before and after delivery versus time in weeks. *Source* Own

From Figs. 18.6, 18.7, 18.8, 18.9 and 18.10: the maximum of new born babies from the infected women with Covid-19 is reported negative. Further the rate of infection in new born is comparatively less due to that the corresponding rate of recovery and mortality changes accordingly.

From Figs. 18.11, 18.12, 18.13, 18.14 and 18.15: After delivery, both mother and born baby are separated upto 10 days under the supervision of medical staff to improve the immune system and to reduce the rate of infection transmission. Hence

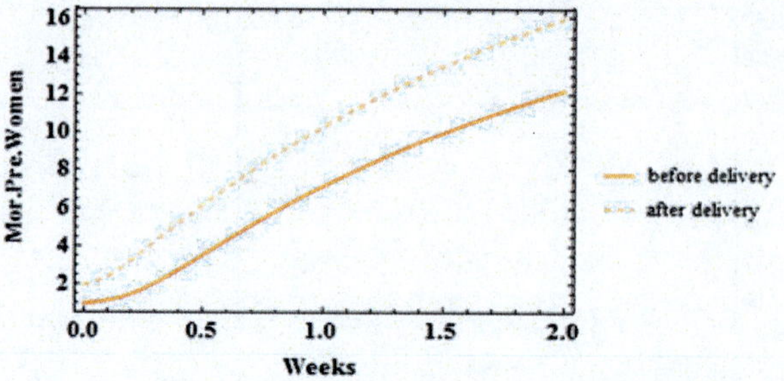


**Fig. 18.4** Variation of mortality in pregnant women before and after delivery versus time in weeks. *Source* Own

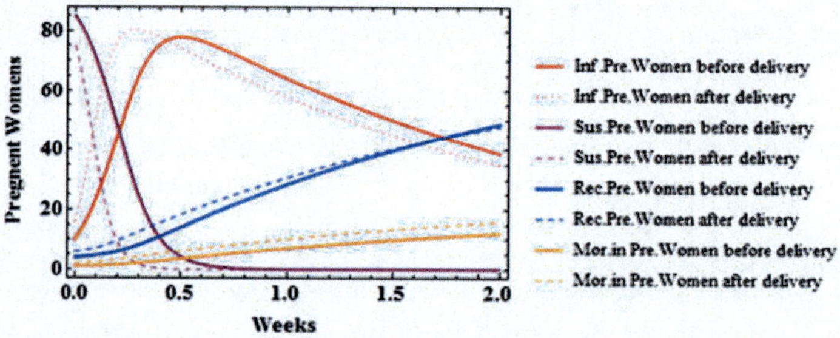


**Fig. 18.5** Variation of infection, recovery and mortality in pregnant women before and after delivery versus time in weeks. *Source* Own

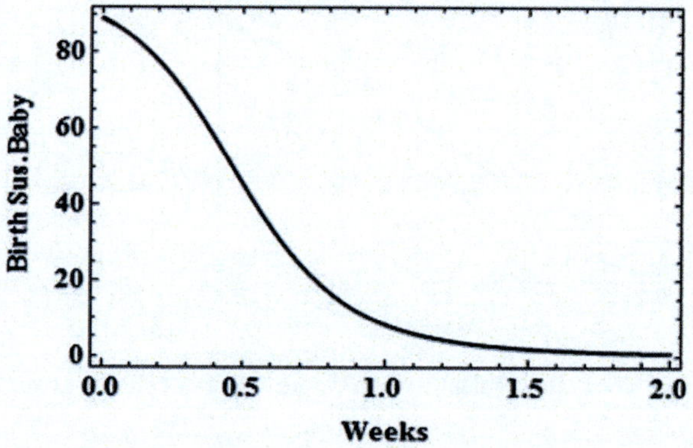


**Fig. 18.6** Variation of suspectibles in newly born babies versus time in weeks. *Source* Own

due to which the rate of recovery in both mother and baby goes on increasing and corresponding rate of mortality is found very less.

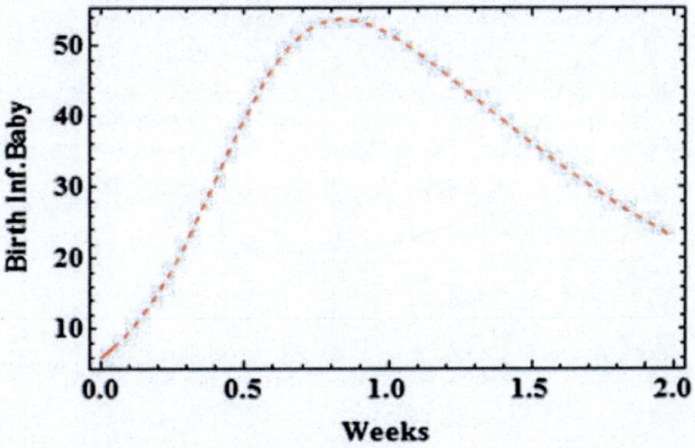


**Fig. 18.7** Variation of infection in newly born babies versus time in weeks. *Source* Own

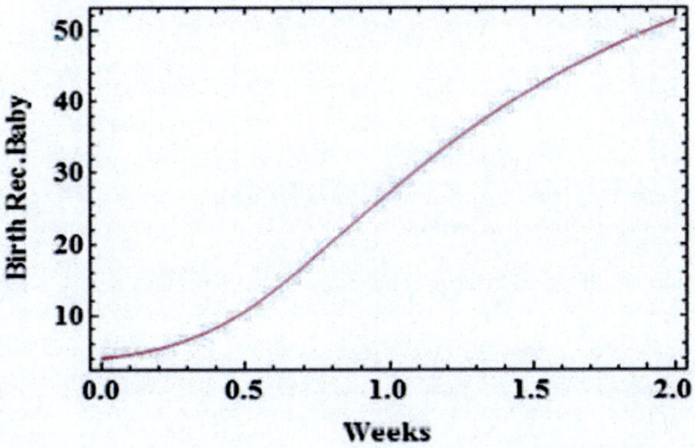


**Fig. 18.8** Variation of recovery in newly born babies versus time in weeks. *Source* Own

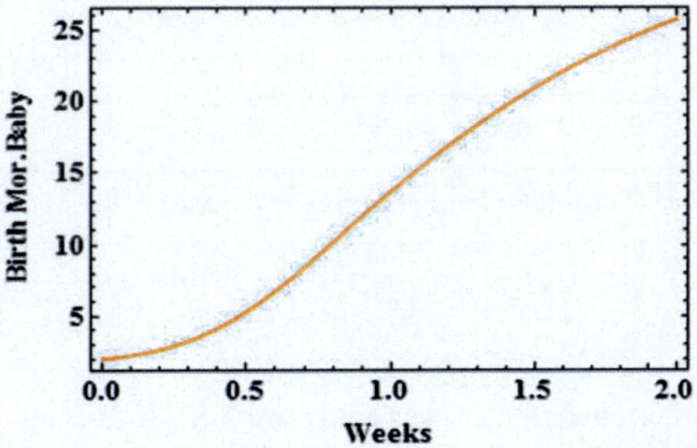


**Fig. 18.9** Variation of mortality in newly born babies versus time in weeks. *Source* Own

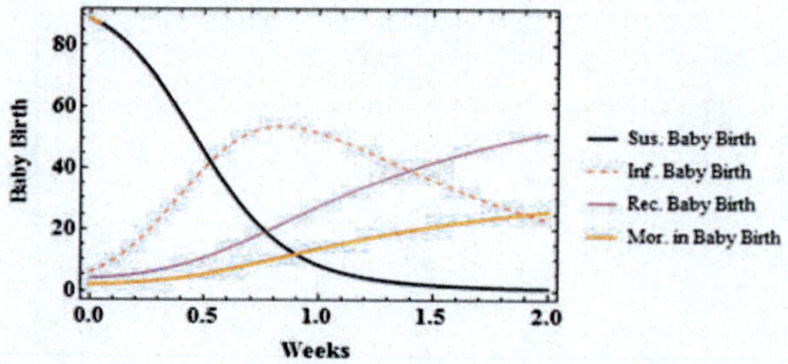


**Fig. 18.10** Variation of infection, recovery and mortality in in newly born babies versus time in weeks. *Source* Own

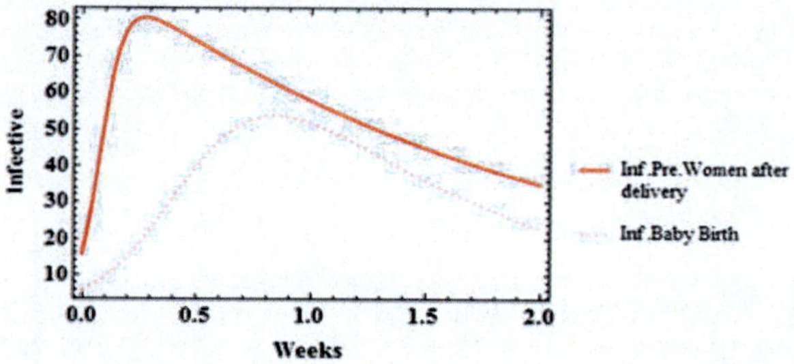


**Fig. 18.11** Correlation of infection after pregnanency in women with newly born versus time in weeks. *Source* Own

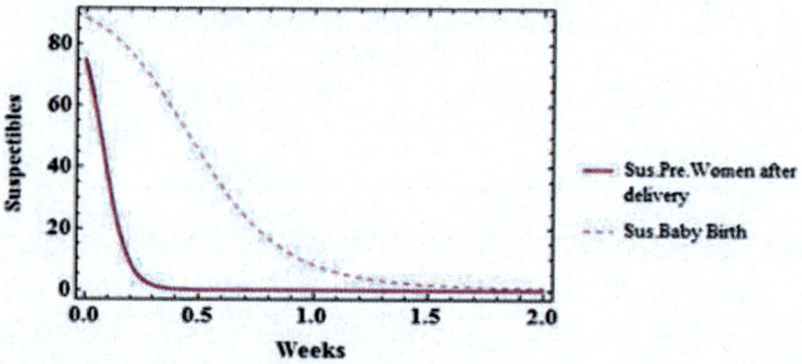


**Fig. 18.12** Correlation of suspectibles after pregnanency in women with newly born versus time in weeks. *Source* Own

## Conclusions

In this research article, the mathematical equations of S-I-R-M model for pregnant women and the born babies have been constructed. The infection rates in the pregnant women before and after delivery have been increased due to transmission rate and the carrying capacity. The infected women are recovered before and after delivery due to the isolation and the quarantine, whereas the newly infected babies are recovered due to the incubation. The mortality in the pregnant women before and after delivery

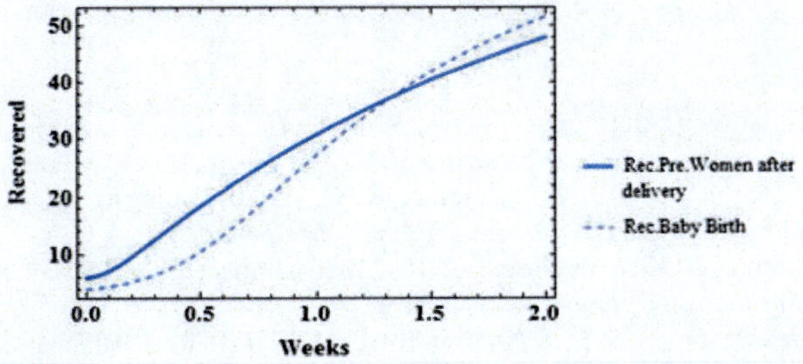


**Fig. 18.13** Correlation of recovery after pregnanency in women with newly born versus time in weeks. *Source* Own

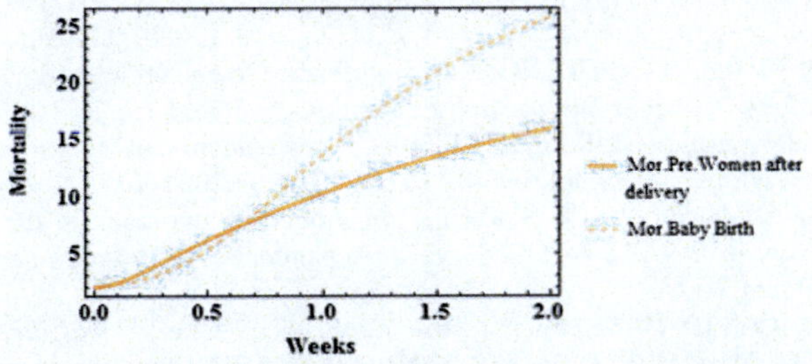


**Fig. 18.14** Correlation of mortality after pregnanency in women with newly born versus time in weeks. *Source* Own

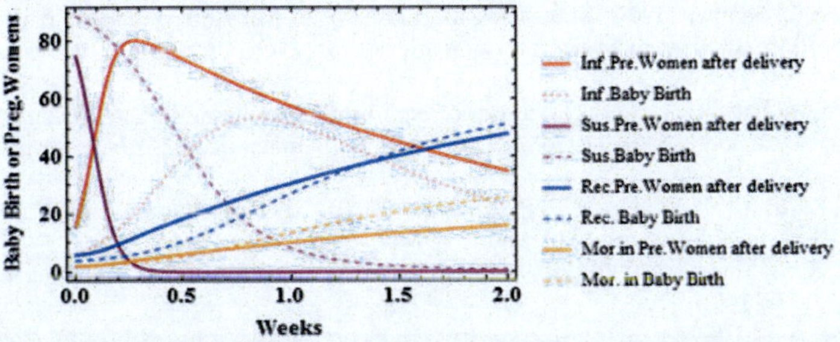


**Fig. 18.15** Correlation of infection, recovery and mortality after pregnanency in women with newly born. *Source* Own

is very slow. Also, the rate of mortality is negligible as compared to the infection and recovery rate which are shown in the figures. This model is presented in hypothetical manners for suitable functions and parameters and it may be feasible for the real working problems.

**Conflicts of Interest** The authors declare that we have no potential conflicts of interest with respect to the research, authorship and publication of this article.

## References


1. https://swachhindia.ndtv.com/pregnancy-in-the-times-of-a-pandemic-what-impact-can-covid-19-have-on-moms-to-be-and-babies-44304/.
2. Lamba, N. K., Warbhe, S. D., & Deshmukh, K. C. (2020). Study of COVID-19 in India: A-mathematical model. *Journal of Interdisciplinary Mathematics.*
3. Warbhe, S. D., Lamba, N. K., & Deshmukh, K. C. (2020). Impact of COVID-19: A mathematical model. *Journal of Interdisciplinary Mathematics.*
4. Ndairou, F., Area, I., Nieto, J. J., & Delfim Torres, F.M. (2020). Mathematical modeling of COVID-19 transmission dynamics with a case study of Wuhan. *Chaos Solitons Fractals, 135*, 109846. https://doi.org/10.1016/j.chaos.2020.109846.
5. Ivorra, B., Ferrandez, M. R., Vela-Perez, M., & Ramos, A. M. (2020). Mathematical modeling of the spread of the coronavirus disease 2019 (COVID-19) taking into account the undetected infections. The case of China (pp.1–29). https://doi.org/10.13140/RG.2.2.21543.29604.
6. Li, Y., Wang, B., Peng, R., Zhou, C., Zhan, Y., Liu, Z., et al. (2020). Mathematical modeling and epidemic prediction of COVID-19 and its significance to epidemic prevention and control measures. *Annals of Infectious Disease and Epidemiology, 5*(1), 1052.
7. Bhola, J., Venkateswaran, V., & Koul, M. (2020). Corona epidemic in India context: Predictive mathematical modelling. https://doi.org/10.1101/2020.04.03.20047175.
8. Cakir, Z., & Savas, H. B. (2020). Mathematical modelling approach in the spread of the Novel 2019 Coronavirus SARS-CoV-2(COVID-19) pandemic. *Electronic Journal of General Medicine, 17*(4), em205.
9. Mbae, N. (2020). COVID-19 in Kenya. *Electronic Journal of General Medicine, 17*(6), em231.
10. Santacroce, L., Charitos, I. A., & Prete, R. D. (2020). COVID-19 in Italy: An overview from the first case to date. *Electronic Journal of General Medicine, 17*(6), em235.
11. Shaikh, A. S., Shaikh, I. N., & Nisar, K. S. (2020). A Mathematical Model of COVID-19 using fractional derivative: Outbreak in India with dynamics of transmission and control. https://doi.org/10.20944/preprints202004.0140.v1.
12. Thevarajan, I., Nguyen, T. H., Koutsakos, M., et al. (2020). Breadth of concomitant immune responses prior to patient recovery: a case report of non-severe COVID-19. *Nature Medicine*, 1–3.
13. Riou, J., & Althaus, C. L. (2020). Pattern of early human-to-human transmission of Wuhan 2019 novel coronavirus (2019-nCoV), December 2019 to January 2020. *Eurosurveillance, 25*(4).
14. Sahin, A. R., Erdogan, A., Agaoglu, P. M., et al. (2020). 2019 novel coronavirus (COVID-19) outbreak: A review of the current literature. *EJMO, 4*(1), 1–7.
15. Cheng, Z. J., & Shan, J. (2020). 2019 Novel coronavirus: Where we are and what we know. *Infection, 48*(2), 1–9.
16. Buncombe, A. (2020, March 28). Coronavirus: Baby dies after testing positive for Covid-19 in US. News in *Independent.co.uk*. https://www.independent.co.uk/news/world/americas/coronavirus-latest-baby-dies-first-infant-death-us-chicago-illinois-a9432096.html.
17. Miller, K. (2020, May 29s). What to know about coronavirus if you're pregnant. *whattoexpect.com*. https://www.whattoexpect.com/news/pregnancy/coronavirus-during-pregnancy.
18. Biswas, S. (2020, May 29). Mumbai: India hospital delivers 100 babies from Covid-19 mums. *BBC News*. https://www.bbc.com/news/world-asia-india-52693987.

# Chapter 19
# Sensor and IoT-Based Belt to Detect Distance and Temperature of COVID-19 Suspect

**Rishabh Gautam, Shruti Mishra, Akhilesh Kumar Pandey, and Jitendra Kumar Singh**

**Abstract** Owing to the pandemic issue of the coronavirus disease 2019 (COVID-19), it is imperative to keep up more than 1-m of social distancing and 37.5 °C temperature to stop the transmission of COVID-19 from human to human. Therefore, it is utmost requirement to make the smart belt installed with ultrasonic and LM35 sensors for distance and temperature measurements to reduce the transmission of COVID-19, respectively. The embedded sensors with NodeMCU show that once anything come in the proximity of 1-m near to the smart belt or helmet fixed to human body, it automatically makes an alarm for distance contact as well as temperature of incoming/outgoing body and sends an email to the controller with the help of Blynk application through Internet of things (IoT). These data can be stored in the cloud for the future purpose. However, the distance sensor has detected the movement of a person from 3 cm up to around 240 cm. The LM35 temperature sensor measures the actual temperature of the host body, i.e., 35.4 °C with time. With the help of this research, it is possible to interface a camera module which can detect the suspects. It could be interfaced with global positioning system (GPS) which can give location-wise data and help us to obtain the probability of suspects at a particular region. It is cost effective, i.e., $14/belt which can help to control the transmission of coronavirus from human to human.

**Keywords** COVID-19 · Social distancing · Sensors · Internet of things · Wearable electronics

---

Rishabh Gautam and Shruti Mishra—Both authors have contributed equally.

---

R. Gautam · S. Mishra · A. K. Pandey (✉)
Department of Electronic Engineering, Institute of Engineering and Rural Technology, Uttar Pradesh, 26, Chatham Line, Dharhariya, Prayagraj 211002, India
e-mail: akhileshpandey@iert.ac.in

J. K. Singh
Innovative Durable Building and Infrastructure Research Center, Department of Architectural Engineering, Hanyang University, 1271 Sa 3-dong, Sangrok-gu, Ansan 15588, Republic of Korea

 
N. H. Shah and M. Mittal (eds.), *Mathematical Analysis for Transmission of COVID-19*, Mathematical Engineering, https://doi.org/10.1007/978-981-33-6264-2_19

## Introduction

Nowadays, coronavirus disease 2019 (COVID-19) pandemic has become worldwide sensation and is the biggest problem in 2020. It has not only locked down the speed of human life but also economy of the world. This pandemic is found to be caused by severe acute respiratory syndrome coronavirus 2 (SARS–CoV-2) [1, 2]. The epicenter of this virus was firstly found in Wuhan, Hubei, China, in December 2019. Since it was not only restricted to China but also spread worldwide, it was considered as a pandemic by World Health Organization [WHO] on March 11, 2020 [3, 4]. The origin of the virus was said to be a new member of the betacoronavirus genus, closely related to the bats or possibly pangolins [5, 6]. It proliferates when people come in close contact to each other which approximately more than 1-m radius of a person or it also spreads with the respiratory droplets discharged when infected person coughs or sneezes. But the virus is not regarded as air borne [7]. It can only be transmitted by touching, close contact by small droplets of cough, sneezing and talking of the infected person where he/she can infect the uninfected person [8–10]. The symptoms do not appear immediately after coming in contact with the infected person rather it takes 2–14 days sometimes up to 21 days' incubation periods to get the indication of this disease [11, 12]. Ordinary symptoms which are observed due to COVID-19 are fever, cough, headache, nasal congestion, sore throat, etc. The emergency symptoms of COVID-19 such as shortening of breath, persistent pain in chest, bluish face are required immediate medical attention. There is no proper medicine or vaccine made for COVID-19. But, there are some preventive measures which have been recommended by the Centers for Disease Control and Prevention (CDC) to control its transmission. These are often washing hands at regular interval of time or after touching anything, covering one's mouth with mask while going outside, covering one's mouth while coughing, by keeping an eye on suspected people and by keeping social distancing [13]. Efforts which are being taken worldwide to prevent the spread of virus are restrictions on travel, quarantines, curfews, hazard controls, event postpone, closure of schools and universities, border closure, or restrictions of passengers' travel, screening at airports, etc. Till September 3, 2020, COVID-19 has infected 26,192,041 patients out of which 18,454,873 are recovered and 867,542 are dead worldwide [14]. Until this date, 6,869,626 are active patients. However, among all the suffered countries, India is at number 3 in total cases of COVID-19. In India, first corona patient was found on 30th January in Kerala's Thrissur district. The patient was student in China and had returned home on an eve [15].

Nowadays, technologies have grown so far. Everything is growing in the shade of Internet and technology. Internet is not only connecting the people globally but also connecting the things with peoples, places and emergence of Internet of things (IoT). Internet of things or Internet of objects is wireless connection which connects electrical or electronic components through various things on a same platform through Internet [16].

These days IoT and sensors are covering almost every area such as home automation to control home appliances, security systems, health monitoring systems, waste management systems, vast medical applications, education system and in many other applications. IoT can also be viewed as technology which is used to automate the things. As we are moving in smart world, we need a smart city and smart system making IoT a more useful platform [17]. As we are moving and focusing on digital world, IoT has substantial importance. Through IoT, we can connect oneself digitally with their things and person. It is helping us to observe all our daily things and works digitally. IoT is helping us in protecting our house from any place.

By considering and keeping in mind of these things, it is our prudent thought to assemble a belt embedded with temperature and distance sensors to detect the temperature and distance of a person in COVID-19 pandemic. These embedded sensors can detect the distance and temperature of people via IoT which can help the government in detection of number of patients. The people coming near to the infected person, our assembled belt with sensor can send an email to the controller with the help of Blynk application through Internet of things (IoT). These data can be stored in the cloud for the future purpose. Thus, this belt can control the transmission of COVID-19.

## Experimental Details

### *Node MCU*

NodeMCU 1.0 (ESP-12E ESP8266 Wi-Fi Module) was a 32-bit microcontroller (Fig. 19.1) with 32 K RAM, 4 Mb flash memory, 16 GPIO pins. Out of 16 pins, only one pin was dedicated for analog signal, whereas rest was digital purpose [18, 19]. Arduino IDE software was used for programming to upload the code via a micro-USB cable. This microcontroller was mainly designed for IoT purpose as NodeMCU interfaced with the board which enables us to connect with the Wi-Fi network. NodeMCU requires 5 V power supply.

### *Programming Software*

NodeMCU uses Arduino 1.8.10 IDE software. The two main constituents of coding part are setup and loop where our program will be coded. Furthermore, for recursive task, we can code the program and drop under the loop. Our programing language was very much similar to the C/C$^{++}$ language [20] which were used in Arduino IDE.

**Fig. 19.1** Image of NodeMCU. *Source* Own

## *Buzzer*

This device takes input in form of electrical signal and provides the output as an audio signal. We can calibrate the loudness as well as frequency with the help of NodeMCU microcontroller.

## *Smart Phone Application Software*

Blynk (Robo India) is a smart phone software used to control many microcontroller boards like Arduino, NodeMCU, Raspberry Pi, etc. Blynk provides many useful widgets for the project. Blynk is a highly secured application with secured network as it requires certain unique token code for the establishment of the connection.

## *Ultrasonic Sensor*

The ultrasonic sensor (HC-SR04, Robokits India, Gandhi Nagar, Gujrat, India) was used for distance measurement (Fig. 19.2). It contains four pins; among them, two

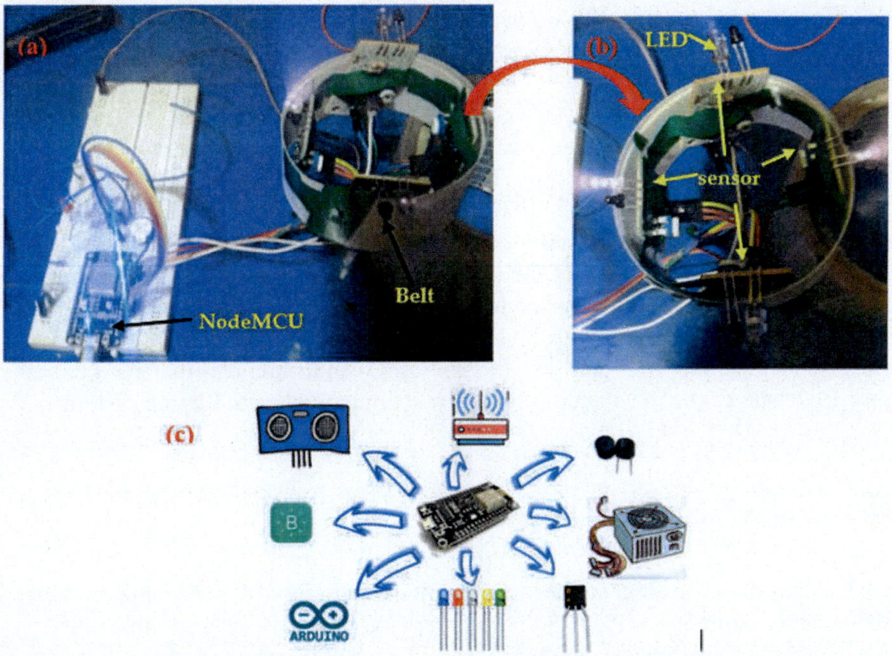


**Fig. 19.2** Project setup **a** all overview, **b** installed belt and **c** raw data of different components. *Source* Own

were used for power supply and the remaining for trigger and echo. It works on 5 V DC power supply, 15 mA and 40 Hz. The maximum sensing capacity of this sensor is 4-m (400 cm) and minimum 2 cm. The dimension of the sensor was 45 mm × 20 mm × 15 mm (Fig. 19.2). The working of this sensor is sending and receiving the sound wave. Basically, this sensor generates the pulse for the fixed duration of time with the help of trigger pin and then stops the pulse for receiving which has been reflected from the object. After receiving process, the data can be collected from the echo pin.

## *Temperature Sensor*

We have chosen LM35 temperature sensor (Texas instrument Pvt. Ltd. Bangalore, India) with ±0.5 °C accuracy which consists of three pins where two pins were used for power supply and the remaining one for data output. LM35 can work with 4–20 V of power supply. It can give the reading from −55 to 150 °C. This sensor was installed separately in wrist which requires direct contact with person body/skin. However, in place of LM35, IR sensor can be used to detect the temperature which does not require body contact. Moreover, LM35 was used in present study to know the body temperature of host person.

### *Hardware Implementation*

As our project is based on IoT, firstly we require Internet access and then connect it with NodeMCU. The overview of the system is shown in Fig. 19.2a. Furthermore, four ultrasonic sensor module was connected with digital pins D0, D1, D2 and D3 of the microcontroller (Fig. 19.2b). Now pins D4, D5, D6, D7 and D8 have been connected to the four light-emitting diodes (LED) indicator and one for buzzer. Figure 19.2c shows the raw data of the different components used in the system.

## Results and Discussion

The basic purpose of present studies is to maintain a proper distance and restrict the further transmission of COVID-19. Therefore, consequently, we have assembled a belt with a remarkable identification number which can be worn on the neck, wrist, headband or helmet. Since, we have built the belt by installing temperature and distance sensor which can verify a person's movement and detect the temperature and distance of incoming person, respectively, whether the person is suspect of COVID 19 or not. With the help of the received data, we can maintain the record for how many times imaginary line, i.e., 1-m (100 cm) has been breached as well as breached location in present study. However, the imaginary line can be altered accordingly. With the help of received data, we can prevent the spread of COVID-19.

### *Distance Measurement*

The belt consists of a microcontroller NodeMCU with four ultrasonic sensors (left, right, front and back) placed at the peripheral of the belt as shown above in Fig. 19.2 where all the major processing commands are taking place (Table 19.1). These ultrasonic sensors are at equidistance from the axis of the belt and also placed at 90° with adjacent. In this command "0" and "1" implies for beyond and within 100 cm distance, respectively (Table 19.1). As soon as someone tries to breach out the imaginary line, i.e., 1-m, a LED was used to state the direction of the object/person which alarms us with the help of implanted buzzer as "line has been breached" (Fig. 19.3a). Furthermore, it also gives the notification on smart phone device and alerts the respective area via email with the help of IoT (Fig. 19.3b) for further enquiry. The message notification can be seen on mobile (Fig. 19.3a) as well as in the email (Fig. 19.3b) to the selected persons. This data can be stored in the hard drive or cloud and shared

**Table 19.1** Command of the ultrasonic sensor

| Front ultrasonic sensor | Left ultrasonic sensor | Back ultrasonic sensor | Right ultrasonic sensor | Front LED indicator | Left LED indicator | Back LED indicator | Right LED indicator | Buzzer |
|---|---|---|---|---|---|---|---|---|
| 0 | 0 | 0 | 0 | OFF | OFF | OFF | OFF | OFF |
| 0 | 0 | 0 | 1 | OFF | OFF | OFF | ON | ON |
| 0 | 0 | 1 | 0 | OFF | OFF | ON | OFF | ON |
| 0 | 0 | 1 | 1 | OFF | OFF | ON | ON | ON |
| 0 | 1 | 0 | 0 | OFF | ON | OFF | OFF | ON |
| 0 | 1 | 0 | 1 | OFF | ON | OFF | ON | ON |
| 0 | 1 | 1 | 0 | OFF | ON | ON | OFF | ON |
| 0 | 1 | 1 | 1 | OFF | ON | ON | ON | ON |
| 1 | 0 | 0 | 0 | ON | OFF | OFF | OFF | ON |
| 1 | 0 | 0 | 1 | ON | OFF | OFF | ON | ON |
| 1 | 0 | 1 | 0 | ON | OFF | ON | OFF | ON |
| 1 | 0 | 1 | 1 | ON | OFF | ON | ON | ON |
| 1 | 1 | 0 | 0 | ON | ON | OFF | OFF | ON |
| 1 | 1 | 0 | 1 | ON | ON | OFF | ON | ON |
| 1 | 1 | 1 | 0 | ON | ON | ON | OFF | ON |
| 1 | 1 | 1 | 1 | ON | ON | ON | ON | ON |

*Source* Own

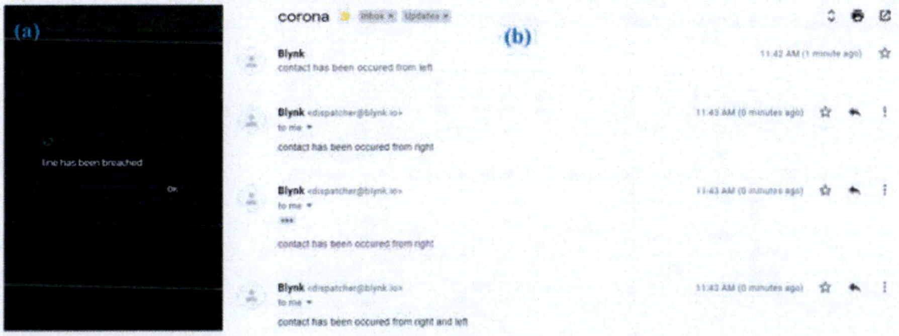


**Fig. 19.3** Notification recived on **a** mobile, i.e., line has been breached and **b** email using IoT. *Source* Own

to the government for record. With the help of this data, government can find the details of infected persons.

The results of ultrasonic sensors for distance measurement are shown in Figs. 19.4 and 19.5. Three consecutive data were collected and the average value is shown in results with ±5% error. The installation of four equidistance ultrasonic sensors at 90° each has been made. This sensor coverage angle was 90° at periphery. The sensors

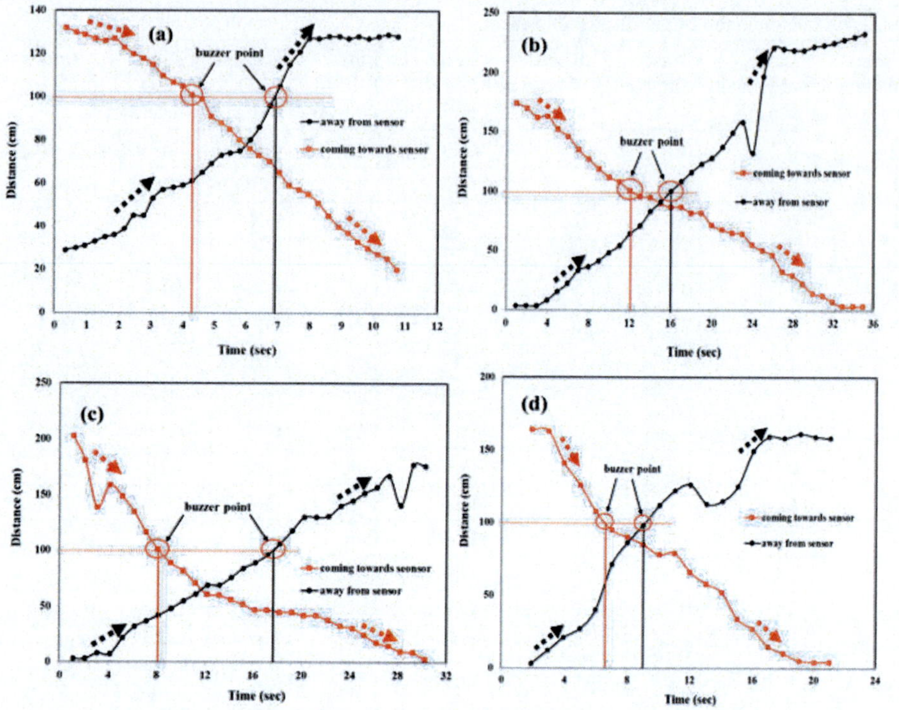


**Fig. 19.4** Results of **a** front **b** back, **c** left and **d** right ultrasonic sensor installed in belt for distance measurement of incoming or away object/person from the sensor with time (sensor wore by person are static while suspect is dynamic). *Source* Own

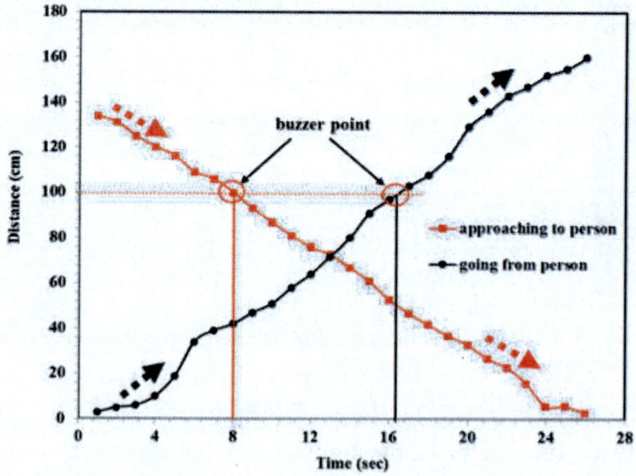


**Fig. 19.5** Result of ultrasonic sensor approaching to person/object and going from the person/object with time (suspect is static while sensor wore by person is dynamic. *Source* Own

have been fixed in a particular direction such as front, back, left and right at 90°. Once any suspected object/person is subjected to come/aproach toward or going back/away from the host, the sensor makes sound by buzzer, notifies by message and sends an email via IoT (Fig. 19.3). The results of front sensor are shown in Fig. 19.4a. If the object/person going back/away (dynamic) from the very close, i.e., 29 cm to the host, the buzzer started to make sound until the dynamic/suspect object/person away from 100 cm (1 m). After 100 cm away from the host, there was no message or notification. This result suggests that if any infected or non-infected person coming or going back from host person, the message can be received on mobile. In present study (for front sensor), a person who was going back from host (static), i.e., 29 cm, it makes sound until he/she away from 100 cm; thereafter, there was no notification on mobile. It means suspected person is not close to range, i.e., 100 cm (1 m). However, if any suspected coming/approaching from 132 cm toward host person (static), it has not shown any notification until unless he/she comes in the range of 100 cm.

Once the suspect comes within 100 cm range, it started to make notification on mobile. Moreover, if he/she has symptoms of COVID-19, the sensor notifies the temperature as well as distance by every second of interval. On the basis of obtained results, we can predict the speed of person. In this study, the front sensor found that incoming and outgoing/back person has 10.75 cm/s and 9.6 cm/s speed, respectively. On the basis of speed, we can stop the person movement.

The back, left and right side sensor results are shown in Figs. 19.4b–d, respectively. It can be seen from these figures, that if any body/person/object coming or going away from the host body (with sensor), immediately it detects the actual distance of the movement. However, it is observed that the incoming or going person has not consistent speed and direction movement which show fluctuation in distance with time. Moreover, if wearer/host is passing through the objects such as wall, table, etc. which are non-living things, it can detect the distance where these objects are located as well as direction. Through LED and buzzer, the surrounding people would also get aware about the contact made. Therefore, the transmission of COVID-19 can be reduced/mitigated or detected through this smart belt. It has been made with the IoT technology where it has its own IP address. The obtained data can be stored in cloud which can be used for further research. The ultrasonic sensor can detect greater dynamic range, i.e., 400 cm (4-m) and more precise data of the location. Moreover, ultrasonic sensor is immune to other light sources which tends to give accurate reading. Thus, it is suggested to use this sensor instead of infra-red. It is possible to interface a camera module which not only protectss us from COVID-19 but also helps us to detect the suspects. It can also be interfaced with global positioning system (GPS) which can give location-wise data and help us to obtain the probability of suspects at a particular region.

It is important to verify the sensor capability; thus, we have moved the person who wore the sensor. The person who wore the belt is moving from 134 cm and approaching toward suspected person near to 100 cm, automatically buzzer started to make sound as well as LED emits the light and sends the notification on mobile via IoT (Fig. 19.5). This person has reached at 3 cm and immediately returned from the suspected person. The host person was going back (away) from the suspected person and reached around 160 cm. The buzzer started to make sound until host reached beyond 100 cm. Once it reached greater 100 cm, there was no buzzer sound and notification on mobile. This result verifies that this sensor is still working and detecting the distance. However, this sensor sentivity up to 400 cm (4 m) but we have studied until 160 cm distance only.

### *Temperature Measurement*

Moreover, to detect the temperature of the person, it was set to be 37.5 °C as it is recommended for COVID-19. If a suspect comes under 100 cm (1-m) periphery and the temperature is above 37.5 °C, then this person will be tagged as suspected. When any object enters in the range of imaginary line, i.e., 100 cm, it detects and sends the message to the NodeMCU; then immediately, it transmits wirelessly to Blynk software and collects all the data and gives us the results. If some wall or non-living object comes in front of the wearer/host, it cannot detect the temperature until unless attach to the object. The LM35 temperature sensor needs direct contact. The results of LM35 temperature sensor are shown in Figs. 19.6 and 19.7. Since, LM35 sensor was wore on the wrist of the wearer/host and measure the temperature with time. Figure 19.6 shows the results of a person having actual body temperature around 35.4 °C with the time. This sensor has ±1 °C deviation attributed to improper contact with body. However, this sensor requires body contact to measure the temperature. The body temperature started to measure from 0.6 s until 4.1 s. It can be seen from Fig. 19.6 that temperature is consistent with marginal fluctuation owing to the limitation of LM35 sensor.

However, it is necessary to know the durability and actual performance of sensor at different temperature. Therefore, we have performed an experiment of drinking

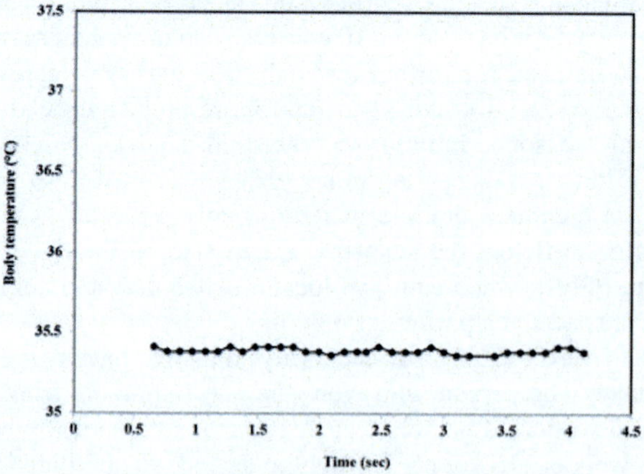


**Fig. 19.6** Temperature measurement of a person with time. *Source* Own

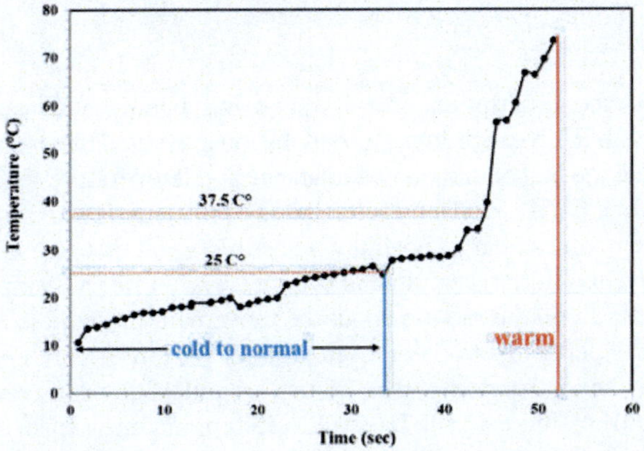


**Fig. 19.7** Temperature measurement of drinking water from cold to warm region with time. *Source* Own

water from cold to hot condition by attaching the sensor on steel pot. We have started to boil 200 ml drinking water from cold, i.e., 10 °C to hot, i.e., 73.5 °C and measured the temperature with time. Once the water was heated at high flame, the temperature is gradually increased up to 73.5 °C with time as shown in Fig. 19.7. In this figure, there are two regions, i.e., cold to normal and worm condition which suggest that if any COVID-19 suspect having more than 37.5 °C, he/she avoids to go outside from home.

## *Cost of the Setup*

We have calculated total cost of the setup. Following break up can calculate the total cost (INR).

PCB plate: Rs 20/-
Four ultrasonic sensors (Distance): 4 × Rs 120 = Rs 480/-
LM35 sensor (temperature): Rs 25/-
NodeMCU: Rs 350/-
Buzzer: Rs 15/-
LED: Rs 10/-
Jumper wire: Rs 45/-
Belt: Rs 70/-
Total: Rs 1015/-

The total cost is around Rs 1015/-, i.e., equal to approximately $14. From the cost calculation, it can be suggested that it is not expensive. Thus, it can be worn by a people.

## Conclusions

In present study, the description of smart belt has been made using sensor as paramount tool with IoT's execution. The conclusion presented by results and discussion manifests that the belt equipped with distance and temperature sensors can limit the transmission of COVID-19. The results shown by distance sensor impart that once an infected person comes near to host, it detects the actual distance and movement of incoming body, and in the meantime at every second of interregnum, it sends the email and notification on the mobile about the movement of the person. It detected the movement of a person from 3 cm up to around 240 cm. It commences a sound through buzzer if any suspected person comes around 100 cm in periphery and it also sends the notification using IoT. The LM35 temperature sensor measures the actual temperature of the host body, i.e., 35.4 °C with time. The measurement of temperature is shown by taking water at different temperature. Once the cold water was boiled, it shown the variation in actual temperature with time and detected the temperature which embarked variation in different regions of water. The assembled device is not so expensive and it costs around $14/belt. Consequently, it is proposed that this technology can be accustomed to control the transmission of COVID-19 pandemic as well as in many medical and environmental issues. With the help of this research, it is possible to interface a camera module which not only protects us from COVID-19 but also helps us to detect the suspects. It can also be interfaced with global positioning system (GPS) which can give location-wise data and help us to obtain the probability of suspects at a particular region.

**Conflict of Interest** The authors declare no conflict of interest.

**Author Contributions** Funding acquisition, Jitendra Kumar Singh; Investigation, Rishabh Gautam and Shruti Mishra; Methodology, Rishabh Gautam and Shruti Mishra; Supervision, Akhilesh Kumar Pandey and Jitendra Kumar Singh; Validation, Rishabh Gautam and Shruti Mishra; Writing—original draft, Rishabh Gautam, Shruti Mishra, Akhilesh Kumar Pandey and Jitendra Kumar Singh; Writing—review and editing, Rishabh Gautam, Shruti Mishra, Akhilesh Kumar Pandey and Jitendra Kumar Singh. All authors have read and agreed to the published version of the manuscript.Funding acquisition, Jitendra Kumar Singh; Investigation, Rishabh Gautam and Shruti Mishra; Methodology, Rishabh Gautam and Shruti Mishra; Supervision, Akhilesh Kumar Pandey and Jitendra Kumar Singh; Validation, Rishabh Gautam and Shruti Mishra; Writing—original draft, Rishabh Gautam, Shruti Mishra, Akhilesh Kumar Pandey and Jitendra Kumar Singh; Writing—review and editing, Rishabh Gautam, Shruti Mishra, Akhilesh Kumar Pandey and Jitendra Kumar Singh. All authors have read and agreed to the published version of the manuscript.

**Funding** This research received no external funding.

## References

1. Chan, J. F. W., Yuan, S., Kok, K. H., To, K. K. W., Chu, H., Yang, J., et al. (2020). A familial cluster of pneumonia associated with the 2019 novel coronavirus indicating person-to-person transmission: A study of a family cluster. *The Lancet, 395*(10223), 514–523.